NEURO-BEHAVIORAL DETERMINANTS OF INTERLIMB COORDINATION

NEURO-BEHAVIORAL DETERMINANTS OF INTERLIMB COORDINATION
A multidisciplinary approach

Edited by

Stephan P. Swinnen
K.U. Leuven, Belgium

Jacques Duysens
K. U. Nijmegen, The Netherlands

Kluwer Academic Publishers
Boston/Dordrecht/London

Distributors for North, Central and South America:
Kluwer Academic Publishers
101 Philip Drive
Assinippi Park
Norwell, Massachusetts 02061 USA
Telephone (781) 871-6600
Fax (781) 681-9045
E-Mail: kluwer@wkap.com

Distributors for all other countries:
Kluwer Academic Publishers Group
Post Office Box 322
3300 AH Dordrecht, THE NETHERLANDS
Telephone 31 786 576 000
Fax 31 786 576 254
E-Mail: services@wkap.nl

 Electronic Services <http://www.wkap.nl>

Library of Congress Cataloging-in-Publication Data

A C.I.P. Catalogue record for this book is available
from the Library of Congress.

Neuro-Behavioral Determinants of Interlimb Coordination: A multidisciplinary approach
edited by Stephan P. Swinnen and Jacques Duysens
ISBN 1-4020-7778-5

The Publisher offers discounts on this book for course use and bulk purchases. For further information, send email to <Melissa Ramondetta@wkap.com>.

This book is dedicated to our previous and current research collaborators for their admirable energy to explore the mysteries of life

Contents

Explanatory boxes

Contributing Authors

Peter J. Beek, Institute for Fundamental and Clinical Human Movement Sciences, Faculty of Human Movement Sciences, Vrije Universiteit, Amsterdam, The Netherlands

Allison J. Bigbee, Neurobiology, University of California at Los Angeles, USA

Winston D. Byblow, Human Motor Control Laboratory, Department of Sport and Exercise Science, University of Auckland, Auckland, New Zealand

Simone Cardoso de Oliveira, Institut für Arbeitsphysiologie an der Universität Dortmund, Ardeystr. 67, 44139 Dortmund, Germany

Richard G. Carson, Perception and Motor Systems Laboratory, School of Human Movement Studies, The University of Queensland, Brisbane, Australia

Francois Clarac, CNRS – INPC, Développement et Dysfonctionnement des Réseaux Locomoteurs, 31 Chemin Joseph Aiguier, 13402 Cedex 20, Marseille, France

Andreas Daffertshofer, Institute for Fundamental and Clinical Human Movement Sciences, Faculty of Human Movement Sciences, Vrije Universiteit, Amsterdam, The Netherlands

Filiep Debaere, Motor Control Laboratory, Department of Kinesiology, K.U.Leuven, Group Biomedical Sciences, Tervuurse Vest 101,3001 Leuven, Belgium

Ray D. de Leon, Kinesiology and Nutritional Science, California State University, Los Angeles, USA

Jörn Diedrichsen, Department of Psychology, University of California, Berkeley, USA

Volker Dietz, University Hospital Balgrist, Zürich, Switzerland

Opher Donchin, Dept. Biomedical Engineering,720 Rutland Ave., Traylor 416, Johns Hopkins University, Baltimore, MD 21205, USA

Stella Donker, SMK-Research, Sint Maartenskliniek, Hengstdal 3, 6522 JV Nijmegen, the Netherlands

Jacques Duysens, SMK-Research, Sint Maartenskliniek, Hengstdal 3, 6522 JV Nijmegen, the Netherlands, and Department of Biophysics, University of Nijmegen, Geert Grooteplein 21, 6525 EZ, Nijmegen, the Netherlands

V. Reggie Edgerton, Physiological Science, Neurobiology, Brain Research Institute, University of California at Los Angeles, USA

Eliot Hazeltine, Department of Psychology, University of Iowa, USA

Herbert Heuer, Institut für Arbeitsphysiologie an der Universität Dortmund, Ardeystraße 67, 44139, Dortmund, Germany

Richard Ivry, Department of Psychology & Helen Wills Neuroscience Institute, University of California, Berkeley, CA 94720, USA

Gwyn N. Lewis, Human Motor Control Laboratory, Department of Sport and Exercise Science, University of Auckland, Auckland, New Zealand & Sensory Motor Performance Program Rehabilitation Institute of Chicago, 345 E. Superior Street, Chicago, Illinois 60611, USA

Ichiro Miyai, Neurorehabilitation Research Institute, Bobath Memorial Hospital, Osaka, 536-0023, Japan

Edouard Pearlstein, CNRS – INPC, Développement et Dysfonctionnement des Réseaux Locomoteurs, 31 Chemin Joseph Aiguier, 13402 Cedex 20, Marseille, France

C. (Lieke) E. Peper, Institute for Fundamental and Clinical Human Movement Sciences, Faculty of Human Movement Sciences, Vrije Universiteit, Amsterdam, The Netherlands

Andras Semjen, Centre de Recherche en Neurosciences Cognitives, CNRS, 31 Chemin Joseph Aiguier, 13402 Cedex 20, Marseille, France

Jean Francois Pflieger, CNRS – INPC, Développement et Dysfonctionnement des Réseaux Locomoteurs, 31 Chemin Joseph Aiguier, 13402 Cedex 20, Marseille, France

Roland R. Roy, Physiological Science, Brain Research Institute, University of California at Los Angeles, USA

Bouwien C.M. Smits-Engelsman, SMK-Research, Sint Maartenskliniek, Hengstdal 3, 6522 JV Nijmegen, the Netherlands, and Nijmegen Institute for Cognition and Information (NICI), University of Nijmegen, P.O. Box 9104, 6500 HE, Nijmegen, the Netherlands

Rebecca Spencer, Department of Psychology, University of California, Berkeley, USA

Will Spijkers, Institut für Psychologie der RWTH, Aachen, Germany

James W. Stinear, Human Motor Control Laboratory, Department of Sport and Exercise Science, University of Auckland, Auckland, New Zealand

Stephan P. Swinnen, Motor Control Laboratory, Department of Kinesiology, Group Biomedical Sciences, K.U.Leuven,Tervuurse Vest 101,3001 Leuven, Belgium

Niranjala J.K. Tillakaratne, Physiological Science, Brain Research Institute, University of California at Los Angeles, USA

Sabine M.P. Verschueren, Motor Control Laboratory, Department of Kinesiology, Group Biomedical Sciences, K.U.Leuven,Tervuurse Vest 101,3001 Leuven, Belgium

Laurent Vinay, CNRS – INPC, Développement et Dysfonctionnement des Réseaux Locomoteurs, 31 Chemin Joseph Aiguier, 13402 Cedex 20, Marseille, France

Nicole Wenderoth, Motor Control Laboratory, Department of Kinesiology, Group Biomedical Sciences, K.U.Leuven,Tervuurse Vest 101,3001 Leuven, Belgium

Preface

When we were involved in publication of a book on interlimb coordination about ten years ago, we had never expected that this area of research would make such a tremendous evolution across the past decade. With the development of new technologies to study brain function, we have witnessed a rapid progress in the discovery of the brain areas associated with interlimb coordination and the role of reflex pathways that serve to optimize the coordination patterns in the face of perturbations or environmental contingencies. We can possibly not do justice to this expanding area of research in a small book such as this one. Accordingly, it is limited in the sense that it focuses on bimanual coordination against the broader context of the coordination between the upper and lower limbs. However, it is also broad in scope in that it reviews recent developments in the study of coordination by means of the latest technologies for the study of brain function, such as functional magnetic resonance imaging, near-infrared spectroscopy, magneto-encephalography, and transcranial magnetic stimulation. In addition, new developments in recovery of interlimb coordination following spinal cord injury and other insults of the central nervous system, such as stroke, are reviewed. No wonder that this area of research has progressed with major strides in the past years.

Whereas conducting science requires rigorous specialization in a narrow area of research, we currently witness a strong evolution towards an interdisciplinary approach whereby the walls between scientific disciplines are taken down progressively. The present book is testimony of this evolution whereby the merging of the neurosciences and the behavioral sciences is put into practice. Readers who are interested in behavioral and kinematic descriptions of skillful behavior and their central representations but also wonder about the neurophysiological substrate underlying the control of these skills, will find their taste in this book, from the single neuron to the systems level. It has not been our intention, however, to provide the reader with a full coverage of the control of patterns of interlimb coordination. We rather wish to introduce the reader to a variety of approaches that provides convergent information about human movement control in general and bimanual movement in particular. In this endeavor, the study of normal and pathological conditions serves as two sides of the same coin.

The book is intended to be a helpful source of information for scientists in basic research as well as practitioners involved in clinical settings. Those who will benefit most are neuroscientists, neurologists, neuropsychologists, cognitive neuroscientists, kinesiologists, motor and rehabilitation scientists, physical therapists etc. Special efforts have been made to make the contents accessible to graduate students by means of review chapters that contain explanatory boxes. We hope to convey our excitement and enthusiasm about the field of interlimb coordination and what it has to offer as a prototypical vehicle for a cognitive neuroscience approach to movement control. We nourish the hope that students will become seduced to the interdisciplinary approach as a result of these efforts. However, it will be necessary to climb high enough to be able to overlook many terrains of scientific endeavor and to become familiar with each of them.

S. P. Swinnen & J. Duysens   *September 25, 2003*

Acknowledgments

Editing a book requires a major commitment of time and control of effort and emotion. We would like to thank many people who have directly or indirectly contributed to completion of the present book. First and foremost, we would like to thank the authors who have made a major effort to review the most recent information and discuss the latest developments in the field of neural control of movement as related to interlimb coordination. They have gone a long way to complete their chapters and to please the senses by means of visually appealing figures and boxes. Secondly, a number of experts were involved in reviewing parts of the contents: Fausto Baldissera, Mario Wiesendanger, Deborah Serrien, Ivan Toni, Elisabeth Franz, Charles Walter, and the authors of the present chapters who also served as peer reviewers. Third, we would like to thank our wives and children for their love, patience, and understanding. It will be impossible to ever refund them for their contribution to our course of life.

We hope that these efforts have resulted in a book that will help promote the multidisciplinary study of movement control in general and interlimb coordination in particular.

S. P. Swinnen & J. Duysens   *Septembert 25, 2003*

Prologue

UNDERSTANDING COORDINATION: A COALESCENCE OF CONSTRAINTS AT MULTIPLE LEVELS

Stephan P. Swinnen[1] & Jacques Duysens[2,3]

[1]*Motor Control Laboratory, Department of Kinesiology, Group Biomedical Sciences, K.U.Leuven, Belgium,* [2]*SMK-Research, Sint Maartenskliniek, Nijmegen, and* [3]*Department of Biophysics, University of Nijmegen, the Netherlands*

Abstract: To better understand the neural architecture underlying the broad repertoire of movements that humans perform with the upper limbs, it is necessary to start with the basic spinal networks that have been investigated in the context of locomotion in animals (Part 1). From a phylogenetic perspective, the arms are controlled by spinal cervical networks that exhibit flexible associations with the lumbar networks of the lower limbs. With the evolution towards upright stance and gait, more direct cortico-motoneuronal control pathways, promoting hand dexterity, have emerged and these are primarily addressed in Part 2. The resulting multilayered neural architecture allows the involvement of arm functions in automated control tasks, such as in gait, as well as in more complex tasks, characterized by increased involvement of higher cognitive functions. To come to grips with the principles inherent to interlimb coordination, it is mandatory to obtain a deeper understanding of the constraints that emerge at multiple systems levels, from behavioral to neural.

Key words: motor control, interlimb coordination, spinal cord injury, functional recovery, bimanual skill, medical imaging technologies, central pattern generators, fMRI, TMS, MEG, NIRS

1. PART 1: CENTRAL PATTERN GENERATORS FOR LOCOMOTION: A STARTING POINT FOR UNDERSTANDING BIMANUAL COORDINATION

The present book intends to address interlimb coordination constraints at multiple systems levels, i.e. from behavioral to neural. In Chapter 1, Duysens and coworkers provide a framework for the notion of a central pattern generator (CPG), i.e., a network of neurons located in the spinal cord that is capable of generating rhythmic bursts of activity autonomously. Each limb is controlled by a spinal CPG and the CPGs of the various limbs are interconnected at low CNS levels, allowing various coupling modes among the limbs. Depending on the nature of these interconnections, different forms of locomotion such as pacing, trotting or galloping can be generated. In addition, these connections may help to explain why certain types of interlimb coordination patterns are favored, even when tasks unrelated to locomotion are produced (see Chapters 6-11). Duysens et al. discuss evidence for the existence of similarities between low-level networks in animals and humans (see also Chapter 4). They point to the important role of afferent feedback and reflexes in the regulation of interlimb coordination in animals and humans. The afferent input plays a major role in shaping the resulting coordination, induced by these CPGs. This is a basic entry point for exploring the recovery of function following spinal cord injury (Chapter 3). The role of proprioceptive afferents bas been studied by muscle vibration (selective stimulus for Ia afferents) and by adding load (activating mainly Ib afferents). Whereas tendon vibration has little effect on intra- and interlimb coordination, load manipulations induce more profound changes. During gait, the loading of one of the limbs induces adaptations in the 3 remaining limbs, thereby providing rhythm constancy. This is in line with other evidence indicating that the coordination between arm and leg movements is quite robust across various types of locomotion, suggesting a strong coupling between both homologous and non-homologous limbs. This coupling among effectors is also addressed in the subsequent chapters. Low-level coupling is predominantly addressed in Part 1 and high level coupling in Part 2.

The early development of the locomotor CPGs is discussed in Chapter 2 by Clarac and coworkers. They focus on the spinal neuronal networks during rat development before and after birth. They find early expressions of rhythmic locomotor activity at birth in vitro: When administering neuroactive substances to the entire brainstem/spinal cord preparation, rhythmical activity can be induced in the cervical and lumbar generators. This coordination results from a mutual interaction between these generators and provides an example of interlimb interactions at spinal cord level in the early stages of life. Indeed, when functionally isolated, the cervical network is slower than the lumbar burst generator. When both are in operation, the faster lumbar rhythm accelerates the slower cervical one whereas the slower reduces the frequency of the faster toward a common intermediate frequency. It is reminiscent of the magnet effect, as described by Von Holst (1973). Thus, propriospinal pathways that link the rostral and caudal locomotor generators, appear to be already functional in the neonatal rat and may account for coordinated forelimb-hindlimb locomotion.

Even though the limb generators develop before birth, voluntary locomotion is not evident at birth because the postural regulatory mechanisms are not yet in place. However, when postural constraints are reduced and/or a strong stimulus is provided, the neonatal rat is able to swim or to produce "air stepping" with an ipsilateral and a contralateral alternation between the different limbs during the first two weeks. Adult rat walking only emerges during the third postnatal week. This is associated with the recruitment of proprioceptive peripheral loops whose information is fed into the central network, allowing adaptations in a coordinated manner. Thus, a basic interlimb coordination network is capable of generating alternating patterns of limb coordination at birth and is gradually subjected to control from supraspinal structures. Evidence suggests that similar networks are also present in humans, as discussed by Edgerton et al. (Chapter 3) and Dietz (Chapter 4). In this context, recovery of function following insults of the central nervous system is also elaborated upon. Chapter 3 and 4 focus on plasticity at the spinal cord level and Chapter 5 and 7 at the supraspinal level.

Edgerton and coworkers (Chapter 3) discuss the recovery of posture and locomotion in mammals following a complete spinal cord lesion and reveal the degree of adaptability of the spinal cord. They review a broad spectrum of intervention strategies that may enhance motor function after a spinal cord lesion: (1) motor training and spinal cord learning, (2) electrical stimulation of nerves, muscles, and the spinal cord below the level of the lesion, (3) pharmacological facilitation of motor function, (4) growth/neurotrophic factor administration, and (5) several types of cell and tissue implants to induce axonal growth. Their general message is that a significant level of motor function can be recovered after a complete spinal cord lesion. This is a direct consequence of the fact that many details of the neural control of locomotion and posture are embedded in the spinal cord itself and this can be exploited in the absence of supraspinal-spinal connectivity. This is because some peripheral inputs that project to the spinal circuitry can activate or modulate the CPGs (Chapter 1). Although motor training has been shown to be one of the most effective strategies to date (see also Chapter 4 and 5), a number of other interventions appear to induce significant improvements in motor function. Some of these improvements can be attributed solely to functional reorganization of the spinal cord circuitry distal to the lesion, whereas other interventions appear to have been mediated by renewed connectivity between the neural elements above and below the lesion site. A general consensus appears to emerge that motor recovery is particularly enhanced by a combination of the aforementioned therapeutic strategies.

Dietz (Chapter 4) addresses the search for a central pattern generator in humans. He discusses existing evidence that neuronal networks for cyclical upper and lower limb coordination within the spinal cord and the associated patterns of reflex modulation, bear many similarities in human and animal gait (notably in the cat) (see also Chapter 1). This suggests that bipedal gait on the one hand and the use of the upper limbs on the other hand, may both have evolved from pathways that were initially involved in quadrupedal locomotion? However, the evolution towards upright stance and gait has paved the way for the emergence of differentiated hand functions with dedicated neural structures (Wiesendanger & Serrien, 2001). Across time, the phylogenetically older propriospinal control of arm movements may have gradually given way to the direct cortico-motoneuronal system that is associated with dexterity. However, the former may still be in place during locomotor types of movements in humans. This suggests a flexible neuronal linkage of cervical and thoraco-lumbar circuits that is task-dependent. As such, the phylogenetically newer components are integrated with the more archaic pre-existing neuronal circuits. Dietz explains how the upper limb system can be involved in both automatic types of activities (such as locomotion) as well as in the more complex bimanual functions, discussed in the later chapters. He also points to the potentially important clinical implications of knowledge about regenerating pathways in animals that may be transferable to humans. For example, the observed interactions between cervical and thoraco-lumbar neuronal circuits in humans may imply that it is meaningful to include arm movements during rehabilitation of patients with an incomplete spinal cord lesion because of its positive effects on their locomotor capacity.

Miyai (Chapter 5) builds further on these clinical implications. He moves away from spinal neuronal networks for interlimb coordination and reviews the supraspinal control of locomotion in humans, as revealed by means of near-infrared spectroscopy (NIRS). This optical imaging system allows real-time monitoring of cortical activation during various locomotor tasks. Accordingly, head movements are much less restrained than during positron emission tomography (PET) or functional magnetic resonance imaging (fMRI) and this is a distinct advantage. The limitation is that only the cortical areas can be revealed with restrained spatial resolution. This cortical activation is assessed by means of changes in levels of regional oxygenated hemoglobin. Miyae has applied this technique to both healthy subjects and patients recovering from a cerebro-vascular accident. In healthy subjects, he observes activation in the medial sensorimotor cortices and supplementary motor areas during walking at low speeds and reduced activation during walking at higher speeds (3 km/hr or 5 km/hr). In unilateral stroke patients, cortical activation patterns are characterized by asymmetrical activation in the sensorimotor cortices and recruitment of other motor-related areas, such as the premotor and prefrontal regions. These activation patterns can be modified as a result of rehabilitative intervention techniques. The restoration of gait in these hemiparetic patients appears to be associated with increased asymmetry in sensorimotor cortex activation and with recruitment of other motor related areas, such as the ipsi-lesional premotor cortex. However, more research is necessary to establish which of these changes are directly associated with optimal functional improvement in locomotion and other daily tasks. Nevertheless, it is clear from Miyai's work that NIRS may contribute to the development of brain-based strategies for neurorehabilitation. These can complement the earlier discussed strategies (Chapters 3-4) that promote the training of the spinal pattern generators for locomotion.

2.   PART 2: NEURAL AND BEHAVIORAL DETERMINANTS OF BIMANUAL COORDINATION

Part 2 addresses the neural basis of bimanual coordination, from the single neuron to the system's level. In addition, the behavioral principles that become evident when facing performers with tasks requiring the simultaneous execution of different movements with both arms, are also discussed. In Chapter 6, Donchin and Cardoso de Oliveira discuss the relevance of studying bimanual movements in primates. They review research on single- and multi-unit activity and focus on two fundamental questions. The first addresses 'which parts of the brain mediate bimanual coordination' (see also Chapter 8). They arrive at the conclusion that a widespread network of multiple brain areas is involved in this function. This marks a deviation from traditional thinking in which the supplementary motor area was assigned an important role in bimanual coordination. They list a number of candidate brain areas for which electrophysiological evidence has been generated to support their role in bimanual coordination. The authors' conclusion is largely convergent upon that made by Wenderoth et al. (Chapter 8) which is mostly based on evidence from medical imaging studies. The second question refers to 'how the activity of different parts of the brain is coordinated to achieve bimanual coordination'. In this context, they refer to the binding phenomenon as previously discussed within the context of resolving ambiguities in visual scenes. Similarly, temporal synchronization of activity in the cortical hemispheres may bind the movements of both arms. They suggest that neuronal assemblies exploit the temporal relations between neuronal activities as a coding dimension that can be used in addition to and parallel with the firing rates of individual neurons. They hypothesize that bimanual coordination may be reflected both in bimanual-specific changes in single unit firing rate and in temporal coupling because both the firing rate code and the temporal code rely on independent dimensions of neuronal activity. Binding in the context of motor performance has recently received increasing attention and appears to offer an interesting avenue for exploring the role of neural synchronies in functional brain organization (Sanes & Truccolo, 2003).

Byblow and coworkers (Chapter 7) take a different approach to bimanual coordination by focusing on the corticospinal pathway using transcranial magnetic stimulation (TMS). During the past decade, TMS has developed into a useful technique to address a variety of neuroscientific questions, such as changes in corticospinal excitability as a result of experimental interventions. The authors first discuss the contribution of peripheral afferent feedback to the crossed facilitation/inhibition of motor pathways that arises during active and passive rhythmic movements of the upper limbs. They observe a cyclic potentiation and inhibition of corticospinal excitability and suggest that movement elicited afference modulates corticomotor excitability. They subsequently turn their attention to factors that influence the degree of coupling between the limbs by examining the changes in corticomotor excitability that result from (voluntary) movements on the opposite side of the body. When homologous forearm muscles are shortened and lengthened synchronously, there appears to be suppression of intracortical inhibition, whereas less suppression of intracortical inhibition is evident when the movement pattern is asynchronous. Hoffman-reflex modulation does not appear to distinguish between the two patterns. Therefore, they hypothesize that the modulation of afference-driven corticomotor excitability is likely a cortical phenomenon. Alterations in the composition of muscle synergies impact directly upon between-limb neural coupling, i.e., cortical input to spinal motoneurons is strongly modulated by changes in the functional context of opposite limb muscles. It appears that there is an excitatory relationship between homologous cortical representations when a prime mover is engaged in a task. This relationship is largely inhibitory if the prime mover is not engaged. More generally, cortical input to spinal motoneurons appears to be strongly modulated by changes in the functional context of opposite limb muscles. This may be an interesting entry point for therapy, as discussed by the authors within the context of unilateral hemispheric stroke. They provide evidence for an association between the down-regulation of intracortical inhibition, cortical reorganisation and functional recovery after stroke. Moreover, they suggest that the use of bilateral rhythmical movement in a rehabilitation setting may be a promising avenue for rehabilitation and is worthy of further investigation. This approach deviates from those discussed earlier (Chapter 3-5) in that the focus is primarily on the upper limbs and on modulation of intra- and inter-hemispheric excitability.

Wenderoth et al. (Chapter 8) address bimanual coordination from the perspective of systems neuroscience. They review studies that have made use of medical imaging techniques (PET, fMRI) to reveal basic information about the bimanual coordination network during the production of rhythmical movement. They discuss two approaches that have been used to tackle this issue. On the one hand, studies have compared brain activations during bimanual movement with the summed activations observed during each of the unilateral movements. This is presumed to directly reflect the coordination effort. On the other hand, studies have also compared the brain activations observed during a variety of bimanual tasks that differ with respect to degree of intermanual compatibility, i.e., from the preferred 1:1 in-phase coordination mode to less preferred modes with different relative phasing features and/or frequency ratios. They show that differences in activation among these conditions emerge, particularly when tasks are performed at higher cycling frequencies. This has resulted in the identification of activity in a network, consisting of the supplementary motor area, the premotor cortex, cerebellum, Broca's area/Insula and the superior temporal gyrus. The ultimate network that becomes activated during the production of bimanual tasks is more extensive however (see also Chapter 6), because the latter tentative list primarily refers to activity related to the coordination effort. Moreover, the interplay between the areas involved in this network is not yet clear, neither is the specific contribution of each of these areas to bimanual coordination. Nevertheless, a preliminary attempt is made to provide a general framework for task division. Additional work is necessary to verify the hypothesized functional assignments.

The next two chapters shift the focus further towards a neuro-behavioral account of coordination. When moving both limbs across different distances and/or directions, interference between limbs emerges and this point to limitations in neural resources. The reason for studying the combined performance of different movements with the upper limbs is to stress the human control system sufficiently in order to reveal its basic or default modes of operation (Swinnen, 2002). One hypothesis that has received considerable attention is the notion of 'neural crosstalk' that refers to neural leakage, possibly occurring at different levels, i.e., from cortical to spinal. Whereas initial experimental work focused upon interference during movement execution (Marteniuk et al., 1984; for review see Swinnen, 2002), more recent models by Heuer and coworkers have made a division between the programming and execution level.

In Chapter 9, Spijkers and Heuer review their experimental project on coupling or bimanual interference at the programming level. They provide extensive evidence for their model across a range of experimental paradigms. A distinction is made between static and phasic (i.e., transient) crosstalk. Static crosstalk persists during motor preparation whereas phasic crosstalk fades with increasing preparation time. Transient crosstalk refers to the capability to decouple concurrent processes of movement specification during motor preparation. The de-coupling is adaptive because it allows the concurrent production of different movements. To the extent that it is incomplete (i.e., static crosstalk remains), the concurrent production of different movements will result in interference. The notion of adaptive crosstalk modulation during motor programming implies that the observable crosstalk effects depend on the state of motor programming. It suggests that investigators should not only search for the softer or harder constraints on coordination, but also for the dynamics of change and for the conditions that support adaptive modulations. Spijkers and Heuer acknowledge that their decomposition into crosstalk at the programming and execution level may be insufficient and that a higher "cognitive" level may have to be added. Potential avenues for these higher levels of constraint are outlined in more detail in Chapter 10.

Ivry and coworkers argue that bimanual coordination and interference depends critically on how these actions are represented at a cognitive level (Chapter 10). Even though they admit that it is not always possible to divide space and time, they discuss constraints with respect to both aspects of movement. In relation to *spatial interactions*, they have identified differences in degree of interlimb interference between movements directed at visual targets (via target illumination) and movements cued symbolically (a letter indicates whether a short or long amplitude should be made, see Chapter 9). They argue that interactions manifest during response programming are limited to the latter condition and that interactions in the formation of the trajectories of the two hands are associated with processes involved in response selection, rather than interactions at the motor programming level, as Spijkers and Heuer propose. The question emerges to what extent a strict distinction between response selection, planning, and execution can be maintained or whether these stages reflect a continuum of motor control. Nevertheless, it appears that when performers are visually cued towards the targets they have to reach for with each limb, interference is absent or at least smaller than when a transformation is required from a symbolic code to a motor program for movement. They discuss a cognitive neuroscience account for these observations and refer to the degree to which the processing is dealt with within the dorsal versus ventral stream pathways. These pathways may be prone to a differential degree of interference. An alternative account may be the differential networks involved in movements that are internally (without visual information) versus externally cued (with visual information) (see Debaere et al., 2003). Further work is necessary to reveal more detailed information about the locus of bimanual interference.

In the second part of their chapter, Ivry et al. focus upon *temporal constraints*, suggesting that discrete bimanual movements are marked by a common event structure (e.g., the finger tap on the table), i.e., an explicit representation that ensures temporal coordination of the movements. When rhythmic movements do not entail an event structure (e.g., continuous circle drawing), timing may be an emergent property and the event-based representation may not be essential. The emphasis on task conceptualization leads them to the conclusion that many of the constraints underlying bimanual coordination arise at an abstract level that can be divorced from processes devoted to motor execution. However, evidence in favor of interference at a cognitive level does not invalidate its well-documented existence at the programming and execution levels. It rather implies that interference can originate at multiple levels, i.e., from action conceptualization to programming and on-line control of movement. Unraveling this coalescence of constraints will become an important avenue for future investigation and holds promise for better understanding and appreciating the extraordinary flexibility with which humans use their two hands (Kelso et al., 2001).

Dynamic pattern theory (DPT) and its recent evolutions are treated in the chapter by Peper et al. (Chapter 11). DPT has had an enormous impact in the past decades, leading to a proliferation of behavioral research on interlimb coordination as witnessed nowadays. Within DPT, biological systems are formally described in terms of their time-dependent changes. Such systems are composed of many subcomponents that organize themselves into coherent global patterns as a result of local interactions. The marvellous collective behavior of a school of fish, as often shown on National Geographic Channel, is a pertinent example. Order in these systems is not prescribed by a superior command structure but emerges in a self-organized fashion as a consequence of local cooperation among the subcomponents. Similarly, motor coordination is an example of orderly behaviour in a system with many degrees of freedom at multiple levels. DPT has highlighted the importance of stability and loss of stability of coordination patterns. The empirically observed stability characteristics of rhythmic interlimb coordination have been modeled in terms of gradient dynamics of the (order) parameter that defines the coordination modes. Relative phase (ϕ) is such an order parameter quantifying the relation between two signals. Cycling frequency has typically been used as a control parameter that probes the stability of the coordination patterns and that triggers transitions from less stable (e.g., anti-phase or 3:2 performance) to more stable patterns (i.e., 1:1 frequency ratio in-phase). A mathematical description of the coordination dynamics as a function of cycling frequency, was originally provided by Haken, Kelso, and Bunz, known as the HKB model (Haken et al., 1985). The produced ϕ is represented by a ball moving within a potential landscape V $(\phi) = - a \cos(\phi) - b \cos(2\phi)$, with b/a representing the control parameter (e.g.,cycling frequency). The HKB model is a landmark in the study of coordination dynamics and has undergone a number of modifications over the years as a result of new experimental evidence. Peper et al. provide an overview of these developments. In addition to modeling interlimb coordination in terms of gradient dynamics, the authors suggest that sorting out how interlimb interactions lead to pattern (in)stability and how changes in stability are modulated by cognitive interventions, such as intention, attentional strategies, and learning, will provide vital information about behavioral coordination dynamics. According to the authors, this will ultimately result in formulating an encompassing dynamical model of rhythmic interlimb coordination that reflects essential features of the underlying system properties and processes, in spite of its relatively high level of abstraction. They propose an appealing dynamical model of interlimb coordination that is divided into the neural and effector level. They admit that the neural level may have to be further decomposed into additional functional levels. This stems from the fact that interactions may range from spinal entrainment (Chapters 1-5) to cortical (and transcortical) levels (Chapter 6-10).

They provide an example of their recent work using magneto-encephalography during polyrhythm performance and show similar dynamics of frequency and phase locking at the cortical and behavioral levels. Recordings of the cortical activity during bimanual coordination show qualitative properties that are consistent with behavioral (in)stabilities. However, the cortical dynamics appears to be much more complex because of nontrivial spatially distributed activation and a broad range of contributing spectral components. Various research groups have started to use experimental recordings with sufficient temporal resolution by means of non-invasive techniques such as electro- or magneto-encephalograms. This will help to unravel the cerebral concomitants of dynamical interlimb interactions by means of analysis of patterns of temporal encoding across cortical areas during the course of limb movements.

3. SUMMARY

With the present book, it has been our attempt to diminish the gap between different approaches towards the study of interlimb coordination. In doing so, we wanted to inform the reader implicitly about recent experimental technologies that have been used to study brain function in general and movement control in particular.

The study of central pattern generators in various animal systems dates back a long time. Sherrington was already involved in the study of locomotion in animals. More recently, Grillner (1975) developed the notion of the CPG which has subsequently been further elaborated. Nevertheless, it is only recently that stronger claims for the existence of these neural substrates in humans have been made. The message of the first chapters is that a better understanding of the principles of bimanual coordination ultimately rests on basic knowledge about its role during quadrupedal locomotion across evolution. Even though the cervical and thoracic girdles can be probed independently, their interaction secures successful and adaptive gait and posture in a variety of animal species. In association with the study of the basic operation of the CPGs, researchers have also addressed the role of afferent information and reflex pathways in the modulation of the output of these neural substrates to meet environmental contingencies and to build in adaptability and flexibility. More recent work has focused on the potential role of the supraspinal structures in interlimb coordination (including locomotion) by means of imaging technologies.

Behavioral studies on interlimb (and particularly bimanual) coordination have developed relatively independently from the aforementioned studies on locomotor CPGs. This is understandable because humans perform a huge variety of complex bimanual movements that go far beyond their role in postural control and propulsion within a gravitational field. In spite of the gap between both subfields of research, it is of critical importance to communicate and inform each other about the different systems levels at which interactions between limbs emerge or constraints become evident. We realize that this is a formidable task but one that is worth exploring further to better understand the great variety of movements we produce every day.

From the perspective of research technology associated with the study of human movement control, the chapters in the present book address a battery of technologies and experimental approaches, some of which are very recent. Each technique has its own strengths and weaknesses in terms of spatial or temporal resolution but it is evident that progress in fundamental and clinical neuroscience will be secured if scientists will use and appreciate them as complementary technologies.

We hope the present book has convinced the reader that interlimb and particularly bimanual coordination has become an active, energetic, and highly dynamic forum of interdisciplinary scientific inquiry that serves as an important vehicle for exploring the (cognitive) neuroscience of motor control but also digs into the essence of our everyday movements.

REFERENCES

Debaere F, Wenderoth N, Sunaert S, Van Hecke P, Swinnen SP (2003) Internal versus external generation of movements: differential neural pathways involved in bimanual coordination performed in the presence or absence of augmented visual feedback. *Neuroimage* 19: 764-776

Grillner S (1975) Locomotion in vertrebrates: central mechanisms and reflex interaction. Physiol. Rev. 55, 247-306

Haken H, Kelso JAS, Bunz H (1985) A theoretical model of phase transitions in human hand movements. Biol Cybern 51:347-356

Holst, Von E (1973) The behavioural physiology of animals and man : The collected papers of Erich Von Holst (Vol 1) (R. martin, trans.). London: Methuen.

Kelso, JAS. *et al.* (2001) Haptic information stabilizes and destabilizes coordination dynamics. *Proc R Soc Lond B Biol Sci.* 268: 1207-1213.

Marteniuk RG, MacKenzie CL, Baba DM (1984) Bimanual movement control: Information processing and interaction effects. Q J Exp Psychol A 36A:335-365

Sanes, J. N. and Truccolo, W. (2003) Motor "binding:" Do functional assemblies in primary motor cortex have a role? *Neuron* 13, 669-673

Swinnen, SP (2002) Intermanual coordination: from behavioural principles to neural-network interactions. *Nat. Rev. Neurosci.* 3: 348-359

Wiesendanger, M, Serrien, D.J. (2001) Toward a physiological understanding of human dexterity. *News Physiol Sci.* 16: 228-233.

PART 1

INTERLIMB COORDINATION

NEURAL NETWORKS FOR LOCOMOTION AND PROCESSES OF FUNCTIONAL RECOVERY

Chapter 1

SENSORY INFLUENCES ON INTERLIMB COORDINATION DURING GAIT

Jacques Duysens[1,2], Stella Donker [1], Sabine M.P. Verschueren[3] , Bouwien C.M. Smits-Engelsman,[4] and Stephan P. Swinnen[3]

[1]*SMK-Research, Sint Maartenskliniek, Nijmegen, the Netherlands, [2]Department of Biophysics, University of Nijmegen, the Netherlands, [3]Motor Control Laboratory, Department of Kinesiology, K.U.Leuven, Belgium. [4]Nijmegen Institute for Cognition and Information (NICI), University of Nijmegen, the Netherlands*

Abstract: This chapter addresses the role of afferent feedback and reflexes in the regulation of interlimb coordination in animals and humans with a focus on locomotion. From the work on cats it is known that the rhythmic muscle activities during gait are generated by specialized neural circuits located in the spinal cord (the so-called central pattern generator, "CPGs"). These CPGs are coordinated by neurons, which interconnect both sides or which transmit information between the cervical and lumbar spine. It is argued that afferent input, especially load-related information, plays a major role in shaping the resulting coordination of these CPGs. Induced changes are seen not only with a general loading of the animal but also with the selective loading of a given limb. Such principles also apply to human locomotion. Studies on infants have shown that basic coordination patterns exist, very similar to those found in the cat. The effects of afferents (notably those related to load and to hip position) play an important role in phase transitions, much as was described in feline models. In adults, the role of proprioceptive afferents was studied by muscle vibration (selective stimulus for Ia afferents) and by adding load (activating mainly Ib afferents). When applied during gait, tendon vibration has little effect on intra – and interlimb coordination. In contrast, load manipulations produce more profound changes. During gait, the loading of one of the limbs induces adaptations in inter-limb coordination in the 3 remaining limbs, thereby providing rhythm constancy (stable cadence). This is in line with other evidence indicating that the coordination between arm and leg movements is quite robust across various types of locomotion, suggesting a strong coupling between both homologous and non-homologous limbs.

Key words: locomotion, gait, interlimb coordination, load afferents, central pattern generator, humans, cats, infants, arm sway

1.   INTRODUCTION: GENERAL PRINCIPLES

This chapter addresses the role of afferent feedback and reflexes in the regulation of interlimb coordination in cats and humans with a focus on locomotion. The theme is that some aspects of human interlimb coordination can be understood from knowledge of interlimb coordination during gait in animals. The consequences of load are of particular importance in this respect. The present review will focus on the following proposals.

a. To understand interlimb coordination in humans it is essential to realize that each limb is controlled by a spinal CPG (Central Pattern Generator) much as in animal models.

b. These CPGs of the various limbs are interconnected at a low level in the CNS as can be demonstrated in locomotor studies. These interconnections help to understand how different forms of locomotion such as pacing, trotting or galloping are generated (Collins and Stewart, 1993). In addition, they may help to explain why certain types of interlimb coordination patterns are favored, even when tasks are involved which are not related to locomotion. In humans, it is now thought that CPGs can control arm movements under certain conditions.

c. During gait the coordination between the limbs depends on sensory input, in particular afferent input related to load. The latter input not only profoundly influences the CPG actions of individual limbs but also the type of coordination between CPGs.

Some introduction on CPGs is required first.

2.   SPINAL CPGS (CENTRAL PATTERN GENERATORS)

For many species the cyclical patterns needed for walking, respiration, mastication or other rhythmical activities, are generated by special neural networks (see Clarac, Chapter 2). For locomotion one usually refers to the term central pattern generator (CPG) to indicate a set of neurons responsible for creating a basic rhythm and for shaping the pattern of the bursts of motoneurons (Grillner and Wallen, 1985; Grillner, 1985). For the cat it is assumed that there is at least one such CPG for each limb and that these CPGs are located in the spinal cord. The term "pattern" is used here in a broad sense to indicate alternating activity in groups of flexors and extensors.

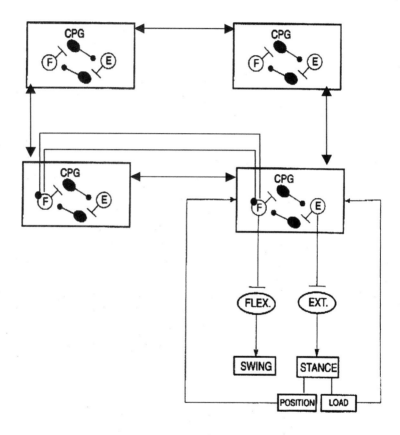

Figure 1-1. Schematic representation of the current hypothesis about the spinal organization of locomotion and its reflex regulation. The Central Pattern Generator (CPG) contains F and E: flexor and extensor half-center, controlling flexors (Flex.) and extensors (Ext.) respectively. Each limb is controlled by such a CPG (top: two arms, bottom two legs). Furthermore there are interconnections between these various CPGs. Some of these are schematically shown for the legs. The main connections are such that activity in a flexor center on one side inhibits activity in the corresponding contralateral flexor center. Each leg

separately also contains feedback loops as shown for the left leg only. These loops allow
automated switching between the stance and swing phases (see text).

How are these CPGs coupled to produce coordinated locomotor activity?
In the spinal turtle, it has been proposed that the central pattern generators
for the hind limbs are part of a bilateral core in which some of the neurons
involved in pattern generation of one leg also participate in producing the
normal motor rhythm for the contralateral leg (Stein and Smith 1997; Stein et
al. 1995). In various species, neurons have been described which connect
CPGs either at the same or at different segmental levels. These can either be
inhibitory or excitatory. They can provide out-of-phase and in-phase
synchronization. The main difference between mammalian and lower
vertebrates is that the former have commissural interneurons making
polysynaptic rather than monosynaptic connections (Butt et al., 2002). It is
thought that the extra synapses allow for a greater integration of information.
In the neonatal rat a distinction is made between short-range inhibitory and
excitatory commissural interneurons (short-range interneurons for left-right
coordination and long-range ones for intersegmental coordination ; Butt et
al., 2002). In the cat, commissural neurons were recently studied by
Jankowska et al. (2003). They claim that these neurons can contribute to
spinal coordination of the two sides and that they also mediate reticulospinal
signals. At the lumbar level the commissural neurons are antidromically
activated from contralateral motor nuclei and monosynaptically from the
ipsilateral reticular formation. They are located in Rexed's lamina VIII.
They can be considered to underlie the close coupling between spinal
interneuronal systems such as CPGs. In the next section the role of
supraspinal input in interlimb coordination will be treated in more detail.
Later the role of sensory input will be discussed. In each case a comparison
will be made between cat and human.

3.   DOES INTERLIMB COORDINATION DEPEND
ON SUPRASPINAL STRUCTURES?

Since the cat has traditionally been a favorite model for the study of
interlimb coordination, it is appropriate to briefly describe recent advances in
this field. An excellent review of the earlier work can be found in Rossignol
et al. (1993; 1996).

3.1   Cat

Studies in the cat have shown that the spinal cord itself contains mechanisms for interlimb coupling. One clear indication for the strong coupling between hind leg CPGs is that rhythmic activity in one limb is influenced by manipulations of another limb. In spinal cats (cats with transected spinal cord), the rhythm can either be facilitated or suppressed by manipulations on one side. For example, continuous imposed flexion on one hind limb blocks the rhythm on both sides (Rossignol et al., 1993). Hence sensory input from one side affects the other. In agreement with this, the reduction of input on one side produces clear deficits on the contralateral side. For example, unilateral deafferentation in spinal cats disrupts both ipsi- and contralateral stepping (Giuliani and Smith 1987).

Coupling of the movements between fore- and hind limbs is also a feature of gait, both in spinal and in intact cats (Cruse and Warnecke, 1992; Miller et al., 1975; English and Lennard, 1982). In alternating gait of the hind limbs, there are two basic patterns. One pattern is related to the pacing gait where flexion of the forelimb precedes extension of the hind limb (as measured at the elbow and knee, respectively). Another pattern is typically found in the trot where flexion of the forelimb follows extension of the hind limb. These patterns of coupling occur symmetrically on both sides. For in-phase stepping of the hind limbs, one distinguishes gallop and half-bounding. In the rotatory and transverse gallop the coupling is asymmetrical. On one side it is comparable to pacing (forelimb flexion precedes hind limb extension), and on the other side it resembles trotting (forelimb flexion follows extension). These distinct patterns of coordination of forelimb-hind limb step cycles were described in most detail by English and Lennard (1982). These authors emphasized that transitions between the coordination patterns were usually gradual. In contrast, changes in hind limb-hind limb phase relationships during transitions were nearly always abrupt. They concluded that "the 4 limbs are coordinated during in-phase stepping according to a few patterns, but that the variability about these patterns makes their association with simple neural circuitry rather speculative". This is in contrast to earlier suggestions that interlimb coupling could be understood on the basis of strict coupling by long propriospinal pathways (Miller and Van der Meché 1976). The latter suggestion was made because high lesioned spinal cats show typical locomotor patterns of interlimb coordination between fore-and hind limbs. Forelimbs in cats are controlled by CPGs and these are in turn coupled to the hind leg CPGs through neurons, which transmit information between the cervical and lumbar spine. Not only in the cat, but also in the rabbit (Viala, 1978) such interlimb coordination has been attributed to long intersegmental propriospinal circuits.

The question remains however what the contribution of these spinal connections is in animals with more intact central nervous systems. Nowadays it is thought that spinal coupling between girdles is relatively weak and that the brainstem is essential for appropriate interlimb coordination. When some connections from the reticular formation to the spinal cord are left intact (such as in some types of decerebrate cats), the interlimb coordination is much better than the one seen in spinal cats. Kato (1994) for example, argues that for interlimb coordination in cat gait, the brainstem locomotor centers are more important than interlimb reflex systems. One strong argument for this idea comes from his observations that coordination between hind limbs is well preserved in chronic spinal cats, even when the lumbar cord is longitudinally separated into halves. Furthermore, interlimb coupling is fairly robust when the brainstem connections to the spinal cord are intact. In cats with transected brainstem ("premammillary cat") it was shown that when one hind limb is prevented from moving, the other hind limb continues its stepping behavior while rhythmic alternating burst persist in the motionless leg (Duysens and Pearson 1980; Grillner and Rossignol 1978). Hence, coupling of the CPGs is probably so strong that activations persist when one of the hind limb CPGs is active and this even in the absence of sensory input in the retained hind limb. Similarly, in decerebrate cats both pacing and trotting patterns remain even after bilateral deafferentation of the hind limbs (Miller and Van der Meché,1976). Furthermore such decerebrate cats can learn to adapt their interlimb coordination. Yanagihara et al. (1993) showed that such cats could learn to walk on a split belt . In their study, the treadmill belt for the left forelimb was driven at twice the speed of the other limbs. After about 50 steps, the animals achieved stable locomotion. They shortened step cycle durations, and adjusted durations of double support phases asymmetrically in the left and right forelimbs. It is unclear however, how much of this adjustment is due to brainstem mechanisms. Indeed spinal cats can also adjust to some extent to split belt conditions with different speeds (Forssberg et al. 1980).

It may be expected, however, that in decerebrate cats the adaptations are more robust, primarily because the changes will also involve cerebellar plasticity. Sensory information from the limbs reaches the spinocerebellum, which has an important role in coordinating the limbs during locomotion. In the past 10 years, our view on this spinocerebellar input has changed considerably. Neurons of the ventral spinocerebellar (VSCT) tract have long been known to be involved since they carry information about the timing of step cycles from more than one limb (Arshavsky 1984). In contrast, until recently it was thought that the dorsal spinocerebellar tract (DSCT) neurons encoded localized sensory information from a few muscles or from a portion of an ipsilateral limb (see Bloedel and Courville, 1981). This was because these early investigators believed that only short latency inputs were significant for interlimb coordination. In agreement with these initial observations, anatomical tracing of primary muscle afferents had failed to show contralateral projections to DSCT neurons (e.g., Walmsley and Nicol 1990). Contralateral projections were found but they ended on presumed interneurons in the intermediate zone of the spinal cord (Scheibel and Scheibel 1969) and were therefore missed by the electrophysiological studies, mentioned above. Nowadays, it is increasingly clear that DSCT neurons encode global, whole limb parameters (Bosco et al. 1996; Bosco and Poppele 2001). Furthermore, Fedirchuk et al. (1995) showed that DSCT cells are modulated during fictive locomotion when locomotor related sensory input is absent. Most recently, Poppele et al. (2003) found that DSCT-neurons respond to contralateral limb stepping. The DSCT neurons are also modulated by passive step-like movements of either hind limb, implying that they receive a bilateral sensory input. On the output side it is primarily the reticulospinal tract, which is thought to be important for interlimb coordination (Matsuyama and Drew, 2000). During gait, neurons of this tract provide a signal related to the activity of groups of flexor and extensor muscles, in either a single or in multiple limbs.

In summary, in the cat the basic wiring for interlimb coordination is present in the spinal cord but for a full expression of this coordination the brainstem and cerebellum are essential.

3.2 Human

In humans there are no coordinated locomotor movements present after a complete spinal cord lesion. Nevertheless there are some data from patients with incomplete lesions, which indicate that interconnections between spinal CPGs may be important. One question for example, is whether there are facilitatory connections between the arm and leg CPGs while walking? Barbeau and colleagues showed that patients with incomplete spinal cord injury (SCI) have greater gait symmetry and more normal EMG while walking on a treadmill when body weight support (BWS) was provided and arm swings were allowed than when arm swings were absent because of the use of parallel bars for the same amount of weight support (Visintin & Barbeau, 1994). More generally, in the case of neurologic pathology, the impaired arm function during gait using a walking aid is likely to add to the neural limitations that already are present (Behrman, 1998). The neural substrate for these interlimb interconnections in humans has long been studied in reflex studies (Delwaide et al., 1977). The more recent work on this topic (e.g the work of Zehr and others; Zehr et al., 2001a,b) is described elsewhere by Dietz (Chapter 4).

4.   HOW MUCH DOES COORDINATION BETWEEN LIMBS DEPEND ON PERIPHERAL AFFERENT INPUT?

4.1   Cat

In the cat this question about the need for afferent input for interlimb coordination during gait can be best studied in fictive locomotion (see Box 1-1). Under these conditions the locomotor output is present in a motionless animal. If coordination persists it must be due to structures within the central nervous system. The older studies, reviewed by Rossignol et al., (1993; see also Swinnen et al., 1994), found that interlimb couplings, consistent with walking, trotting or pacing can be present in cat fictive locomotion. In the study of Orsal et al. (1990), efferent discharges in muscle nerves of the four limbs were recorded simultaneously during spontaneous fictive locomotion in cats with a high level decerebration The onset of the bursts of activity in the nerve of a given flexor muscle in each limb allowed the temporal relationships between the fictive step cycle of a pair of limbs to be determined. The fictive step cycles of the two forelimbs were always strictly alternated whereas the phasing of the step cycles of either the two hind limbs or pairs of homolateral or diagonal limbs was more variable. The time interval between the onsets of the flexor bursts of one of the two pairs of diagonal limbs was independent of the step cycle duration. Distinct patterns of interlimb coordination, corresponding to the walking and the trotting gaits in the intact cat, were seen very frequently. This supports the view that interlimb coordination essentially results from diagonal interaction between a fore- and hind limb CPG for locomotion. These results further show that the central nervous system has the ability to generate most of the patterns of interlimb coordination that occur during real locomotion even when phasic afferent inputs from the periphery are lacking.

4.2   Human

In humans, there is no equivalent to "fictive locomotion". Nevertheless there are some experimental data available with respect to the effects of a reduction of sensory input on one side on the locomotor output of the other side. Two experimental paradigms were used, one involving gait, the other cycling. As in the cat, the question is whether there is evidence that coupling between leg CPGs is so strong that rhythmic output can persist in a leg even if sensory input is reduced but the contralateral side continues locomotion.

Box 1-1 Central Pattern Generator (CPG)

Central pattern generators (CPGs) are networks that generate the rhythm and form the pattern of the locomotor bursts of motoneurons (Grillner and Wallen, 1985; Grillner, 1985). For the cat it is thought that the CPGs are located in the spinal cord and that there is at least one such CPG for each limb. The history of CPGs started with Sherrington (1910), showing that stepping movements could be elicited in spinal cats (transected spinal cord) with hanging legs provided the tail was pinched. He noted the similarity between this reflex stepping and the activation of the same muscles in the flexor and extensor reflex synergies. Graham Brown (1912) showed that alternating activity in antagonist muscles persists after cutting spinal cord, even after deafferentation (cutting the dorsal roots blocks the afferent input). In the 1960s, Anders Lundberg and his colleagues took up the idea of Sherrington about the similarity between flexor reflex and swing phase. They demonstrated in cats that particular neurotransmitters (L-dopa) could induce not only flexor reflexes but also sustained alternating activity in flexors and extensors, presumably by activating the spinal circuitry involved in these reflexes and locomotion (Jankowska et al. 1967). Since the animal is motionless, one refers to this activity as "fictive locomotion". These preparations are of interest since they allow studying the activity of the locomotor central networks in the absence of the sensory inputs related to the ongoing movement. One of Lundbergs' students, Sten Grillner, has extensively used this fictive locomotion, both in cat and in dogfish. Since ventral root afferents are left intact, they additionally curarized the animals and showed that the full pattern of motor output remained essentially identical to that seen in normal locomotion. With his colleagues he could demonstrate that hip position-related afferent activity can affect the output of fictive locomotion (Andersson et al., 1978, 1981, 1983). They entrained the locomotor rhythm in low spinal cats with L-Dopa-induced fictive locomotion using small-amplitude sinusoidal hip movement of a partially denervated hind limb. Around the same period a second type of direct input to the CPG was demonstrated (Duysens and Pearson, 1980; this chapter). It was shown that unloading the ankle extensors in the leg at the end of the stance phase is essential to induce the onset of the swing phase. It is thought that the CPGs consist of functionally grouped interneurons, which activate alpha and gamma motoneurons of flexor or extensor half-centers, which in turn rhythmically activate muscle extensors and flexors (Brown, 1912). Knowledge on the spinal generators for locomotion (CPGs) and their interconnections came from studies on the cat (see Burke et al., 2001), and more recently on the neonatal rat (see Clarac, Chapter 2). Lower vertebrates such as tadpoles and lamprey have also been used extensively. In all these species there are neurons that connect CPGs either at the same or at different segmental levels. Pattern-generating networks are very flexible and can produce different motor behaviors (reviewed in Grillner, 1981; Clarac, 1984, 1990; Pearson, 1993). In cats the different forms of walking (walk, trot, gallop, upslope, down slope, forward and backward) are likely to be generated by the same locomotor CPGs (Miller et al. 1975a,b; English, 1979; Buford & Smith, 1990; Buford et al., 1993; Perell et al., 1993; Carlson-Kuhta et al., 1998). In the lamprey, different directions of swimming are produced by the same central pattern generator (Matsushima and Grillner, 1992). Likewise in humans, both forward and backward gait are controlled in the same way with respect to phase-dependent modulation of reflexes (Duysens et al., 1996).

A first example is the use of so-called "reduced gait" (Van de Crommert et al. 1996; Faist et al. 1999). This consists of limping with one leg on a moving split belt while the other leg is held as stiff as possible on a stationary belt to reduce sensory feedback. It was shown that some form of rhythmic activity persists in the stiff leg, much as was observed in cats walking on one leg (see above; Duysens, 1977). Furthermore, in the human studies, it was shown that the basic features of the phase-dependent modulation of biceps femoris tendon reflexes were preserved in the "stiff" leg provided the other leg kept walking (Van de Crommert et al. 1996; Faist et al. 1999).

In recent years, several authors have used cycling instead of walking as a paradigm to address this issue of CPG coupling in humans. H-reflexes in the ipsilateral stimulated leg were shown to be modulated by passive contralateral movement and this modulation depended on the phase of active ipsilateral movement (reviewed in Brooke et al., 1997, see also Chapter 7). Another group has used split-crank pedaling to study interlimb interactions. This method allows decoupling of the two legs and control of the mechanical loading (for a detailed review, see Ting et al., 1998). Subjects pedaled a modified bicycle ergometer in a two-legged (bilateral) and a one-legged (unilateral) pedaling condition. In the unipedal condition the subject pedaled actively only on one side while the other leg pedaled passively (hence the so-called unipedal condition was also bipedal but the "unipedal" refers to the subjects using only one leg actively). The loading on the leg during unilateral pedaling was adjusted to be nearly identical to the loading experienced by the leg during bilateral pedaling. This was achieved by having a trained second subject pedal along with the experimental subject and exert on the opposite crank (on the side of the "passive" leg) the torque that the subject's contralateral leg generated in bilateral pedaling. The question was whether this would lead to exactly the same muscle coordination pattern, as predicted by totally independent "generators" for both legs, or whether active contralateral cycling facilitated the ipsilateral cycling, as one would predict if the generator actions were coupled.

The latter prediction would yield higher EMG activity in unipedal cycling, as was indeed observed.

It was concluded that unilateral pedaling is not performed with the same muscle coordination utilized in a bipedal condition, even if such coordination would be equally effective in the execution of the unilateral task. These data can be interpreted as indicating that during bipedal pedaling there is a facilitation of the output on one side due to the pedaling of the other side. Further support for such ideas came from another study from the same group (Kautz et al, 2002). They were able to show that pedaling is not completely essential to obtain the contralateral facilitation. Instead rhythmic isometric contractions in a motionless leg can produce the same type of effect. In some of the sessions, the mechanically decoupled contralateral leg was first relaxed and then produced rhythmic isometric force trajectories. With contralateral force production in the extension-to-flexion transition (predominantly by the hamstrings), rectus femoris activity and work output increased in the pedaling leg during its flexion-to-extension transition, which occurs simultaneously with contralateral extension-to-flexion in conventional pedaling. Similarly, with contralateral force production in the other transition (i.e., flexion-to-extension; predominantly by rectus femoris), hamstrings activity and work output increased in the pedaling leg during its extension-to-flexion transition. The rhythmic isometric force generation in the contralateral leg supported the ongoing bifunctional muscle activity and resulting work output in the pedaling leg. Hence these various pieces of evidence support the idea that in humans the rhythmic locomotor output on one side facilitates the rhythm on the other side. This conclusion has important clinical implications since contralateral facilitation can be used in treadmill training of patients with Spinal Cord Injury (SCI) (see also Dietz, Chapter 4).

5.   COORDINATION BETWEEN THE LIMBS DEPENDS ON SENSORY INPUT, IN PARTICULAR AFFERENT INPUT RELATED TO LOAD.

In the previous sections, it was proposed that the locomotor coupling between the various limbs is fairly robust and survives a reduction in sensory input. It would be misleading however to think that this implies that sensory input is not important for interlimb coordination. In the cat the various interlimb coordination patterns are much more stable when normal sensory feedback is allowed during gait (see Rossignol et al., 1993, for review). In this section the emphasis is on interactions of sensory inputs with central structures such as CPGs. The main sensory inputs involved are those derived from receptors monitoring muscle length (muscle spindles) and those involved in measuring load (muscle Golgi Tendon Organs, GTOs and mechanoreceptors from the foot).

5.1   Muscle Spindles

Spindles are very sensitive to changes in muscle length. During gait there are large length changes and therefore one may expect that this afferent input can have a large impact, especially since this input normally results in stretch reflexes in the lengthened muscles and related synergists. A global way to study the impact of sensory input (including those from spindles) is by comparing the CPG locomotor output in fictive and in real locomotion. The EMG pattern seen in various leg muscles during over ground walking can differ considerably from the output of the CPG as measured in fictive locomotion. This discrepancy emerges from the fact that during normal over ground walking, some parts of the muscle activation patterns are not centrally generated but are induced by reflexes, most likely through stretch reflexes (Perret and Cabelguen, 1980, Prochazka, 1979; Wisleder et al. 1990; Van de Crommert et al., 1996). These stretch reflexes originate in muscle spindles and they provide activation of the stretched muscles as well as inhibition of antagonists. In other words, the input from muscle spindles contributes to the EMG amplitude during locomotion.

How important is this contribution? During human gait there is a general suppression of H-reflexes (which depend on spindle afferents; Morin et al., 1982; Capaday and Stein, 1986). Nevertheless there are clear indications from experiments using brief taps on muscle tendons that the burst of activity seen in hamstrings at end swing is partly due to stretch reflexes (Van de Crommert et al., 1996); Faist et al.1999). Similarly, the activity in quadriceps muscles at the onset of the stance phase is likely due to stretch reflexes (Dietz et al., 1990a,b). Further support for such reflex contributions comes from experiments in which muscle vibration is used to activate spindle afferents selectively. It was found that such stimuli cause significant increases in hamstrings and quadriceps activity, while largely leaving unaffected the activity in other muscles such as the triceps surae (Verschueren et al., 2003). Hence, these actions of stretch reflexes are restricted to particular muscles and there are only small effects at a more general level. Interlimb coordination is largely unaffected (Courtine et al., 2001). Nevertheless, in the study of Verschueren et al. (2002) the vibration of biceps femoris induced small but significant phase shifts in the movements of the 2 legs. One study found a small tendency to speed up gait in subjects receiving continuous bilateral vibration of given muscles (Ivanenko 2000), whereas in another study with unilateral continuous vibration there was a tendency to slow down (Verschueren 2003). Overall, the effects of vibration and thus of Ia afferent activity on locomotion, described in the few studies that have been performed in humans, are rather small (Courtine et al. 2001, Ivanenko et al. 2000, Verschueren et al.2002, Verschueren et al. 2003).

5.2   Load receptors

5.2.1   Animals

The second important type of afferent signal is related to loading. Load receptors, such as muscle Golgi Tendon Organs (GTO) in extensors, are important in shaping the motor output during locomotion since they do so in a more global fashion, by affecting the timing of phase transitions, rather than at the level of individual muscles (as do spindle afferents, see above). After gait initiation, GTO afferents deliver load-related information, which acts directly on the CPG to aid the phase transitions during the step cycle, thus providing the possible induction of variations to meet the environmental demands (Duysens et al., 2000; Pearson, 2000). The consequences of load are of particular importance, not only for the structure of the step cycle of the loaded limb but also more broadly for the coordination between the limbs. In fact, the type of interlimb coordination critically depends on load in a variety of species (Duysens et al., 2000). A few examples from the latter review can illustrate this point. The insect hemipterus Nepa rubra normally walks according to a classical alternating tripod coordination pattern (at least 3 of the 6 legs on the support surface at any time), with alternation of the legs of the same segment. When swimming, however, all legs are in phase and the forward movement (equivalent to the swing phase) is significantly longer than the backward movement (equivalent to stance phase) while during walking the stance phase is normally longer than the swing phase. This shows that load, which is important during stance, is crucial in controlling the inter-leg pattern. The fishing spider Dolomedes, which rows on the water surface and walks on land, shows a similar load-dependent switch in interlimb coordination. Other typical examples are found in amphibious animals, which can move equally well in and out of water. The crab, Carcinus maenas, walks laterally over ground with four pairs of legs. In seawater, the crab's weight is several times less than on land and, correspondingly, the cadence under water is faster than on land. In the rock lobster, the removal of load receptors makes the remaining stump to switch from an alternating pattern, such as seen during walking, to an activity profile in phase with the other legs, as is seen during swimming, when load receptors are minimally activated. In the cockroach it has been demonstrated that there are some interneurons (INs), which can induce flight if load-related tarsus information is absent and walking if this information is present.

In mammals, load also affects interlimb coordination during gait. For example, in adult rats walking on a treadmill at moderate speed, the effects of 14 days of hind limb unloading on interlimb coordination were quite extensive (Canu et al., 1998). The unloading was achieved by simply suspending the animal by their tail, taking care that only the forelimbs could touch the ground (see also Canu et al., 2001). Following the hind limb unloading, the general organization of locomotion was preserved since the two hind limbs were always strictly alternating, suggesting a normal basic coupling between the hind limb CPGs. However, the locomotor pattern was found to be very irregular and a lateral instability was observed. There were frequent hyperextensions of the ankle when walking. The EMG analysis showed an increase in step cycle duration and in coactivation duration of the soleus muscles (i.e. in the double stance duration). These changes point to the important role of load in shaping locomotor output. In each leg separately, the periodic loading and unloading can affect the automated switching between phases. In the cat or the rat, unloading the leg at the end of the stance phase is essential to provoke the onset of the swing phase (Duysens and Pearson 1980; Fouad and Pearson 1997, Pearson et al. 1998; Whelan et al 1995). However, contralateral effects occur as well. For example, an unexpected loss of ground contact during the stance phase, resulting in a sudden unloading, induces a premature onset of the next swing phase and prolongs the ensuing contralateral stance phase (Hiebert et al., 1994). The perturbation induces a short latency reflex activation of the extensors on the contralateral side. This allows the animal to support its body and keep upright. Similarly, in rats the changes in load on one side induced effects on the contralateral side as well (Timoszyk et al., 2002). Changes in load were achieved through robotic arms, which applied a perturbation to the lower shanks of the legs during the stance phase. The effects on stepping on a treadmill were studied in adult rats that received complete, midthoracic spinal cord transections as neonates The perturbations resulted in a decrease not only of the stance duration of the ipsilateral but also in the swing duration in the contralateral leg, thereby preserving interlimb coordination in these spinal animals. Overall, these various experiments in mammals indicate that changes in load in one leg can clearly affect gait on the contralateral side as well.

5.2.2   Humans

How do these findings relate to human locomotion? It took quite some time before the idea became accepted that human leg movements during walking can be controlled by spinal CPGs, just as in the cat (Duysens et al., 1996; Duysens and Tax, 1994). In intact humans the automated switching effects are hard to demonstrate in isolation (Stephens and Yang, 1999) but the principles were shown to operate in cases where supraspinal control is limited so that spinal mechanisms prevail. Examples are subjects with a spinal cord injury (Harkema et al., 1997) and infants which all have immature supraspinal pathways (Yang et al 1998). The data on these two groups will be discussed separately.

5.2.2.1   Infants

For infants, the view that CPGs control leg and arm movements has been supported by findings from the laboratory of Yang. In this group, the view was taken that infants are very appropriate subjects because they allow examination of the very basic mechanisms of locomotor control, in the near absence of superimposed cortical control (Yang et al., 1998). Infant stepping is possible, provided weight support is given. Walking is then largely controlled by the spinal and brainstem circuitry. Such infants (ages 2-11 mo) show well organized phase-dependent and location-specific reflex responses to mechanical disturbances during walking in various directions, requiring very different interlimb coordination (Pang and Yang, 2000,2002; Pang et al., 2003; Lam and Yang, 2000) (see Box 1-2).

So far the group of Yang has focused on intra- and interlimb coordination of the legs. It was shown that information about limb loading and hip position are powerful signals for regulating the stepping pattern in human infants (Pang and Yang 2000), in much the same way as that reported in decerebrate and spinal cats (Grillner & Rossignol 1978; Duysens & Pearson 1980; Hiebert et al. 1996). Manual loading during the stance phase of gait delayed the onset of the swing phase (Pang and Yang, 2000), much as was seen in the cat. Inversely, limb unloading was shown to be important for the stance to swing transition, not only for forward walking (Pang and Yang, 2000, 2001) but also for walking sideways or backward (Pang and Yang, 2002). Unloading effects were studied by placing a piece of cardboard under the foot and by pulling the cardboard. As expected on the basis of the animal experiments described above, these changes also affected contralateral stepping but these effects have not been described in detail yet (Yang, personal communication). Loading and unloading during the swing phase was studied as well (Lam et al., 2003). The infants adapted to additional load

---

Box 1-2 Development of interlimb coordination in humans

Interlimb coordination between arm and leg movements in infants during unperturbed locomotion has been described in detail by Ledebt (2000). In infants, basic coordination patterns exist, very similar to those found in the cat. For example, infants crawling on hands-and-knees preferably use a diagonal interlimb pattern (diagonally opposite limbs move synchronously; Freedland and Bertenthal, 1994). When they start to walk upright, infants first hold their arms in a high guard position (arm position at onset step cycle is with the shoulder in exorotation and with the elbow in flexion). This coordination is seen in the first 10 weeks of independent walking (Ledebt, 2000). It is linked to the characteristic instability of this period of walking (Bril and Breniere, 1998; Bril and Ledebt, 1998). Later the infants lower the arms and produce normal arm swings as gait becomes more stable (as evidenced by a decrease in step width). This is usually around 1.5 years of age (Sutherland et al. (1980). Full maturation takes up to six years or longer after the onset of independent walking (Berger, Altenmueller, & Dietz, 1984; Stolze et al., 1997; Sutherland, Olshen, Cooper, & Woo, 1980; Berger, Quintern, & Dietz, 1987; Okamoto & Kumamoto, 1972). During gait of older children and adults, the movements of the upper limbs are often thought to counteract the rotation of the pelvis about the longitudinal body axis. Recently there has been revived interest in alternative theories as proposed earlier by Jackson (1983). It was suggested that upper limb movements during gait are produced by CPG centers within the spinal cord (see Dietz 2002; Dietz, Chapter 4).

---

on the leg by immediately increasing the generation of hip and knee flexor muscle torques. When the weight was removed, some infants exhibited an after-effect (high stepping) in the first step after removal of the weight. These after-effects were shown to be unrelated to a sudden change in cutaneous input with removal of the weight. The after-effect suggests that some infants made a continuous adaptation to their stepping pattern.

5.2.2.2 Adults: whole body loading

In adults, temporary whole body loading increased the amplitude of the extensor burst but had little effect on timing of the step cycle (Stephens and Yang, 1999; Misiaszek et al., 2000). In the study of Misiaszek et al (2000) transient loading of the leg at the end of stancewas applied along the long axis of the leg. This resulted in an unexpectedly complex reaction, involving rapid co-contraction of antagonist pairs of muscles around the ankle and knee (reminiscent of startle responses during gait; see Nieuwenhuijzen et al., 2000). There was also a prolongation of the stance phase. The loading produced much less changes in phase switching than in infants. This is thought to be related to the role of matured supraspinal influences overruling the spinal mechanisms. This is consistent with the finding that intact cats also fail to show large phase-delaying effects with stimuli that have powerful effects in decerebrate cats (Duysens and Stein, 1978). In contrast, the evidence for a role for spinal load reinforcing reflexes in extensors remains robust. Sinkjaer and coworkers, for example found that loading (unloading) of subjects during the stance phase increased (decreased) soleus EMG, suggesting a role of afferent feedback in the generation of soleus EMG activity (Sinkjaer et al., 2000). Similarly, Misiaszek et al (2000) found that loading of the leg at the end of the stance phase enhanced the ongoing extensor-muscle activity. Apparently an increased cortical control can overrule spinal switching automatisms but not the loading and unloading reflexes.

The loading experiments described so far involved transient loading. In adults the effects of continuous general body loading on phase switching have been studied as well (LaFiandra et al., 2003). Subjects walked on a treadmill with a backpack containing 40% of their body mass. It was found that the additional loading resulted in a shorter stride length and higher stride frequency. In addition, there was decreased pelvic rotation during load carriage requiring an increased excursion in the hip joint to compensate. However, the increase in hip excursion was insufficient to fully compensate for the observed decrease in pelvis rotation, presumably requiring an increase in stride frequency during load carriage to maintain a constant walking speed.

What are the effects of loading on the arm movements? In bipedal adults the arms play no role in load bearing. Nevertheless, the arm swings are important in gait. When subjects walk while their arm movements are prevented, there are usually relatively minor effects but subjects report discomfort (Tang et all., 1998). During load carriage, there is reduced transverse plane pelvic rotation, which in turn may reduce the angular momentum of the lower body (LaFiandra et al., 2003). Hence, one might expect that arm swing and counter rotation between the pelvis and thorax may be less essential as a means of reducing the net angular momentum of the body. However, this is not the case and there is clear reciprocal motion between ipsilateral arm and leg swings during load carriage (LaFiandra et al., 2003). These arm swings could be important in counterbalancing the angular momentum of the lower body (for EMG studies see Elftman, 1939; Fernandez-Ballesteros et al., 1965). At preferred walking velocities the upper limbs swing in alternation, with each limb swinging forward and backward in phase with the diagonal lower limb, whereas at lower walking velocities the upper limbs swing in phase at a frequency twice as high as the stride frequency of the legs (Craik, Herman, & Finley, 1976; Webb & Tuttle, 1989). These observations prompted the hypothesis that the stride frequency at which "single swinging" changes to "double swinging" occurs at or slightly below the natural pendular frequency of the upper limbs. Webb et al. (1994) explained the occurrence of observed coordination modes (i.e., "single swinging" or "double swinging") in terms of the biomechanical properties of one of the two components (legs and arms) participating in the coordination. Biomechanical properties are important for coordinative phenomena, but coordinative structures are functional units of organization that are based more on neuronal rather than on biomechanical principles. Experiments with added load on single limbs illustrate this.

5.2.2.3   Adults: Loading a single limb

Although the addition of mass to one leg leads to changes in the duration of swing and double support phase, the cadence remains virtually unchanged (c.f., Bonnard & Pailhous, 1991). Similarly, adding mass to one leg or arm does not change the coordinative walking patterns between arm and leg movements (Donker, 2002). However, such load manipulations do affect the motor output. For example, adding mass to a wrist results in an increase in arm muscle activity and in a decrease of movement amplitude of the perturbed arm. Notably, this manipulation also results in increased movements in the non-perturbed arm Additionally, adding mass to the ankle causes adaptive changes in both arms in that both muscle activity and arm movement increases (Donker, Mulder, Nienhuis, & Duysens, 2002; see Fig. 1-2).

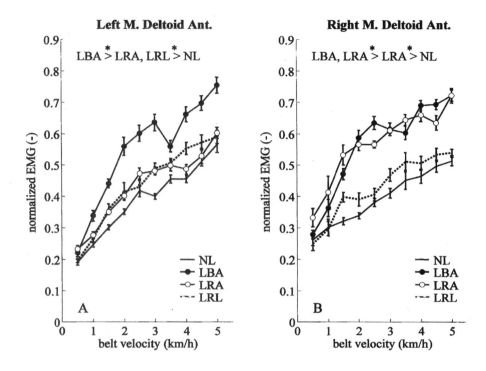

Left M. Deltoid Ant.   **Right M. Deltoid Ant.**

LBA $\overset{*}{>}$ LRA, LRL $\overset{*}{>}$ NL   LBA, LRA $\overset{*}{>}$ LRA $\overset{*}{>}$ NL

normalized EMG (-)   belt velocity (km/h)

A   B

— NL
—●— LBA
—○— LRA
--- LRL

Figure 1-2. Averaged over all participants the averaged amount of left (a) and right (b) M. deltoid anterior are depicted for each experimental condition as a function of walking velocity. In each panel the significant differences ($\alpha = .05$ level) between the four experimental conditions are given, which were analyzed by means of pair-wise comparisons (Bonferroni adjustment for multiple comparisons) based on the estimated marginal means of all walking velocities. (NL: no mass added; LBA: mass added to both arms; LRA: mass added to the right leg; LRL: mass added to the right leg). [Adapted with permission from Donker, Mulder, Nienhuis, & Duysens, 2002].

These observations indicate that during gait the changes in one limb induce adaptations not only in the loaded limb but also in the 3 remaining limbs, presumably to provide (rhythm) constancy of inter-limb coordination. This explanation is in line with other evidence indicating that the coordination between arm and leg movements is quite robust across various types of locomotion (Wannier, Bastiaanse, Colombo, Dietz, 2001). These findings and the findings of studies on the effect of load on cyclical arm and leg movements during non-locomotor tasks (e.g., Baldissera & Cavallari, 2001; Serrien & Swinnen, 1998; Swinnen, Dounskaia, Verschueren, Serrien, & Daelman, 1995) suggest a strong neural coupling between both homologous and non-homologous limbs.

Although more and more research is done on the existence of central pattern generators in humans and the neural coupling between the different limbs (for an overview see Duysens & van de Crommert, 1998), human locomotion has 'traditionally' been approached from a biomechanical point of view. Often the observed stepping patterns have been explained by the physical properties of the legs (e.g., Holt et al., 1991). In a similar vein the motions of the arms during walking have been explained in biomechanical terms. During walking, two different coordinative patterns between arm and leg movements can be observed in humans. At customary walking velocities the arms swing alternately and in phase with the diagonal lower leg, whereas at very low walking speeds the arms often swing in unison and at twice the frequency of the legs (cf. Craik, Herman, & Finley, 1976; Webb & Tuttle, 1989). From a biomechanical perspective, the moving limbs can be viewed as force-driven pendulums and the observed movement patterns can be explained in terms of the physical properties of the pendulums and their muscular forcing (cf. Holt, Hamill, & Andres, 1990; Schot & Decker, 1998). Following this line of reasoning, the different frequency ratios between arms and legs during walking can be explained in terms of limbs preferring to move at or near their natural frequency (cf. Holt, Hamill, & Andres, 1991). The fact that the natural frequency of an arm is higher than that of a leg is thought to account for the double swinging of the arms (i.e., 2:1 frequency coordination) at low walking frequencies and the 1:1 frequency coordination at customary walking frequencies. Specifically, in this context, adding mass to an arm or leg is an interesting manipulation. Adding mass to the distal end of the arm is predicted to decrease the natural frequency of the arm and, as a result, to decrease the difference between the natural frequencies of the arm and leg. Consequently, these biomechanical principles would predict that the 2:1 frequency coordination will disappear in favor of the 1:1 frequency coordination after adding a load to the arm. This hypothesis was tested (Donker et al., 2002) but in only one out of seven participants the 2:1 frequency coordination pattern between arm and leg movements disappeared in favor of 1:1 frequency coordination at low walking frequencies. Hence, a pure biomechanical approach to this coordinative phenomenon is of limited use for its understanding. Instead, it is suggested that although biomechanical properties of the limbs play a role during locomotion, the coordinative structures are functional units of organization that are based foremost on neuronal principles. In this light, a comparison of load manipulation during walking and non-walking tasks is of interest. Whereas during walking the coupling modes between the different limbs and their stability are not basically altered by the load perturbation (Donker, 2002), previous studies on interlimb coordination during non-walking tasks (Jeka & Kelso, 1995; Serrien & Swinnen, 1998) have shown somewhat larger effects.

In agreement with the gait data, it has been shown that homologous limbs (i.e. both legs) are more strongly coupled than limbs with dissimilar eigenfrequencies (non-homologous limbs; i.e., right arm and right leg) (see also Swinnen et al., 1995, 1997). However, the coupling of non-homologous limbs may be stronger in walking than in non-walking conditions. Apparently, the interaction between central control and peripheral feedback is modified due to altered sensory information (e.g., the mechanical interaction with the support surface). Put differently, the nervous system is organized according to a task-related rather than an anatomy-related fashion. As a result, the control structure for a given movement strongly depends on the nature of the task that has to be performed.

These findings have implications for the coordination between arm and leg movements during pathological walking where the task is to maintain a stable walking pattern and to resist gravity. For example, in prosthetic walking the limb load depends on the prosthesis, which introduces asymmetry between the legs and, as a result, asymmetry in the gait pattern. Based on a series of experiments on patients with an above-knee prosthesis, it was found that the gait asymmetry brought about by the prosthesis leads to a decrease in the stability of the coordination between the legs as compared to a control group (Donker & Beek, 2002). In addition, the 2:1 frequency coordination between arm and leg movements that is generally observed in healthy walkers at low walking velocities is basically absent in prosthetic walkers. Given the strong tendency of the human movement system to move its limbs in synchrony (Beek, Peper & Stegeman, 1995; Serrien & Swinnen, 1997; Swinnen, 2002; Swinnen et al., 1991; Von Holst, 1939/1973), it is conceivable that the destabilizing effect of the prosthesis results in a preference for the (most stable) 1:1 frequency coordination between arm and leg movements at all walking velocities. Likely, the in-phase relation between contralateral arm and ipsilateral leg movements at the moments of footfall provide stability and, as such, dominate the coordination during locomotion. Hence, the function of the adjustments in the arm movements as a result of mass perturbation to a single arm or leg (as described earlier) or as a result of leg prosthesis is probably to preserve the specific coordination between arm and leg movements and to enhance the overall dynamic stability of walking.

All in al, the above described capacity of adaptation without influencing the overall coordination is likely to have implications for studies on gait in general and pathological gait in particular. It emphasizes a basic characteristic of motor control, namely the capacity of the motor system to reorganize its output more or less immediately as a result of peripheral disturbances, allowing for maintenance of motor constancy.

5.2.2.4   Adults: Unloading one leg

Unloading of a limb occurs for example when making a turn. Walking along a curve is accompanied by displacement of the body center of mass (COM) toward the inner foot, thereby leading to the loading of the inner and the unloading of the outer leg (Patla et al., 1999; Hollands et al., 2001; Vallis et al., 2001). Such changes induce adaptations in the stance swing proportions but they do not basically affect cadence. The re-distribution of loading during veering is expected to cause changes in the EMG pattern, especially with respect to the activation of extensors during stance and the initiation of flexor activity just prior to swing (Stephens & Yang, 1999; Pang & Yang, 2000; Fouad et al., 2001). This type of questions has been addressed in studies in which turning is simulated by walking on a treadmill with split belts adjusted at different speeds. Dietz and collaborators (Dietz et al., 1994; Jensen et al., 1998; Dietz, Chapter 4) have shown that the stride length is increased on the fast side (equivalent to the outer side) as compared to the slow side (inner side). Alternating gait is achieved using a step cycle duration, which is intermediate to the values found during normal walking with both legs moving on a belt, either at the high or low speed. The duration of the stance phase is reduced on the fast side and prolonged on the slow side.

Brief unloading also affects interlimb coordination. In several species, an unexpected perturbation of a loaded leg often induces a reflex unloading due to a suppression of extensor activity and an activation of flexors (see introduction). In the contralateral leg, the Sherringtonian prediction is that crossed extensor activation prepares the leg for load bearing. Whether this is achieved through CPG circuitry is not known but a link is likely (Jankowska et al, 1967). Can one find this basic crossed reflex pattern in humans as well? Recently unexpected unloading of the stance leg has been studied, for example in experimental "slips" (Tang et al., 1998, Tang et al. 1999). In the slips described by Tang and Woollacott (1999) subjects walked across a movable force plate with its movement timed to midstance. Postural responses were found in the contralateral non-perturbed leg, which seemed to prepare the leg for an earlier and safe landing and thereby assist in foot liftoff of the perturbed leg. In such cases, the contralateral leg aids substantially in step recovery with a latency, which is only a few ms longer than the ipsilateral reactions. Unfortunately, in slips only the first trial is truly unexpected and the first reactions differ basically from those observed in subsequent trials (Marigold and Patla, 2002). In fact, some authors choose to study only the first trial (Brady et al, 2000). Following a slip, the reactions in the ipsilateral leg occur mainly as a flexor synergy with activations of biceps femoris (BF) and tibialis anterior (TA). The interpretation of some authors is that this brings the foot back near the body and decelerates the sliding heel (Redfern et al., 2001). However, the activation pattern is also compatible with a basic flexor reflex synergy, which leads to unloading of the perturbed leg and loading of the contralateral leg. This is reminiscent of the condition following stumbling, where there is a fast activation of contralateral extensors, presumably to prepare this stance leg to accept load (Schillings et al., 2000). Similarly, in responses to unexpected slips, the muscles of the unperturbed limb are recruited at almost the same latency as those for the perturbed limb (Marigold et al., 2003). In summary, these data on single leg unloading show that humans, much like cats, produce a bilateral reaction in which interlimb coordination aims at recovering stability.

6.   SUMMARY

During activities such as walking, running, and swimming, the rhythmical movements exhibit a specific coordination of arms and legs. The choice of this pattern and the organization of the phases of these movements depends on how much the animal as a whole, as well as each limb separately, is loaded. Load receptors in the limbs act upon the neural structures that generate the locomotor output of each limb separately (Central Pattern Generator, CPG). In this way, load can induce changes in the timing of phase switching during gait. However, these changes can only be observed when these basic effects are not overruled by supraspinal influences (e.g. in human infants or in cats with transected brainstem). The CPGs consist of flexor or extensor half-centers, which rhythmically activate muscle extensors and flexors (Brown, 1912). Through coupling of these CPGs, different forms of locomotion such as pacing, walking or galloping can be performed (Collins and Stewart, 1993). This organization is preserved in human infants. The limb CPGs are coupled to provide alternating movements of the limbs during walking. In humans the arm swings are probably also generated by CPGs , as was also suggested by others (Eke-Okoro, 1991;Yang and Pang, 2000). Furthermore, there is a linkage between the CPGs in the lumbar and cervical spinal cord as seen in cats. In man, these various couplings are quite robust. Adding mass to one leg or arm does not change the overall coordinative walking patterns between arm and leg movements (Donker, 2002). However, such load manipulations do affect the motor output. Adding mass to a wrist results in an increase in arm muscle activity and in a decrease of movement amplitude of the perturbed arm. Such manipulations also result in alterations in the non-perturbed limbs, presumably to provide (rhythm) constancy of inter-limb coordination. This explanation is in line with other evidence indicating that the coordination between arm and leg movements is quite robust across various types of locomotion (Wannier, Bastiaanse, Colombo, Dietz, 2001) as it is in non-walking tasks (e.g., Baldissera & Cavallari, 2001; Serrien & Swinnen, 1998; Swinnen, Dounskaia, Verschueren, Serrien, & Daelman, 1995).

REFERENCES

Andersson OS, Forssberg, Grillner S, Lindquist M (1978) Phasic gain control of the transmission in cutaneous reflex pathways to motoneurones during "fictive" locomotion. Brain Res 149: 503-507

Andersson O, Grillner S (1981) Peripheral control of the cat's step cycle I. Phase dependent effects of ramp-movements of the hip during "fictive locomotion". Acta Physiol Scand 113: 89-101

Andersson O, Grillner S (1983) Peripheral control of the cat's step cycle. II. Entrainment of the central pattern generators for locomotion by sinusoidal hip movements during "fictive locomotion". Acta Physiol Scand 118: 229-239

Arshavsky YuI, Gelfand IM, Orlovsky GN, Pavlova GA, Popova LB. (1984) Origin of signals conveyed by the ventral spino-cerebellar tract and spino-reticulo-cerebellar pathway. Exp Brain Res.;54(3):426-31.

Baldissera F, Cavallari P (2001). Neural compensation for mechanical loading of the hand during coupled oscillations of the hand and foot. Exp Brain Res 139: 18-29

Beek PJ, Peper CE, Stegeman DF (1995) Dynamical models of movement coordination. Hum Mov Sci 14: 573-608

Behrman ATP, Cauraugh JH (1998) Verbal instructional sets to normalise the temporal and spatial gait variables in Parkinson's disease. J Neurol Neurosurg Psychiatry 65: 580-582

Berger W, Altenmueller E, Dietz V (1984) Normal and impaired development of children's gait. Human Neurobiology 3: 163-170

Berger W, Quintern J, Dietz V (1987) Afferent and efferent control of stance and gait: developmental changes in children. Electroencephalography & Clinical Neurophysiology 66: 244-52

Bloedel JR, Courville J (1981) Cerebellar afferent systems. In: Brooks VB (ed) Handbook of physiology, sect 1. The nervous system, motor control, vol 2:2. Am Physiol Soc, Bethesda, MD, pp 735–829

Bonnard M, Pailhous J (1991). Intentional compensation for selective loading affecting human gait phases. Journal of Motor Behavior 23: 4-12

Bosco G, Rankin A, Poppele R (1996) Representation of passive hindlimb postures in cat spinocerebellar activity. J Neurophysiol 76: 715–726

Bosco G, Poppele RE (2001) Proprioception from a spinocerebellar perspective. Physiol Rev 81: 539–568

Brady RA, Pavol MJ, Owings TM, Grabiner MD (2000) Foot displacement but not velocity predicts the outcome of a slip induced in young subjects while walking. J Biomech 33 (7): 803-8

Bril B, Breniere Y (1998) Development of postural control of gravity forces in children during the first 5 years of walking. Exp Brain Res 121(3): 255-62

Bril B, Ledebt A, Breniere Y *(1998) The build-up of anticipatory behaviour. An analysis of the development of gait initiation in children. Exp Brain Res 120 (1): 9-17

Brooke JD, Cheng J, Collins DF, Mcilroy WE, Misiaszek JE, Staines, WR (1997) Sensori-sensory afferent conditioning with leg movement: gain control in spinal reflex and ascending paths. Progress in Neurobiology 51: 393-421

Brown TG (1912) The factors in rhythmic activity of the nervous system. Proc R Soc London Ser B84: 278-289

Buford JA, Smith JL (1990) Adaptive control of backward quadrupedal walking. II. Hindlimb muscle synergies. J Neurophysiol 64:756-766.

Buford JA Smith JL (1993) Adaptive control for backward quadrupedal walking. III. Stumbling corrective reactions and cutaneous reflex sensitivity. J Neurophysiol 70:1102-1114.

Burke RE, Degtyarenko AM, Simon ES. (2001) Patterns of locomotor drive to motoneurons and last-order interneurons: clues to the structure of the CPG.J Neurophysiol.;86(1):447-62.

Butt SJB, Lebret JM, Kiehn O (2002) Organization of left-right coordination in the mammalian locomotor network. Brain Research Reviews 40 (1-3): 107-117

Canu MH, Falempin M, Orsal D. (2001) Fictive motor activity in rat after 14 days of hindlimb unloading. Experimental Brain Research 139 (1): 30-38

Canu MH, Falempin M (1998) Effect of hindlimb unloading on interlimb coordination during treadmill
locomotion in the rat. Eur J Appl Physiol Occup Physiol 78(6): 509-15

Capaday C AND Stein RB. (1986) Amplitude modulation of the soleus H-reflex in the human during walking and standing. J. Neurosci. 6: 1308-1313

Carlson-Kuhta P, Trank TV, Smith JL (1998) Forms of forward quadrupedal locomotion. II. A comparison of posture, hindlimb kinematics, and motor patterns for upslope and level walking. J Neurophysiol 79:1687-1701.

Clarac F (1984). Spatial and temporal co-ordination during walking in Crustacea. Trends Neurosci 7:3293-3298.

Clarac F (1990) Introduction to a comparative neurobiological approach to locomotion. In: Gravity, posture and locomotion in primate, edited by F.J. Jouffrouy and C. Niemitz Stock. Florence: II Sedicesimo,.33-44.

Collins JJ, Stewart IN (1993) Coupled nonlinear oscillators and the symmetries of animal gaits. J. Nonlinear Sci 3: 349-392

Courtine G, Pozzo T, Lucas B, Schieppati M. (2001) Continuous, bilateral Achilles' tendon vibration is not detrimental to human walk.Brain Res Bull. 55(1):107-15.

Craik RL, Herman RM, Finley FR (1976) The human solutions for locomotion: Interlimb coordination. In: Herman RM, Grillner S, Stein PS (eds) Neural control of locomotion. Plenum Press, New York, pp 51-63

Cruse H, Warnecke H. (1992) Coordination of the legs of a slow-walking cat. Exp Brain Res. 89(1):147-56.

Delwaide PFC, Richelle C (1977) Effects of postural changes of the upper limb on reflex transmission in the lower limb. Journal of Neurology, Neuosurgery, and Psychiatry 40: 616-621

Dietz V, Discher M, Faist M and Trippel M. (1990a) Amplitude modulation of the human quadriceps tendon jerk reflex during gait. Exp. Brain Res. 82: 211-213.

Dietz V, Faist M AND Pierrot-Deseilligny E. (1990b) Amplitude modulation of the quadriceps H-reflex in the human during the early stance phase of gait. Exp. Brain Res. 79: 221-224.

Dietz V, Zijlstra W, Duysens J (1994) Human neuronal interlimb coordination during splitbelt locomotion. Exp Brain Res 101: 513-520

Dietz V.(2002) Proprioception and locomotor disorders.Nat Rev Neurosci. 3(10):781-90.

Donker SF, Beek, PJ (2002) Interlimb coordination in prosthetic walking: Effects of asymmetry and walking velocity. Acta Psychologica 110: 265-288

Donker SF, Mulder T, Nienhuis B, Duysens J (2002) Adaptations in arm movements for added mass to wrist or ankle during walking. Exp Brain Res 146: 26-32

Donker SF (2002) Flexibility of Human Walking: A study on Interlimb Coordination. Thesis. RUG, Groningen, The Netherlands

Duysens J (1977a) Fluctuations in sensitivity to rhythm resetting effects during the cat's step cycle Brain Res 133: 190-195

Duysens J (1977b) Reflex control locomotion as revealed by stimulation of cutaneous afferents in spontaneously walking premammillary cats. J. Neurophysiol 40: 737-751

Duysens J. and Stein RB (1978) Reflexes induced by nerve stimulation in walking cats with implanted cuff electrodes. Exp. Brain Res. 32: 213-224

Duysens J, Pearson KG (1980) Inhibition of flexor burst generation by loading ankle extensor muscles in walking cats. Brain Res 187: 321-332

Duysens J, Tax T (1994) Interlimb reflexes during gait in cats and humans. In: Swinnen SP, Heuer H, Massion J, Casaer P (eds) Interlimb coordination. Neural, dynamical, and cognitive restraints. Academic Press, San Diego, pp 97-119

Duysens J, Tax AAM, Murrer L, Dietz V (1996) Backward and forward walking use different patterns of phase-dependent modulation of cutaneous reflexes in humans. J Neurophysiol 76: 301-310

Duysens J, Van de Crommert HW (1998) Neural control of locomotion: the central pattern generator from cats to humans. Gait and Posture 7: 131-141

Duysens J, Clarac F, Cruse H (2000) Load-regulating mechanisms in gait and posture: comparative aspects. Physiol Rev 80: 83-133

Eke-Okoro ST (1991) Functional dispositions of the spinal stepping generators and their half-centres. Electromyogr Clin Neurophysiol 31:81-83

Elftman H (1939) The functions of the arms in walking. Human Biology 11:529-535

English AW (1979) Interlimb coordination during stepping in the cat: an electromyographic analysis. J Neurophysiol 42:229-243.

English AW, Lennard PR. (1982) Interlimb coordination during stepping in the cat: in-phase stepping and gait transitions. Brain Res. Aug 12;245(2):353-64.

Faist M, Blahak C, Duysens J, Berger W (1999) Modulation of the biceps femoris tendon jerk reflex during human locomotion. Exp Brain Res 125:265-270

Fedirchuk B, Hultborn H, Bennett DJ, Gorassini M (1995) Dorsal spinocerebellar tract neurons can be influenced by the neural circuitry producing fictive locomotion in the cat. Soc Neurosci Abstr 21:1199

Fernandez-Ballesteros ML, Buchthal F, Rosenfalck P (1965) The pattern of muscular activity during the arm swing of natural walking. Acta Physiol Scand 63:296-310

Fouad K, Pearson KG (1997) Effects of extensor muscle afferents on the timing of locomotor activity during walking in adult rats. Brain Res 749:320-329

Fouad K, Bastiaanse CM, Dietz V (2001) Reflex adaptations during treadmill walking with increased body load. Exp Brain Res 137(2):133-40

Forssberg HS, Grillner J, Halbertsma, Rossignol, S (1980) The locomotion of the low spinal cat: II. Interlimb coordination. Acta Physiol Scand 108:283-295

Freedland RL, Bertenthal BI (1994) Developmental changes in interlimb coordination: transition to hands-and-knees crawling. Psychological Science 5:26-32.

Giuliani CA, Smith JL (1987) Stepping behaviors in chronic spinal cats with one hindlimb deafferented. J. Neurosci 7:2537-2546

Grillner S, Rossignol S (1978) On the initiation of the swing phase of locomotion in chronic spinal cats. Brain Res 146:269-277

Grillner S (1981) Control of locomotion in bipeds, tetrapods, and fish. In: Handbook of Physiology, section 1, The Nervous System, vol. II (Brookhart JM, Mountcastle VB, ed), pp 1179-1236. Bethesda: American Physiological Society.

Grillner S (1985) Neurobiological bases on rhythmic motor acts in vertebrates. Science 228:143-149

Grillner S, Wallen P (1985) Central pattern generators for locomotion, with special reference to vertebrates, Ann. Rev. Neurosci 8:233-261

Harkema SJ, Hurley SL, Patel UK, Requejo PS, Dobkin BK, Edgerton VR (1997) Human lumbosacral spinal cord interprets loading during stepping. J. Neurophysiol 77:797-811

Hiebert GW, Whelan PJ, Prochazka A, Pearson KG (1996) Contribution of hindlimb flexor muscle afferents to the timing of phase transitions in the cat step cycle. J Neurophysiol 81:758-770

Hiebert GW, Gorassini MA, Jiang W, Prochazka A, Pearson KG (1994) Corrective responses to loss of ground support during walking II. Comparison of intact and chronic spinal cats. J. Neurophysiol 71(2): 611-622

Hollands MA, Sorensen KL, Patla AE (2001) Effects of head immobilization on the coordination and control of head and body reorientation and translation during steering. Exp Brain Res 140:223-233

Holt KG, Hamill J, Andres RO (1990) The force-driven harmonic oscillator as a model for human locomotion. Hum Mov Sci 9:55-68

Holt KG, Hamill J, Andres RO (1991) Predicting the minimal energy costs of human walking. Medicine and science in sports and exercise 23:491-498

Ivanenko YP, Grasso R, Lacquaniti F. (2000) Influence of leg muscle vibration on human walking. J Neurophysiol. 84(4):1737-47.

Jackson KM, Joseph J, Wyard SJ (1983) Upper limbs during human walking part two: function. Electromyogr Clin Neurophysiol 23:435-446

Jankowska E, Jukes MGM, Lund S, Lundberg A (1967) The effects of DOPA on the spinal cord. 5 Reciprocal organization of pathways transmitting excitatory action to alpha motoneuroned of flexors and extensors. Acta Physiol Scand 70:369-388

Jankowska E, Hammar I, Slawinska U, Maleszak K, Edgley SA (2003) Neuronal basis of crossed actions from the reticular formation on feline hindlimb motoneurons. J Neurosci 23(5):1867-78

Jeka JJ, Kelso JAS (1995) Manipulating symmetry in the coordination dynamics of human movement. Journal of Experimental Psychology: Human Perception and Performance 21:360-374

Jensen L, Prokop T, Dietz V (1998) Adaptional effects during human split-belt walking: influence of afferent input. Exp Brain Res 118:126-130

Kato M (1994) Interlimb coordination during locomotor activities. Spinal-intact cats and chronic cats with horizontal and longitudinal seperation of the spinal cord. In: Swinnen SP, Heuer H, Massion J, Casaer P (eds) Interlimb coordination. Neural, dynamical, and cognitive restraints. Academic Press, San Diego, pp 76-94

Kautz SA, Brown DA, Van der Loos HF, Zajac FE (2002) Mutability of bifunctional thigh muscle activity in pedaling due to contralateral leg force generation. J Neurophysiol 88(3):1308-17

La Fiandra M, Wagenaar RC, Holt KG, Obusek JP (2003) How do load carriage and walking speed inlfuence trunk coordination and stride parameters? Journal of Biomechanics 26:87-95

Lam T, Wolstenholme C, Yang JF.How do infants adapt to loading of the limb during the swing phase of stepping? J Neurophysiol. 2003 Apr;89(4):1920-8.

Lam T, Yang JF (2000) Could different directions of infant stepping be controlled by the same locomotor central pattern generator? J Neurophysiol 83:2814-2824

Ledebt A (2000) Changes in arm posture during the early acquisition of walking. Infant Behavior & Development 23:79-89

Marigold DS, Patla AE (2002) Strategies for dynamic stability during locomotion on a slippery surface: effects of prior experience and knowledge. J. Neurophysiol 88 (1):339-353

Marigold DS, Bethune AJ, Patla AE. (2003) Role of the unperturbed limb and arms in the reactive recovery response to an unexpected slip during locomotion.J Neurophysiol. 89(4):1727-37

Matsushima T, Grillner S (1992) Neural mechanisms of intersegmental coordination in lamprey: local excitability changes modify the phase coupling along the spinal cord. J Neurophysiol 67:373-388.

Matsuyama K, Drew T. (2000) Vestibulospinal and reticulospinal neuronal activity during locomotion in the intact cat. I. Walking on a level surface. J.Neurophysiol. 84(5):2237-56.

Miller S, Van Der Burg J, Van der Meché FGA (1975a). Locomotion in the cat: basic programmes of movement. Brain Res 91:239-253.

Miller S, Van Der Burg J, Van der Meché (1975b) Coordination of movements of the hindlimbs and forelimbs in different forms of locomotion in normal and decerebrate cats. Brain Res 91:217-237.

Miller S, Van der Meché FGA (1976) Coordinated stepping of all four limbs in the high sinal cat. Brain Res 109: 395-398

Misiaszek JE, Stephens MJ, Yang JF, Pearson KG.(2000) Early corrective reactions of the leg to perturbations at the torso during walking in humans. Exp Brain Res.131(4):511-23.

Morin C, Katz R, Mazieres L, and Pierrot-Deseilligny E. (1982) Comparison of soleus H reflex facilitation at the onset of soleus contractions produced voluntarily and during the stance phase of human gait. Neurosci. Lett. 33: 47-53.

Nieuwenhuijzen PHJA, Schillings AM, Galen G P van and Duysens J. (2000) Modulation of the Startle Response During Human Gait. J. Neurophysiol. 84, pp 65-74

Okamoto T, Kumamoto M (1972) Electromyographic study of the learning process of walking in infants. Electromyography and Clinical Neurophysiology 12:149-158

Orsal D, Cabelguen JM, Perret C (1990) Interlimb coordination during fictive locomotion in the thalamic cat. Exp Brain Res 82(3):536-46

Pang MYC, Yang JF (2000) The initiation of the swing phase in human infant stepping; the importance of hip position and leg loading. J Physiol 528:389-404

Pang MYC, Yang JF (2002) Sensory gating for the initiation of the swing phase in different directions of human infant stepping. J of Neurosci 22(13):5734-40

Pang MYC, Yang JF (2001) Interlimb coordination in human infant stepping. J Physiol (Lond) 533:617-625

Pang MY, Lam T, Yang JF. (2003) Infants adapt their stepping to repeated trip-inducing stimuli. J Neurophysiol. (in press)

Patla AE, Adkin A, Ballard T (1999) Online steering: coordination and control of body center of mass, head and body reorientation. Exp Brain Res 129:629-634

Pearson KG (1993) Common principles of motor control in vertebrates and invertebrates. Annu Rev Neurosci 16:265-297.

Pearson KG, Misiaszek JE, Fouad K (1998) Enhancement and resetting of locomotor activity by muscle afferents. Ann N Y Acad Sci 860:203-15

Pearson KG. (2000) Neural adaptation in the generation of rhythmic behavior. Annu Rev Physiol. 62:723-53.

Perell KI, Gregor RJ, Buford JA, Smith JL (1993) Adaptive control for backward quadrupedal walking. IV. Hindlimb kinetics during stance and swing. J Neurophysiol 70:2226-2240.

Perret C, Cabelguen, JM (1980) Main characteristics of the hindlimb locomotor cycle in the decorticate cat with special reference to bifunctional muscles. Brain Res 187:333-352

Poppele RE, Rankin A, Eian J (2003) Dorsal spinocerebellar tract neurons respond to contralateral limb stepping Exp Brain Res 149:361–370

Prochazka A, Stephens, JA, Wand P (1979) Muscle spindle discharge in normal and obstructed movements. J Physiol 287:57-66

Redfern MS, Cham R, Gielo-Perczak K, Gronqvist R, Hirvonen M, Lanshammar H, Marpet M, Pai CY, Powers C (2001) Biomechanics of slips. Ergonomics 44(13):1138-66

Rossignol SP, Saltiel MC, Perreault T, Drew K, Pearson KG, Belanger, M (1993) Intralimb and interlimb coordination in the cat during real and fictive rhythmic motor programs. Semin Neurosci 5:67-75

Rossignol S (1996) Neural control of stereotypic limb movement. Handbook of physiology. Exercise: regulation and integration of multiple systems. Am Physiol Sco, Bethesda, MD, pp 173-216

Scheibel ME, Scheibel AB (1969) Terminal patterns in cat spinal cord. 3. Primary afferent collaterals. Brain Res 13:417–443

Schillings AM, Wezel BMH van, Mulder Th and Duysens J. (2000) Muscular responses and movement strategies during stumbling over obstacles. J. Neurophysiol. 83, pp 2093-2102

Schot PK, Decker MJ (1998) The force driven harmonic oscillator model accurately predicts the preferred stride frequency for backward walking. Hum Mov Sci 17:67-76

Serrien DJ, Swinnen SP (1997) Coordination constraints induced by effector combination under isofrequency and multifrequency conditions. Journal of Experimental Psychology: Human Perception and Performance 23:1493-1510

Serrien DJ, Swinnen SP (1998) Load compensation during homologous and non-homologous coordination. Exp Brain Res 121:223-229

Sherrington CS (1910) Flexorreflex of the limb, crossed extension reflex, and reflex stepping and standing. J Physiol 40:28-121

Sinkjaer T, Andersen JB, Ladouceur M, Christensen LO, Nielsen JB. (2000) Major role for sensory feedback in soleus EMG activity in the stance phase of walking in man. J Physiol. 523 vol 3:817-27.

Stein PSG, Smith JL (1997) Neural and biomechanical control strategies for different forms of vertebrae hindlimb moto tasks. In: Stein PSG, Grillner S, Selverston AI, Stuart DG (eds) Neurons, networks and motor behavior, MIT press, Cambridge, MA

Stein PSG, Victor JC, Field EC and Currie SN (1995) Bilateral control of hindlimb scratching in the spinal turtle: contralateral spinal circuitry contributes to the normal ipsilateral motor pattern of fictive rostral scratching. J. Neurosci 15:4343-4355

Stephens MJ, Yang JF (1999) Loading during the stance phase of walking in humans increases the extensor EMG amplitude but does not change the duration of the step cycle. Exp Brain Res 124:363-370

Stolze H, Kuhtz-Buschbeck JP, Mondwurf C, Boczek-Funcke A, Johnk K, Deuschl G, Illert M (1997) Gait analysis during treadmill and overground locomotion in children and adults. Electromyography and motor control. Electroencephalography and Clinical Neurophysiology 105:490-497

Sutherland DH, Olshen R, Cooper L, Woo SL (1980) The development of mature gait. Journal of Bone and Joint Surgery 62:336-353

Swinnen SP. Intermanual coordination: from behavioural principles to neural-network interactions. Nat Rev Neurosci. 2002 May;3(5):348-59.

Swinnen SP, Dounskaia N, Verschueren S, Serrien DJ, Daelman A (1995) Relative phase destabilization during interlimb coordination: the disruptive role of kinesthetic afferences induced by passive movement. Exp Brain Res 105:439-454

Swinnen SP, Dounskaia N, Walter CB Serrien DJ (1997) Preferred and induced coordination modes during the acquisition of bimanual movements with a 2:1 frequency ratio. J Exp Psychol [Hum. Percept.] e 23, 1087-1110

Swinnen SP, Heuer H, Massion J, Casaer P (eds) (1994) Interlimb coordination. Neural, dynamical, and cognitive restraints. Academic Press, San Diego

Swinnen SP, Young DE, Walter CB, Serrien DJ (1991) Control of asymmetrical bimanual movements. Exp Brain Res 85:163-173

Tang P, Woollacott MH (1999) J Gerontol A Biol Sci Med Sci (2):M89-102!!!

Tang P, Woollacott MH, Chong RKY (1998) Contol of reactive balance adjustments in perturbed human walking: roles of proximal and distal postural muscle activity. Exp Brain Res 119 (2):141-52

Timoszyk WK, de Leon RD, London N, Roy RR, Edgerton VR, Reinkensmeyer DJ (2002) The rat lumbosacral spinal cord adapts to robotic loading applied during stance. J Neurophysiol 88:3108-3117

Ting LH, Raasch CC, Brown DA, Kautz SA, Zajac FE (1998) Sensorimotor state of the contralateral leg affects ipsilateral muscle coordination of pedaling. J Neurophysiol 80(3):1341-51

Ting LH., Kautz SA, Brown DA and Zajec FE (2000). Contralateral movement and extensor force generation after flexion phase muscle coordiation in pedaling. J. Neurophysiol 83, 3351-3365.

Vallis LA, Patla AE, Adkin AL (2001) Control of steering in the presence of unexpected head yaw movements. Influence on sequencing of subtasks. Exp Brain Res 138:128-134

Van de Crommert HWAA, Faist M, Berger W, Duysens J (1996) Biceps femoris tendon jerk reflexes are enhanced at the end of the swing phase. Brain Res 734: 341-344

Verschueren SMP, Swinnen SP, Desloovere K and Duysens J (2002) The effects of tendon vibration on the spatiotemporal characteristics of human locomotion. Exp. Brain Res., 143, 231-239.

Verschueren SMP, Swinnen SP, Desloovere K and Duysens J (2003) Vibration-induced changes in EMG during human locomotion. J. Neurophysiology, 89, 1299-1307.

Viala DVC (1978) Evidence for distinct spinal locomotion generators supplying respectively fore- and hindlimbs in the rabbit. Brain Res 155:182-186

Visintin M, Barbeau, H (1994) The effects of parallel bars, body weight support and speed on the modulation of the locomotor pattern of spastic paretic gait. A preliminary communication. Paraplegia, 32(8):540-553

Von Holst E (1939/1973) Relative coordination as a phenomenon and as a method of analysis of central nervous function. In: Martin R (ed) The collected papers of Erich von Holst. Vol. 1. The behavioral physiology of animal and man. University of Miami Press, Coral Gables, Fl.

Walmsley B, Nicol MJ (1990) Location and morphology of dorsal spinocerebellar tract neurons that receive monosynaptic afferent input from ankle extensor muscles in cat hindlimb. J Neurophysiol 63:286–293

Wannier T, Bastiaanse C, Colombo G, Dietz V (2001) Arm to leg coordination in humans during walking, creeping and swimming activities. Exp Brain Res 141:375-379

Webb D, Tuttle RH (1989) The effects of stride frequency on the motion of the upper limbs in human walking. American Journal of Physiological Anthropology 78:321-322

Webb D, Tuttle RH, Baksh M (1994) Pendular activity of human upper limbs during slow and normal walking. American Journal of Physical Anthropology 93:477-89

Whelan PJ, Hiebert GW, Pearson KG (1995) Stimulation of the group I extensor afferents prolongs the stance phase in walking cats. Exp Brain Res 103:20-30

Wisleder D, Zernicke RF, Smith JL (1990) Speed-related changes in hindlimb intersegmental dynamics during the swing phase of cat locomotion. Exp Brain Res 79:651-660

Yanagihara D, Udo M, Kondo I, Yoshida T. (1993) A new learning paradigm: adaptive changes in interlimb coordination during perturbed locomotion in decerebrate cats. Neurosci Res. Dec;18(3):241-4.

Yang JF, Pang MY (2000) The initiation of the swing phase in human infant stepping: importance of hip position and leg loading.J Physiol 15(528):389-404

Yang JF, Stephens MJ, Vishram R (1998) Transient disturbances to one limb produce coordinated, bilateral responses during infant stepping. J. Neurophysiol 79:2329-2337

Zehr EP, Collins DF, Chua R (2001a) Human interlimb reflexes evoked by electrical stimulation of cutaneous nerves innervating the hand and foot. Exp Brain Res 140:495-504

Zehr EP, Collins DF, Frigon A, Hoogenboom N (2003) Neural control of rhythmic human arm movement: phase-dependence and task-modulation of Hoffmann reflexes. J Neurophysiol 1:12-21

Zehr EP, Hesketh KL, Chua R (2001b) Differential regulation of cutaneous and H-reflexes during leg cycling in humans. J Neurophysiol 85: 1178-1185

Zehr EP, Kido A Neural control of rhythmic, cyclical human arm movement: task dependency, nerve specificity and phase modulation of cutaneous reflexes. J Physiol (2001) 537:1033-1045

Chapter 2

DEVELOPMENT OF INTERLIMB COORDINATION IN THE NEONATAL RAT

Francois Clarac, Edouard Pearlstein, Jean François Pflieger, Laurent Vinay

INPC, CNRS,Marseille, France

Abstract: Locomotion is a type of motor behaviour that is produced by spinal neuronal networks associated with the different limbs. The rat in which development lasts three weeks *in utero* and continues during the first three post-natal weeks appears to be a very attractive model for the study of interlimb coordination maturational mechanisms. An early expression of rhythmic locomotor activity is observed at birth *in vitro*. Neuroactive substances like excitatory amino acid or amines (5-HT...) applied on the entire brainstem/spinal cord preparation induce a unique rhythmical activity over the cervical and the lumbar generators. This strict coordination results from a mutual interaction between generators since both bursting frequencies are altered significantly after functional isolation, with a slowing down of the isolated cervical network and a significant acceleration of the lumbar burst generator. However the rat does not walk at birth due to an absence of postural regulations. During the first two weeks, the neonatal rat is able to swim or to produce "air stepping" with an ipsilateral and a contralateral alternation between the different limbs. Adult walking occurs only during the third postnatal week. At least two mechanisms are able to control this early interlimb coordination. Numerous studies showed that 5-HT has ubiquitous topic and trophic effects on the early development of neurons and synapses. Studies on "PCPA" treated and "spinal" neonatal rats demonstrated strong deficits in locomotor movements. The second mechanism concerns the sensory afferents after birth; proprioceptive peripheral loops feed the central network and adapt its activation in a coordinated manner.

Key words: Central Pattern Generator (CPG), gait, *in vitro* preparation, proprioceptive control, serotonin (5-HT).

1.   INTRODUCTION

It is widely accepted that a type of motor behaviour, such as locomotion, is produced by spinal neuronal networks ("Central Pattern Generators", CPGs) that are modulated by central and peripheral loops (Orlovsky et al., 1999). All the elementary networks that are associated with the different limbs must be coordinated with each other to produce a given behaviour. In quadrupeds, limb coordination depends on the gait (walk, trot or gallop), which is closely related to locomotor speed. During ontogeny, which results from a complex combination of genetic and epigenetic factors (Clarac et al., 1998), the locomotor behaviour depends on the maturation of both the neural networks and the musculo-skeletal framework.

In mammals, among the different species that have been considered to investigate ontogeny of locomotion (cat, rabbit...), the rat appears to be the most attractive model since it has been extensively studied not only electrophysiologically on *in vitro* preparations but also behaviourally on intact young pups. The development lasts three weeks *in utero* (embryonic days 0 to 21: E0 - E21) and continues during the first three post-natal weeks (post-natal days 0 to 21: P0 - P21); after 3 weeks rats are considered as adults even though several structures are still not fully developed. In rat as in human, the central nervous system (CNS) is very immature at birth compared to other mammals such as ungulates (Clarac et al., 2001).

This chapter is divided into three parts. The first one will summarize the main data obtained recently with the brainstem/spinal cord *in vitro* preparations. Considering the cervical and the lumbar networks, coordinations between contralateral as well as ipsilateral activities will be presented (see also Dietz, Chapter 4). In the second part we will describe different forms of immature locomotion and their related interlimb coordinations during the first post-natal week. The last part will focus on some of the central and peripheral nervous mechanisms, which may account for the gradual maturation and adaptation of the locomotor behaviour to the surrounding medium.

2.   ONTOGENESIS OF THE FOUR CENTRAL PATTERN GENERATORS FOR LOCOMOTION AND THEIR RELATIONSHIPS

An early expression of rhythmic locomotor activities has been observed in totally dissected preparations consisting of the CNS from the brainstem to the sacral spinal cord. The presence of these activities in parallel with the absence of myelin at this age have made the neonate rat one of the most powerful electrophysiological *in vitro* preparations for studying rhythmic behaviours such as respiration and locomotion (Smith and Feldman, 1987 ; Cazalets et al., 1992 ; Kjaerulff and Kiehn, 1997).

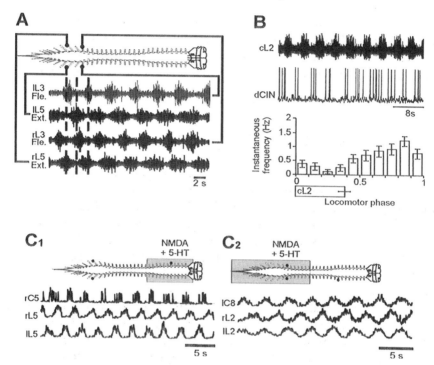

Figure 2-1. In vitro study of fictive locomotion on the neonate rat brainstem – spinal cord preparation. A: In vitro neonate rat brainstem/spinal cord preparation. The brainstem and spinal cord were removed from the animal and placed in a recording chamber. Recording electrodes were placed on the 3rd lumbar ventral roots (L3), that mediate flexor outputs, and on L5 ones that mediate extensor outputs. Bath application of excitatory amino acid agonists (such as NMA) and/or 5-HT induced a patterned rhythmical activity (with both left/right and ipsi- L3/L5 alternation) called fictive locomotion. B: Role of dCIN in interlimb coordination. Here is presented the rhythmic activity of a dCIN that fired out-of-phase with the contralateral L2 (cL2) during pharmacologically induced fictive locomotion. The phase relationships were more visible on the graph that presents the spontaneous firing frequency vs. the cL2 locomotor phase: one can see that the highest frequency was reached during the cL2 silence phase (horizontal bar representing cL2 firing) (adapted from Butt et al., 2002). C: Interactions between cervicothoracic and lumbar CPGs. Simultaneous recordings were made at the cervical and lumbar levels. Bath application of NMDA and 5-HT (gray area) either at the thoracolumbar level (C1) or the cervicothoracic level (C2) induced fictive locomotion like activity at both levels. However the rhythm induced by rostral application was slower than that obtained by caudal application (adapted from Ballion et al., 2001).

Bath application of various pharmacological tools enables to analyse the activities of the neural networks in the foetus and the young pup. In such conditions four CPGs have been identified, two at the cervical level corresponding to the forelimbs and two at the lumbar level corresponding to the hindlimbs.

Although a huge amount of data have been collected, in particular on the lumbar region, little is known about the network structure of these mammalian CPGs. For the cervical region it is only recently that Ballion et al. (2001) have described the anterior rhythms and their relationships with the lumbar centres.

2.1   Lumbar CPGs and left-right coordination modes

The activity recorded from ventral roots in response to bath application of several neurotransmitters or receptor agonists consists of right and left alternating bursts of action potentials (Fig. 2-1). The period ranges from 1 to 5 s (or 0.2 to 1 Hz) depending on the concentration of bath-applied transmitters. In the isolated spinal cord of neonatal rats, locomotor-like activity is induced by activating N-methyl-D-aspartate (NMDA) receptors; excitatory amino-acid action also initiates locomotion via the activation of kainate receptors.

Serotonin (5-HT) alone is also able to activate the locomotor network in a dose-dependent manner. Bath application of a mixture of 5-HT and NMDA is a powerful means to elicit long-lasting (several hours) locomotor-like activity. Other neurotransmitters such as acetylcholine and dopamine have also been demonstrated to activate the locomotor network. The ipsilateral lumbar ventral roots L3 and L5 exhibit out-of-phase bursting activity, suggesting that both flexor and extensor units are active during one step cycle (Fig. 2-1A). Kiehn and Kjaerulff (1996) subsequently identified the burst in L2 as the flexor phase and the burst in L5 as the extensor phase of the cycle, and they also demonstrated that complex motor patterns are generated *in vitro*. Most of the neurochemical agents used, with the exception of acetylcholine, and including the excitatory amino acids, elicit a motor pattern compatible with locomotor activity in terms of period and phase relationship (Nicolopoulos-Stourmaras and Iles, 1984). Not all lumbar segments have the same capacity in inducing rhythmic activity; it seems to be mainly restricted to the rostral lumbar segments (Cazalets et al., 1995; Kjaerulff and Kiehn, 1997). We also know that the alternation between left and right motoneuronal activities is mostly determined by reciprocal inhibition of the networks on the two sides of the cord. This inhibition is mediated mainly by glycinergic commissural interneurons (CINs). Four different populations of CINs have been described on the basis of morphologies and axonal projections: 1 intrasegmental and 3 intersegmental, either descending or ascending or bifurcating INs (Eide et al., 1999). Butt et al. (2002) have investigated the role of the descending CIN (dCIN): their cell bodies are located in the anterior segments (lumbar segment 2 and 3 - L2 and L3) and their axons cross the midline toward more caudal segments (L4-L5). Most of these dCINs are active in one of the different phases of the locomotor cycle (Fig. 2-1B). It is hypothesized that they connect contralaterally the different pools of motoneurons and that they are essential in the rostrocaudal coordination between flexors and extensors. They are responsible for the strict alternation between the flexor phase on one side and the extensor phase on the contralateral side.

Iizuka et al. (1998) have demonstrated that seven days prior to birth (E 14.5) it is possible to record in the lumbar spinal cord synchronous rhythmic activities on both sides but at a frequency ten times lower than that observed at birth. The left/right alternation appears at E18.5; however rhythmic bursts still occur synchronously in flexors and extensors. Nakajima et al. (2002) showed that even at E14.5 CINs are present which are responsible for the coordination. During these very early developmental stages the connections are excitatory and under the control of $GABA_A$ receptors. Changes in the role of GABAergic synaptic transmission from excitation to inhibition during foetal development, is due to the shift of the reversal potential for chloride currents (E_{Cl}) towards further hyperpolarized values (Ben Ari, 2002). This switches the temporal pattern of the pharmacologically-induced rhythmic locomotor activity from left-right synchrony to alternation. Moreover at later foetal stages (after E18.5), near birth, glycine receptors take this role over whereas $GABA_A$ receptors do not contribute significantly to the left-right coordination. At birth, flexor and extensor bursts of a same segment cycle become out of phase. Each CPG becomes completely functional. Lumbar CPGs appear then structurally organised quite early during the foetal period with potent left-right connections.

2.2 Cervical CPGs and coordination with the lumbar networks

The locomotor-like activity recorded in the cervical region can be obtained with the same pharmacological agents as those used to elicit the lumbar rhythm (excitatory amino-acids, 5-HT...). It can be recorded between the cervical roots (C5 to C8) and the first thoracic root (T1). The C5 and C6 ventral roots seem to control the forelimb flexor muscles and C7, C8 and T1 are more related to the forelimb extensor muscles (Ballion et al., 2001). The frequency of the rhythm ranges from 0.2 to 0.3 Hz.

If a neuroactive substance is applied over the entire brainstem/spinal cord preparation, a unique rhythmic activity is recorded from all the ventral roots. This demonstrates that the cervical and the lumbar locomotor generators operate at the same frequency. In that case a detailed analysis shows that forelimb extensor bursts are in phase with hindlimb flexor bursts (Ballion et al. 2001) but sometimes the relationships are opposite. This strict coordination results from a mutual interaction between generators as both cervical and lumbar bursting frequencies are altered significantly after functional isolation. The authors observed a slowing down of the isolated cervical network rhythm and a significant acceleration of the lumbar burst generator. According to various studies on rhythm generator coupling, in general the faster oscillator entrains the slower one. Here the coordination is mutual, there is a double gradient (Fig. 2-1C1, C2); the faster lumbar rhythm accelerates the slower cervical one; but at the same time, the slower reduces the frequency of the faster toward a common intermediate frequency (Von Holst, 1943). Thus, propriospinal pathways that link the rostral and caudal locomotor generators together in the adult appear to be already functional in the neonatal rat and may account for coordinated forelimb-hindlimb locomotion (see also Dietz, Chapter 4).

2.3   Four different rhythms in the neonate brainstem/spinal cord preparation

In this *in vitro* preparation, the situation is even more complex, 4 different rhythmic activities can be recorded: the two locomotor rhythms already presented but also in the rostral region, the respiratory rhythm and in the sacral region, the tail beating. The respiratory rhythm, limited to the cervical region, involves strictly synchronous bursting in left and right ventral roots and thereby can be easily distinguished from the locomotor activation (respiratory frequency = 0.1 Hz). It disappears when the brainstem is removed (Hilaire and Duron, 1999). At first sight it does not seem to be connected with the locomotor rhythm. However, Morin and Viala (2002) have demonstrated that a rhythmic stimulation of a lumbar dorsal root can entrain not only the lumbar locomotor CPGs as already shown by Sqalli-Houssaini et al. (1993) but also the medullary respiratory rhythm. This suggests that low-threshold lumbar afferents project to the medullary area where they can entrain the respiratory rhythm generator.

A sacral isolated rhythm can also be recorded from the different sacral ventral roots (S1 to S4). It is often masked by the lumbar network rhythm; when it is activated, it improves its proper frequency and the same rhythm is observed on the lumbar and sacral ventral roots (Cazalets and Bertrand, 2000). A sacral rhythm can be induced alone by mechanical stimulation of the tail or by electrical stimulation of the S2 ventral root (Lev-Tov et al., 2000).

In conclusion, the CNS at birth can produce a locomotor rhythm composed of different CPG activities coordinated by different potent connections. However, it has been demonstrated that the CNS is still very immature at this age, in particular at the levels of motoneurons (Vinay et al., 2000a), descending pathways (Brocard et al., 1999b) and sensory afferents (Vinay et al., 2000b). Behaviourally, these animals have very restricted movements during the first week due to an absence of postural regulation.

3. INTERLIMB COORDINATION DURING POSTURAL AND LOCOMOTOR ACTIVITY

Just at birth (P0) the rat is able to lift its head and to make some horizontal head movements. Blind and mute the neonate is sensitive to smell and to heat, senses which are useful in the nest. The first locomotor behaviour observed is an alternating crawling when the belly supports most of the animal weight and when the forelegs are mostly used for displacement (Gramsbergen, 2001).

The real walking occurs much later during the second postnatal week when the animal is able to stand up on its 4 legs. It starts by bouts of sequences that are still not well coordinated. Eyes open around P13, then the animal explores its environment walking in a plantigrade fashion. From P15, in one or two days, the immature pattern of locomotion is replaced by a faster and fluent rhythm. The animal is digitigrade, and at this age, is able to stand on its hind paws (rearing). It has been possible to analyse different kinds of locomotor behaviours during the first 3 postnatal weeks:

(i) a very early walking can be obtained if the animal, away from its litter, is stimulated by the odour of its mother. Jamon and Clarac (1998) have studied such a situation between P3 to P10 and characterized the evolution of the different limb movements.

(ii) Postural reactions need some delay to be efficient in walking. However in some particular circumstances this constraint can be greatly reduced and rhythmic limb movements obtained:

- in a fixed pup induced by an olfactory motivational device
- during a swimming behaviour
- during air stepping when the animal is lifted above the ground

3.1   Leg coordination during "early walking"

If a tube filled with nest material is moved ahead of a young rat (at least at P3), away from the litter, the animal is able to follow the stimulus. The mother's odour motivates the pup so strongly that it is able to stand up and walk (Jamon and Clarac, 1998). The movement elicited in this study involved longer stance duration and a lower step frequency than those required for trotting. The gait so induced resembled quadruped locomotion and evolved with age; the swing phase duration decreased slowly with age whereas the stance phase duration varied discontinuously with steep decreases at P5 and P8 and intervening plateaus. The ipsilateral coupling shifted progressively from 220° to 260°. This early walking has some parameters, like speed and period duration, consistent with the walking performances previously described in older rats. However, some fundamental changes such as the disappearance of hindlimb hyperextension, digitigrade walking and the stabilization of swing duration occur only after P15.

3.2   Leg coordination in a behavioural condition with reduced postural constraints

If the pup is placed such that postural requirements are reduced, alternating movements can be obtained, which in general correspond to a stereotyped trotting. Similarly, Jensen et al. (1994) demonstrated that 3-month-old infants lying on their back are able to produce alternating leg movements that result from a CPG activity.

- With a pup fixed with a belt, forelimbs and hindlimbs hanging on each side, Fady et al. (1998) obtained alternating leg movements when mother's odour was presented in a tube in front of the nose of the animal (Fig. 2-2A). In such a situation, long sequences of 30 to 50 cycles of leg beating can be obtained. Rhythmic leg movement cease as soon as the tube is displaced away from the pup's nose. Such a study has been done from P0 to P4. The gait at P0 is consistent with the diagonal gait observed at P5 in swimming and in air stepping even though, the ipsilateral phase is 0.55 and not 0.5. It corresponds to the same value as in "early walking" observed at P3-4. Before

P5, the pup does not use the classical diagonal gait but a coordination characteristic of immature rats. At birth all legs beat at the same cyclic speed (less than 1 s period). Relations between forelimbs and hindlimbs become particular since the forelimbs progressively step faster. At P4 the period is around 600 ms and 700 ms for the forelimbs and hindlimbs, respectively. The pup can correct its phase shift trying to maintain an absolute 1:1 coordination. The coordination shows a tendency for the legs to adjust their step period in response to phase shift (Von Holst, 1943). Double steps exist between ipsi- and contralateral pairs of legs. In most cases, both legs modulate reciprocally their own step period in order to restore an alternating gait as soon as possible. Double steps exist also at P0 when both pairs of limb have a similar rhythm. In that case, it is supposed to compensate for the phase shifts caused by weak relationships (Fig. 2-2A2).

Both air stepping (with the rat suspended; McEwen et al., 1997b; Stehouwer and Van Hartesveldt, 2000) and swimming (when the rat is immersed in water, Cazalets et al., 1990), induce leg beating with alternation at a frequency similar to that obtained *in vitro*. Air stepping can be obtained after a general pharmacological activation with L-Dopa and 5-HT agonists (McEwen et al., 1997a). Both behaviours have a concomitant evolution from about 1 c/s at P0 and 4 c/s at P20. During the first day, only the front legs are swimming. However, 24h later, all four legs are activated with an alternation between ipsi- and contralateral limbs, while the diagonal limbs are in synchrony. The adult swimming behaviour involving only the hindlimbs is present at P12-15. The forelimbs are no more involved in producing propulsive forces but in directing the animal. Cazalets et al. (1990) studied the relationships between the four legs in the postnatal period. Between forelimbs and between hindlimbs, the relationships are always in opposition around 0.5. For both hindlimbs, however, there is a significant decrease in the variability of the phase value with age: the SD decreases from 0.14 at P3 to 0.1 at P8 and 0.05 at P15. By contrast, for the forelimbs, the coupling remains very variable (SD = 0.14 at P3 and 0.13 at P8) (Fig. 2-2B1-B2).

The limitation to the hindlimbs of the swimming propulsion in adult rats is under the control of cortical structures. Vanderwolf et al. (1978) have observed that rats use again both forelimbs and hindlimbs, in a way similar to young animals, after removing the neocortex and the hippocampus.

An EMG study of air stepping behaviour demonstrates a very stereotyped pattern in extensor muscles. At all ages, the biceps femoris, the vastus lateralis and the gastrocnemius have a similar temporal pattern consisting of a short burst which covers only a part of the stance phase (Tucker and Stehouwer, 2000; Iles and Coles, 1991). Such activities resemble those observed during swimming (Gruner and Altman, 1980).

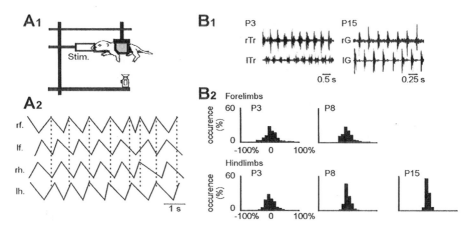

Figure 2-2. In vivo study of locomotor coordinations during postnatal development. A: Induction of air stepping in neonate rat with reduced postural constraints. The experimental device (A1) is composed of a plate supporting the rat's body, linked to a holding system by an adjustable sling. The forelimbs and hindlimbs are hanging on each side of the support without touching the ground. The head is partly inserted in a tube filled with soiled nest bedding materials (Stim.). Two mirrors (not shown) positioned at an angle of 45° and a camera allows simultaneously filming of left, right and ventral views. A2: Stepping patterns of a P0 rat, the stride amplitude was normalized with maxima corresponding to AEP (anterior extreme position) and minima to PEP (posterior EP). The dotted lines highlight the locomotor coordinations between right and left forelimbs (rf., lf.) and hindlimbs (rh., lh.): left-right and ipsilateral anterior–posterior alternations. Even occasional failures in the phase relationships, the generators are able to restore coordination between limbs from the day of birth onwards. [Adapted from Fady et al., 1998]. B: In vivo study of postnatal development of interlimb phase relationships during swimming. B1: EMG recordings of forelimb left and right trapezius muscles (rTr, lTr, on the left) at P3 and of hindlimb left and right gastrocnemius muscles (rG, lG, on the right) at P15 during swimming. B2: Distribution histograms of the difference percentage of successive periods for hindlimbs and forelimbs established at 3 different ages (P3, P8, P15). While there was no improvement in the forelimb beating stability between P3 and P8, the hindlimb paddling became more stable with age. [Adapted from Cazalets et al., 1990].

If we try to compare *in vitro* with *in vivo* analysis, the difficulty is due not only to the differences between preparations but also to the time when experiments are done. *In vitro* studies have been done on the foetus (after E14.5) and very young pup (during the first postnatal week). *In vivo* experiments start at birth and roughly cover the entire developmental period (until P21). Hence, *in vitro*, the evolution of the interlimb coordination has been recorded before birth. It is also present just after birth (Pearlstein et al., in prep.). *In vivo*, when the animal is intact, a rostro-caudal gradient from forelimbs to hindlimbs has been presented during the first week. Subsequently, the diagonal rhythm has been recorded even though only the hindlimbs are beating later in swimming.

In summary, all these results show that the CNS is composed at birth of the main structures of the locomotor CPGs but they become functional later (one or two weeks) when the postural regulations are adapted.

4.   CENTRAL AND PERIPHERAL MECHANISMS OF INTERLIMB COORDINATION

At least two mechanisms enable the control of early interlimb coordination. Numerous studies have shown that 5-HT demonstrates ubiquitous topic and trophic effects on the early development of neurons and synapses (Tanaka et al., 1992). In the neonatal rat experiments have confirmed such an effect. The second mechanism concerns the sensory afferents. After birth, proprioceptive peripheral loops feed the central network and adapt its activation in a coordinated manner.

4.1   Role of the serotonergic system

The contribution of the 5-HT system to the maturation of lumbar motoneurons and networks has been investigated in two different experiments. In the first, a 5-HT synthesis inhibitor, P-chlorophenylalanine (PCPA), was administered daily from P0 onwards; this depleted serotonin in the spinal cord within 3-4 days (Pflieger et al., 2002). PCPA-treated rats exhibited postural changes characterized by reduced flexor activity at the knee and ankle levels and reduced extensor activity at the hip. Posture was asymmetric, suggesting possible deficits in the interlimb coordination (Fig. 2-3 A1-A2). Using the *in vitro* brainstem/spinal cord preparation at P3-5, intracellular recordings from motoneurons innervating different hindlimb muscles, demonstrated that the excitability of motoneurons was decreased (higher rheobase and input conductance) in PCPA-treated rats compared to sham animals. In agreement with postural observations, changes were more pronounced in hip extensor/knee flexor than in ankle extensor motoneurons. In addition, the maturation of repetitive firing properties was stopped by PCPA treatment. Using a similar treatment with PCPA, Myoga et al. (1985) and Nakajima et al. (1998) described in detail the deficits in locomotor movements. In particular, during swimming the neonate is unable to generate coordinated activity. On a transgenic mouse where MAO is absent (thereby leading to high levels of endogenous 5-HT), Cazalets et al. (2000)

demonstrated that the development of swimming behaviour is delayed for at least two days.

The second type of experiment consisted of transecting the spinal cord at a thoracic level on the day of birth in order to suppress the inputs from supraspinal structures. The coordination between hindlimbs has been examined during the postnatal week in two complementary experiments (Norreel et al., 2003). EMGs from the two hindlimb extensor muscles were recorded *in vivo* during air stepping and lumbar ventral roots were recorded *in vitro* during pharmacologically-evoked fictive locomotion. Tail pinching elicited long lasting sequences of alternated cyclic movements of the right and left hindlimbs in the spinal neonate but not in the intact animal demonstrating some degree of disinhibition of the network when the supraspinal pathways are suppressed. *In vitro* experiments on spinal animal confirmed an increased excitability compared to sham experiments.

Similar experiments performed at the end of the first week (P6-7) showed that the left-right alternating pattern is lost in the spinal animals both *in vitro* and *in vivo*. Alternation is restored after activation of 5-HT$_2$ receptors suggesting that the descending pathways and in particular the serotonergic projections control the reciprocal interleg inhibition (Fig. 2-3 B1-B2).

4.2   Peripheral proprioceptive control

The sensory systems are very immature at birth except for olfaction and temperature (Brunjes and Frazier, 1986). The vestibular system and the proprioceptive system develop gradually and they are only fully mature much later (after more than 3 weeks). However, some sensory feedback exists very early; the monosynaptic stretch reflex pathways studied by Kudo and Yamada (1985, 1987) are present at E18.5. Sqalli-Houssaini et al. (1993) have shown that an electrical stimulation of a given dorsal root in the first week induces a phase delay or a phase advance in the ventral root, depending upon the position of the stimulus in the given ongoing rhythm induced by 5-HT/NMA application in the bath. Moreover within a certain range peripheral rhythmic dorsal root stimulation is able to induce the central rhythm (Marchetti et al., 2001). During the first postnatal week, Brocard et al. (1999a) have demonstrated the existence of an antigravity postural reflex induced by a rotation of the whole body in the vertical plane. This protocol, which activates the vestibular system, induces a very weak response in the hindlimb extensor muscles at birth and a strong one at P6. This suggests that afferents are very functional only at the end of the first week to control the posture of each limb. It also plays a role in the coordination between both front and hindlimbs.

All these inputs will adapt the central network to the surrounding medium. Morin and Viala (2002) have shown that sensory afferents from the hindlimb not only control the hindlimb CPG but also are able to act on the respiratory rhythm.

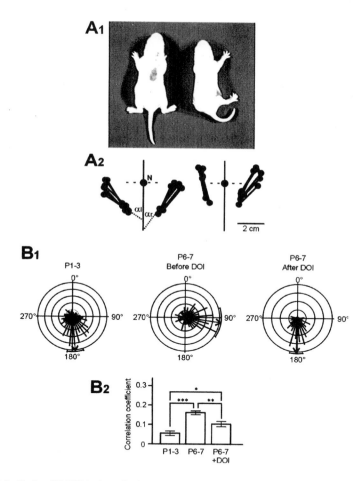

Figure 2-3. Role of 5-HT in interlimb coordinations. A: PCPA-treated rats present postural deficits. A1: Dorsal view of a sham-injected (left) and a PCPA-injected (right) animal (P6), illustrating the difference in posture between the two conditions. A2: Graphs illustrating the positions of hindfeet relative to the navel (N, horizontal dotted line) and the body axis (vertical line). Angles between the body axis and the left (al) or the right (ar) foot axis were measured. The postural asymmetry of the PCPA-treated specimen is due to a hyperextension of the right hindlimb. (adapted from Pflieger et al., 2002) B: DOI-induced reversion of

locomotor pattern disorganization after neonatal spinal cord transection (spinalization). B1: Circular histograms showing the distribution of phase relationships between EMG bursts of activity recorded from left and right ankle extensor muscles in P1–P3 (left) and P6–P7 (middle - Control, right - DOI-treated) spinal rats. While the left–right alternating pattern of air stepping remained at P1–P3 (mean vector tended towards 180°), it was lost at the end of the first postnatal week in the absence of supraspinal inputs. In this latter case, acute injection of DOI restored the left-right alternation. B2: Mean left-right correlation coefficient at P1–P3, P6 –P7 and P6 –P7 after DOI injection. This demonstrated again that, in spinal animals, at the end of the first postnatal week the alternation disappeared (correlation coefficient tending to more positive values than at P1-3) and that DOI acute injection tended to bring back this coefficient towards its initial value. [Adapted from Norreel et al., 2003].

5.   SUMMARY

The previously discussed *in vitro* and *in vivo* experiments have produced a huge amount of data on the locomotor pattern and on the development of interlimb coordination. They have demonstrated a progressive adaptation, which lasts up to two or three weeks after birth. The following conclusions can be presented:

1. Despite the large degree of immaturity of the different nervous structures, the neural locomotor networks develop before birth in the rat and are in principle functional at birth. However the absence of efficient postural regulations delays the development of locomotion for 2 weeks except when a strong emotional stimulus is induced or when postural constraints are reduced. Swimming or air stepping are routinely possible.

2. Interlimb coordination, as studied in the different experimental conditions, is more or less identical and corresponds to an alternating pattern. Gallop occurs later in conjunction with the development of complex behaviours.

3. Sensory afferents operate gradually to stabilize the pattern and to adapt it to the environment.

4. Descending pathways, in particular the 5-HT control, are essential for shaping the motor output and in particular for the interleg reciprocal inhibition. Supraspinal structures like the cerebellum develop after 2 weeks and further complete the development of these motor behaviours.

5. The similarities in many features of the rat ontogenesis with that of human make this animal model very useful for the understanding of human development. Even thought the differential maturation of some functions (sensory, cognitive…), the ontogenesis of motor functions, like posture and locomotion, in rat and human are very comparable.

REFERENCES

Ballion B, Morin D, Viala D. (2001) Forelimb locomotor generators and quadrupedal locomotion in the neonatal rat. Eur J Neurosci 14:1727-1738

Ben Ari Y (2002) Excitatory actions of GABA during development: the nature of the nurture. Nat Rev Neurosci 3:728-739

Brocard F, Vinay L, Clarac F (1999a) Development of hind-limb postural control during the first post-natal week in the rat. Brain Res Dev Brain Res 117:81-89

Brocard F, Vinay L, Clarac F (1999b) Gradual development of the ventral funiculus's input to lumbar motoneurons in the neonatal rat. Neuroscience 90:1543-1554

Brunjes PC, Frazier LL (1986) Maturation and plasticity in the olfactory system of vertebrates. Brain Res 396:1-45

Butt STB, Harris-Warrick RH, Kiehn O (2002) Firing properties of identified interneuron populations in the mammalian hindlimb central pattern generator. J Neurosci 22(22):9961-9971

Cazalets JR, Menard I, Cremieux J, Clarac F (1990) Variability as a characteristic of immature motor systems : an electromyographic study of swimming in the newborn rat. Behav Brain Res 40:215-225

Cazalets JR, Sqalli-Houssaini Y, Clarac F (1992) Activation of the central pattern generators for locomotion by serotonin and excitatory amino-acids in neonatal rats. J Physiol (London) 455:187-204

Cazalets JR, Borde M, Clarac F (1995) Localisation and organization of the central pattern generator for hindlimb locomotion in newborn rat. J Neurosci 15:4943-4951

Cazalets JR, Bertrand S (2000) Coupling between lumbar and sacral motor networks in the neonatal rat spinal cord. Eur J Neurosci 12:2993-3002

Cazalets JR, Gardette M, Hilaire G (2000) Locomotor network maturation is transiently delayed in the MAOA deficient mouse. J Neurophysiol 83:2468-2470

Cazalets JR (2000) Development of the neural correlates of locomotion in rats. In: Kalverboer AF, Gramsbergen A (eds) Handbook of Brain and Behaviour in human development, Kluwer Academic Publishers, Printed in Great Britain, pp. 447-466

Clarac F, Vinay L, Cazalets JR, Fady JC, Jamon M (1998) Role of gravity in the development of posture and locomotion in the neonatal rat. Brain Res Brain Res Rev 28:35-43

Clarac F, Brocard F, Pflieger JF, Vinay L (2001) Maturation of locomotion in the neonate rat, comparisons with another higher vertebrate. In: Gantchev N (Ed.), "From basic motor control to functional recovery II", Bulgarian Academy of Science, Sofia.

Eide AL, Glover J, Kjaerulff O, Kiehn O (1999) Characterization of commissural interneurons in the lumbar region of the neonatal rat spinal cord. J Comp Neurol 403:332-345

Fady JC, Jamon M, Clarac F (1998) Early olfactory-induced rhythmic limb activity in the newborn rat. Brain Res Dev Brain Res 108:111-123

Gramsbergen A (2001) Neuro-ontogeny of motor behaviour in the rat. In: Kalberboer AF, Gramsbergen A (eds), Handbook of Brain and Behaviour in human development, Kluwer Academic Publishers, pp. 467-512

Gruner JA, Altman J (1980) Swimming in the rat : analysis of locomotor performance in comparison to stepping. Exp Brain Res 40:374-382

Hilaire G, Duron B (1999) Maturation of the mammalian respiratory system. Physiol Rev 79:325-360

Iikuza M, Nishimaru H, Kudo N (1998) Development of the spatial pattern of 5-HT induced locomotor rhythm in the lumbar spinal cord of rat fetuses in vitro. Neurosci Res 31:107-111

Iles JF, Coles SK (1991) Effects of loading on muscle activity during locomotion in the rat. In: Armstrong DM, Bush BMH (eds), Locomotor neural mechanisms in arthropods and vertebrates, Manchester University Press, Manchester and New York, pp. 196-201

Jamon M, Clarac F (1998). Early walking in the neonatal rat : a kinematic study. Behav Neurosci 112:1218-1228

Jensen JL, Schneider K, Ulrich BD, Zernicke RF, Thelen E (1994) Adaptive dynamics of the leg movement patterns of human infants: I. The effects of posture on spontaneous kicking. J Mot Behav 26:303-312

Kiehn O, Kjaerulff O (1996) Spatiotemporal characteristics of 5-HT and dopamine-induced rhythmic hindlimb activity in the in vitro neonatal rat. J Neurophysiol 75:1471-1482

Kjaerulff O, Kiehn O (1997) Crossed rhythmic synaptic input to motoneurons during selective activation of the contralateral spinal locomotor network. J Neurosci 17:9433-9447

Kudo N, Yamada T (1985) Development of the monosynaptic stretch reflex in the rat : an in vitro study. J Physiol (London) 369:127-144

Kudo N, Yamada T (1987) Morphological and physiological studies of development of the monosynaptic reflex pathway in the rat lumbar spinal cord. J Physiol (London) 389:441-459

Lev-Tov A, Delvolve I, Kremer E (2000) Sacrocaudal afferents induce rhythmic efferent bursting in isolated spinal cord of neonatal rats. J Neurophysiol 83:888-894

Marchetti C, Beato M, Nistri A (2001) Alternating rhythmic activity induced by dorsal root stimulation in the neonatal rat spinal cord in vitro. J Physiol (London) 530:105-112

McEwen ML, Van Hartesveldt C, Stehouwer DJ (1997a) L-Dopa and Quipazine elicit air stepping in neonatal rats with spinal cord transection. Behav Neurosci 111:825-833

McEwen ML, Van Hartesveldt C, Stehouwer DJ (1997b) A kinematic comparison of L-Dopa-induced air-stepping and swimming in developing rats. Dev Psychobiol 30: 313-327

Morin D, Viala D (2002) Coordination of locomotor and respiratory rhythms in vitro are critically dependent on hindlimb sensory inputs. J Neurosci 22:4756-4765

Myoga H, Nonaka S, Matsuyama K, Mori S (1998) Prenatal development of locomotor movement in normal and parachlorophenylalanine-treated newborn rat. Neurosci Res, 21:211-221

Nakajima K, Matsuyama K, Mori S (1998) Prenatal administration of para-chlorophenylalanine results in suppression of serotonergic system and disturbance of swimming movement in newborn rats. Neurosci Res 31:155-169

Nakajima K, Nishimaru H, Kudo N (2002) Basis of changes in left-right coordination of rhythmic motor activity during development in the rat spinal cord. J Neurosci 22:10388-10398

Nicolopoulos-Stournaras S, Iles JF (1984) Hindlimb muscle activity during locomotion in the rat (Rattus norvegicus) (Rodentia-Muridea). J Zool Lond 203:427-440

Norreel JC, Pflieger JF, Pearlstein E, Simeoni-Alias J, Clarac F, Vinay L (2003) Reversible disorganization of the locomotor pattern after neonatal spinal cord transection in the rat. J Neurosci 23:1924-1932

Orlovsky GN, Deliagina TG, Grillner S (1999) Neuronal control of locomotion - from mollusc to man. Oxford University Press, Oxford.

Pflieger JF, Clarac F, Vinay L (2002) Postural modifications and neuronal excitability changes induced by short-term serotonin depletion during neonatal development in the rat. J Neurosci 22:5108-5117

Smith JC, Feldman JL (1987) In vitro brainstem-spinal cord preparations for study of motor systems for mammalian respiration and locomotion. J Neurosci Methods 21:321-333

Sqalli-Houssaini Y, Cazalets JR, Clarac F (1993) Oscillatory properties of the central pattern generator for locomotion in neonatal rats. J Neurophysiol 70:803-813

Sqalli-Houssaini Y, Cazalets JR (2000) Noradrenergic control of locomotor networks in the in vitro spinal cord of the neonatal rat. Brain Res 852:100-109

Stehouwer DJ, Van Hartesvedlt C (2000) Kinematic analysis of air stepping in normal and decerebrate preweanling rats. Dev Psychobiol 36:1-8

Tanaka H, Mori S, Kimura H (1992) Development changes in the serotoninergic innervation of hindlimb extensor motoneurons in neonatal rats. Dev Brain Res 65:1-12

Tucker LB, Stehouwer DJ (2000) L-Dopa-induced air stepping in the preweanling rat : electromyographic and kinematic analyses. Behav Neurosci 114:1174-1182

Vanderwolf CH, Kolb B, Cooley RK (1978) Behavior of the rat after removal of the neocortex and hippocampal formation. J. Comp Physiol Psychol 92:156-175

Vinay L, Brocard F, Clarac F (2000a) Differential maturation of motoneurons innervating ankle flexor and extensor muscles in the neonatal rat. Eur J of Neurosci 12:4562-4566

Vinay L, Brocard F, Pflieger JF, Simeoni-Alias J, Clarac F (2000b) Perinatal development of lumbar motoneurons and their inputs in the rat. Brain Res Bull 53:635-648

Von Holst E (1943) Uber relative koordination bei arthropoden. Arch Ges Physiol Pfluger's 146:846-865

Chapter 3

LOCOMOTOR RECOVERY POTENTIAL AFTER SPINAL CORD INJURY

V. Reggie Edgerton[1,2,3] Niranjala J.K. Tillakaratne[1,3] Allison J. Bigbee[2] Ray D. de Leon[4], Roland R. Roy[1,3]
[1]Physiological Science, [2] Neurobiology, [3] Brain Research Institute, UCLA, Los Angeles, CA, USA; [4] Kinesiology and Nutritional Science, California State University, Los Angeles, CA, USA

Abstract: In this chapter we examine the recovery of posture and locomotion in mammals following a complete spinal cord lesion. The mechanisms by which the ability to stand and step can be regained following a complete injury is described with respect to central pattern generation, processing of sensory (proprioceptive and cutaneous) input by the spinal cord circuitry and the concept of automaticity. A series of intervention strategies designed to enhance motor function after a complete spinal cord lesion are examined, i.e., motor training and spinal cord learning, electrical stimulation of nerves, muscles, and the spinal cord below the level of the lesion, pharmacological facilitation of motor function, administration of growth/neurotrophic factors, and several types of cell and tissue implants to induce axonal growth. The general conclusion is that a significant level of motor function can be recovered after a complete spinal cord lesion. Although motor training has been shown to be one of the most effective strategies to date, a number of other interventions also are beginning to produce significant improvements in motor function. Some of these improvements can be attributed solely to functional reorganization of the spinal cord circuitry distal to the lesion, whereas other interventions appear to have been mediated by renewed connectivity between the neural elements above and below the lesion site. There seems to be a general consensus that a greater level of motor recovery is likely to be attained by a combination of therapies.

Key words: spinal automaticity; motor training; spinal learning; implants; electric stimulation; spinal transection

1.   INTRODUCTION

A significant level of interlimb coordination and motor control is derived from the spinal cord as well as the brain in the uninjured state. After a spinal cord injury (SCI) these supraspinal and spinal sources of control of movement will differ substantially from that which persisted prior to the injury. It is conceptually useful to think of there being a "new" spinal cord after SCI, i.e., the post-injury spinal cord (as well as brain) will not function as it did before the injury. Recognition of the concept of a "new" spinal cord or perhaps a "different" spinal cord suggests that the movement generated by the spinal cord may differ to some degree as a result of adaptive or plastic processes occurring after an injury (Edgerton et al., 1997a; 2001a). The same perspective also seems applicable to the brain after a SCI in that the "new" brain can adapt to the availability of unique combinations of information from the spinal cord and it responds accordingly to the new input in a novel way.

Historically there has been substantial underestimation of the level of control that the spinal cord can have in performing motor tasks such as stepping and standing. The contribution of the spinal cord to motor output after SCI depends on the severity and specific anatomical location of the injury. For example, after an incomplete SCI, more of the control of movement must be derived from the spinal cord and less from the brain. When the injury is "complete" then all control must be derived from the spinal cord.

We are now beginning to gain some insight into the properties of the spinal cord which can be used to enhance its potential to recover after SCI. This chapter will focus primarily on some of those properties that are involved in motor control after complete a SCI (Fig. 3-1). We will discuss automaticity, spasticity, and the initiation of locomotion, as well as some of the therapeutic strategies that are presently being used to enhance functional recovery after a complete SCI.

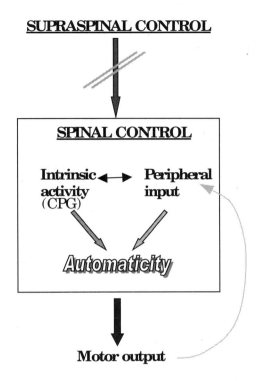

Figure 3-1. Automaticity in the spinal cord. The brain and spinal cord contain a large degree of automaticity, comprised of intrinsic activity (central pattern generation, CPG) and peripheral input, which allow the neural circuitry to perform a motor task, such as walking across a room, without much conscious thought. After complete SCI (see cross-hatch) connections to and from the brain are lost (top arrow), while intrinsic activity (left arrow) and peripheral input (right arrow) at the spinal cord level remain, although in possibly a severely altered state depending on the severity of the injury. Any number of therapeutic strategies may be used to assist in the recovery of locomotor function after SCI, including 1) motor training, 2) electrical stimulation, 3) pharmacological intervention, 4) growth factors, and 5) cell implants. The therapies may act at a number of sites, either by modifying the local circuitry, helping to facilitate re-growth of injured neurons, or both. Most likely a combination of these strategies will be the most efficacious in facilitating the recovery of function after SCI. Finally, as motor output becomes more appropriate for performing motor tasks, the peripheral input will also become more appropriate, thus resulting in a positive feedback loop.

2.   SOME FUNDAMENTAL PRINCIPLES OF SPINAL LOCOMOTION

2.1   Automaticity in posture and locomotion after SCI

There is a strong element of automaticity in the neural control of most movements, both from the brain and spinal cord (Baev, 1998). For example, once the decision to walk across a room is made, very little conscious effort is required to perform the details of that task. It seems that the central nervous system (CNS) is designed so that much of the intricate decision making occurs automatically. Both supraspinal and spinal sources contribute to the pathways responsible for the automaticity. Spinal automaticity, however, remains in the absence of supraspinal input, as occurs after a complete severing of the spinal cord. A large body of data demonstrates that the spinal cord is much more than a static conduit for communication between the brain and muscles.

After a complete transection at the mid-thoracic level of the spinal cord, the automaticity generated from the lumbosacral spinal cord is most likely to be attributable to two primary components. The first is the central pattern generator (CPG), which generates rhythmic, coordinated bursts of motor patterns from flexor and extensor motor pools that occur without any oscillating input from the brain or the peripheral nervous system (PNS) (Grillner, 1985; Grillner et al., 1998). A second extremely important component of automaticity is the sensory input to the spinal cord from the periphery. When sensory input from cutaneous and proprioceptive sources is combined with the intrinsic CPG, the spinal circuitry is capable of executing a much wider range of complex motor tasks, such as locomotion and standing, than when the output is generated by the spinal circuitry alone.

How, then, does the automaticity of posture and locomotion emerge from the interactions between the sensory inputs and the spinal circuits that generate CPG? For these two systems to work in synergy, each system must have intrinsic activation and inhibition patterns that are orderly, sequential and coordinated. For example, the sequence of activation patterns associated with a step cycle can occur only if a critical level of synergy occurs between the sources and modes of peripheral input with the cyclic events that are intrinsic among the neurons that generate CPG. This may be thought of as a complementary pattern of afferent information that interfaces with the CPG circuitry, as do two pieces of a jigsaw puzzle, at any given time "bin" within a step cycle 1) to inform the CPG of the current state of the step cycle; 2) to make the circuitry aware of which "bin(s)" preceded that instant; and 3) to predict which "bin(s)" should occur next. In this scenario a "bin" would consist of that afferent input and the stage of activation-inhibition of the neurons that generate the CPG output at that instant. Again the temporal patterns of ensembles of peripheral inputs must be matched with that of the CPG for locomotion to continue effectively.

The adaptation of the spinal cord to changing peripheral inputs suggests that, contrary to a pervasive perception, the spinal cord is not hard-wired. Rather, the combination of the intrinsic activity and the sensory input to the spinal circuitry allow the spinal cord to readily adjust to parameters such as the speed of stepping, the level of load imposed on the stepping, and a wide range of unpredictable patterns of sensory anomalies (Edgerton et al., 1991). For example, the level of activation of the extensor and flexor muscle groups is enhanced in proportion to the level of loading in spinal cord injured as in non-disabled human subjects when they are stepping on a treadmill (Harkema et al., 1997; Rossignol et al., 2002). An example of the effect of loading on soleus muscle EMG activity is shown in Figure 3-2.

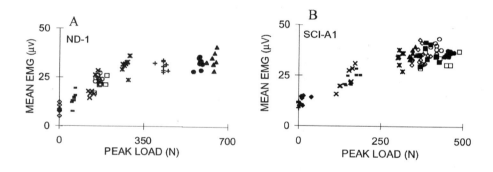

Figure 3-2. Loading effects on activation levels of motor pools during stepping. Relationships between soleus EMG mean amplitude (μV) and limb peak load (N) from a non-disabled subject (A, ND-1) and a subject with a SCI classified on the ASIA scale as an A, i.e. having a complete injury (B, SCI-A1), over a range of loading conditions. Each data point represents one step and each symbol represents a series of consecutive steps at one level of body-weight support. (Reproduced with permission; Edgerton et al 2001, J Physiol 533: 15-22).

In SCI subjects classified as ASIA A (because they have no voluntary control of movement or have no sensation below the spinal lesion), not only is the magnitude of muscle activation for support of the load that is being borne appropriate, but it occurs at the appropriate time within the step cycle to generate extension or flexion (Rossignol et al., 2002). The ensemble of afferent input associated with a greater load manifested through the lower limbs during stepping will generate a complex and temporally precise pattern of excitatory and inhibitory synaptic activity that will result in a highly coordinated motor response, as shown by the observation that increasing load on one side leads to greater flexion during the swing phase of the same leg (Harkema et al., 1997). Similar loading related responses have been described in low-thoracic complete spinal cats (de Guzman, et al 1991). This loading-related response is also readily observed during standing in ASIA A SCI subjects. If a complete or severely injured SCI subject stands with their body weight distributed bilaterally, and then the weight is shifted to one side, the unloaded limb will flex as if the swing phase of a step is being initiated (Susan Harkema, personal communication).

In SCI patients, the level of automaticity is quite striking. While all of the sensory inputs and intrinsic oscillatory patterns will not be appropriately synchronized at all times, the injured spinal cord is highly capable of interpreting and responding to sensory information from the periphery. From a systems standpoint, the specific combination of recruited motor pools and the level of net activation of this combination of motor pools at any given time within a step cycle must be highly precise. That is, the ability to generate successful stepping requires that some mechanism of maintaining balance under highly dynamic conditions, i.e., avoid falling when faced with a perturbation, must be present. Even when the precision of the activation patterns is not impressive at the individual motor pool level, the net effect, e.g., successful stepping, can be remarkably consistent at the systems level. An example of the preciseness of the decision-making capability at the spinal cord level is the response of the complete spinal animal to a tripping stimulation during the swing phase of a step. A series of experiments have demonstrated that a stimulus to the dorsum of the paw of a spinal cat during the swing phase will induce ipsilateral flexion and contralateral extension (Forssberg et al., 1975). When the same stimulus is presented during the stance phase of a step, the opposite reaction occurs. These examples highlight the state-dependent decision-making ability of the sensory and motor circuitry of the lumbosacral spinal cord in the absence of input from the brain.

Two of several interpretations of how the spinal cord senses load "on-line" are that: 1) there are sensory receptors in the limbs (soles of the feet, tendons, muscles, joints) that specifically sense load; and 2) an ensemble of many types of sensory receptors at multiple locations within the limbs generate a highly recognizable "image" to inform the spinal circuitry of the biomechanical status of the weight bearing. The second interpretation is the one we favor and it is consistent with the concept that has been alluded to many times previously, i.e., it is the ensemble of sensory input which has meaning and can be interpreted by the spinal cord circuitry so that an appropriate motor pattern can be generated (Edgerton et al., 2001a; Prochazka and Gorassini, 1998).

Although the purposefulness or teleological nature of this concept may seem inconsistent with fundamental biological principles, this need not be the case. All motor systems have evolved within Earth's gravitational environment and in this evolutionary process the sensory and motor systems have evolved and function within this known environment. Similar sensory and motor components among a wide range of animals with vastly different musculo-skeletal structures have evolved in a manner that enables many postural and locomotor tasks to occur quite automatically within this known gravitational environment (Edgerton et al., 2001b). The automaticity aspect of these functions reflect the successful evolution that enables locomotion to occur more automatically rather than relying on more complicated and delayed decision making by higher neural centers. Relying more on the brain would require additional time and impose disadvantages in the execution of a wide variety of postural and locomotor tasks. This automaticity seems to have been learned in the evolutionary process. In this sense, "evolutionary learning" has played a key role in the automaticity of neural control during the execution of motor tasks.

2.2 Spasticity and recovery of locomotion

The term spasticity itself triggers an array of perceptual and conceptual responses, many of which have their own etiologies. Spasticity can be viewed, in a sense, as automaticity gone awry. Rather than afferent-efferent events being predictable, they become non-predictable and more random as the result of an injury. Our present working hypothesis is that spasticity reflects randomness, or at least a severe deficit in the coordination of the activation of motor pools. During spastic movements, there are extensive contractions but little reciprocal activation of agonistic and antagonistic motor pools in the appropriate temporal patterns. One possible explanation for this more or less randomness of motor pool activation is that the spinal cord has lost the ability to synchronize and associate a coordinated ensemble of afferent input to produce a predictable motor output. This deficit may be due to the absence or rare occurrence of synchronized events that are normally associated with load-bearing stepping. In the absence of these coordinating events the system loses the ability to synchronize them into functional movements of the limbs.

Although some progress has been made in quantifying the level of spasticity in subjects with specific neuromotor disorders, our present understanding of the etiologies for this phenomenon remains meager at best. However, the presence of spasticity is a positive sign of the potential of a

SCI patient to regain locomotor function. A patient who is hypo-responsive to afferent input is less likely to respond to the proprioceptive input associated with load-bearing stepping. Thus, it is to our benefit to understand the physiological mechanisms that underlie spasticity in an effort to enhance locomotor recovery in SCI patients.

2.3   Initiating locomotion

With the increasing success in improving the locomotor potential of laboratory animals as well as humans after a complete SCI, the question has been raised as to the functional significance of this motor recovery since there is no voluntary initiation of stepping or standing. This view, however, is not entirely true. First, both of these motor tasks can be initiated voluntarily, although indirectly, even in ASIA A subjects. For example, when an ASIA A SCI subject is standing with bilateral weight support, stepping can be initiated by shifting the body weight to one leg, extending the hip position of the contralateral leg, and leaning forward (Wernig and Müller, 1991). When stepping on a moving treadmill belt the same events can be used more easily to initiate stepping since the weight-bearing limb will be moving backward into hip extension. A moving treadmill, however, is not required to successfully initiate stepping, i.e., the subject can use the upper body to alternate the loading from one side to the other side at the appropriate time and facilitate the joint actions bilaterally. Simply unloading a limb at the end of the stance phase of a step facilitates the initiation of the swing phase of that leg (Duysens and Pearson, 1980). Although initiating stepping in subjects with a severe, but not complete, SCI also can be challenging, some voluntary control of some motor pools that innervate the lower limbs may persist. But even if the descending pathways that normally initiate stepping are not functional, the subject has the same mechanisms for motor recovery available as described above for the ASIA A subject.

3.   STRATEGIES TO ENHANCE POSTURE AND LOCOMOTION

In the final analysis it is the loss of the ability to activate the appropriate motor pools at an adequate level and with sufficient coordination that limits the ability of an individual to stand or step after a SCI. It is quite clear now that cats with a complete low-thoracic spinal cord transection can regain the ability to activate and coordinate the lower limb motor pools when trained repetitively over a period of weeks (Lovely et al., 1986; 1990). The primary limitation seems to be linked to the proportion of the motor pools that can be activated (Lovely et al., 1986; 1990). The net torque that can be generated is defined by the proportion of motoneurons recruited within a motor pool and the relative level of activation of agonist vs. antagonist motor pools. However, the net torque obviously will be limited as well by the rapid muscle atrophy that occurs after SCI (Roy and Acosta 1986; Roy et al., 1998; 1999). It remains to be determined how much of the loss in the locomotor potential of a SCI subject can be attributed to deficits in the level of 1) coordination; 2) activation of motor pools; and 3) skeletal muscle atrophy.

At the present time there is a clear indication that motor deficits reflecting each of these parameters can be overcome to a significant degree. There is evidence that 1) motor pools can become more effectively coordinated (de Leon et al., 1998a; 1998b); 2) the level of activity in specific motor pools can be elevated (de Leon et al., 1998a,b; Dietz et al., 1998); and 3) the amount of muscle atrophy can be at least partially and perhaps fully recovered (Roy and Acosta 1986; Roy et al., 1984;1991;1998;1999). In spite of these advances, to our knowledge no one to date has successfully rehabilitated an ASIA A subject to the point that they can step independently while fully bearing their body weight, at least to the level demonstrated in the cat.

Currently there is a range of approaches being studied intensely to enhance postural and locomotor performance after SCI. We will examine the state of progress in the following areas: 1) motor training and spinal learning; 2) sensory stimulation; 3) epidural and intradural spinal cord electrical stimulation; 4) pharmacological interventions; 5) growth factor administration; and 6) cellular implants. In many cases, these approaches are used simultaneously.

3.1 Motor training and spinal learning

To date neuromotor training has been the primary intervention that consistently improves the ability to step and stand after SCI (Edgerton et al 1997a; 2001a). There is strong evidence that training of humans on a treadmill belt using a weight-supporting device combined with overground training can increase the levels of motor pool activation, improve coordination of motor pool recruitment, and reduce muscle atrophy (Dietz et al., 1998; Harkema personal communication). However, these improvements occur slowly, i.e., after weeks or even months of training. It appears that neuromotor training provides a means by which the sensory system can become synchronized with the spinal circuitry and with the motor pools that are linked to this circuitry. The evidence to support these general concepts is discussed below.

Early evidence that spinal circuits respond to neuromotor training came from studies of simple hindlimb motor reflexes in complete spinal cord transected animals. Simple hindlimb reflexes, e.g., the hindlimb withdrawal reflex, can be modulated via classical (Durkovic, 1975) or operant (Buerger and Fennessy, 1971) conditioning techniques. An example of operant conditioning within the spinal cord is the prolonged dorsiflexion that occurs in a spinal rat after a series of electrical shocks are presented to the paw. In this paradigm, when the ankle plantarflexes to a threshold position, a shock is delivered to the leg. Using this paradigm a complete spinal rat can be conditioned to avoid the shock by maintaining a more dorsiflexed position, even though there is no perception of the shock (Grau and Joynes, 2001). In contrast, no conditioning occurs in "yoked" spinal rats that get shocked, but the shock is not associated with a specific foot position. All of these findings are consistent with the conclusion that spinal circuits can be trained to perform relatively simple hindlimb motor tasks in the absence of any connection between the spinal cord and the brain.

The work of Nesmeyanova (1977) and Shurrager (1955) also strongly suggested that some complex postural and locomotor tasks could be improved following some training paradigm. It is now clear that hindlimb stepping in spinal cats can be improved with daily practice of walking on a treadmill (Lovely et al., 1986; 1990). Lovely et al. demonstrated that adult spinal cats performed full weight-bearing hindlimb stepping on a treadmill after as little as 2 to 3 weeks of locomotor training. Similar findings have been reported by others (Barbeau and Rossignol, 1987; Belanger et al., 1996). With training, stepping ability as measured by a steady increase in the maximum treadmill speed achieved with full weight-bearing stepping and in the number of plantar surface steps performed progressed (Barbeau and Rossignol, 1987; de Leon et al., 1998a; 1998b). In addition, locomotor training tended to normalize the characteristics of the locomotion based on similarities in electromyographic (EMG) and kinematic patterns of trained spinal and intact cats. Normal flexor and extensor muscle activation relationships (de Guzman et al., 1991), EMG burst waveform shapes (Forssberg et al., 1980a; Forssberg et al., 1980b) and trends in EMG burst duration across speeds of locomotion (Barbeau and Rossignol, 1987; Lovely et al., 1990) were preserved in trained spinal cats. With step training, the overall patterns of joint angle excursions during a step cycle (Belanger et al., 1996), the sequence of flexion and extension movements in hindlimbs joints (Lovely et al., 1986), and the force levels in the soleus muscle (Lovely et al., 1990) were similar, but not identical, during stepping in intact and spinal cats.

Together, these findings indicated that locomotor training in spinal cats improved hindlimb stepping function. However, the most convincing evidence that the spinal cord learned to step initially came from the study performed by Lovely et al. (1986). Unlike most other studies of spinal locomotion at this time, a control group of spinal transected cats that were not trained to step was included in the experimental design. Thus, the extent that hindlimb stepping recovered spontaneously after SCI and the extent that the recovery was associated with the treadmill training could be determined. Surprisingly, a few non-trained spinal cats improved their locomotor ability over the course of the 6-month study. However, at the end of the study, the trained spinal cats stepped at significantly faster treadmill speeds and with more normal kinematics than the non-trained spinal cats. This demonstrated for the first time that treadmill training significantly improved the hindlimb stepping ability of SCI cats beyond the level of recovery that occurred spontaneously. These data provided the most compelling behavioral evidence to date that the circuitry within the lumbosacral spinal cord that generated hindlimb stepping could be modified by the experience of treadmill training.

Since then, we have performed a more detailed comparison of hindlimb locomotion between a group of spinal animals that had been trained to step and a group that had not been trained to step (de Leon et al., 1998a). We were particularly interested in identifying elements of the locomotor pattern that were changed as a result of training. We compared the stepping performance in trained vs. non-trained adult cats for 12 weeks after a complete SCI. Behaviorally, we found that non-trained spinal cats were able to consistently step at slow treadmill speeds with full weight bearing on the hindlimbs. However, when the non-trained spinal cats were tested at faster treadmill speeds, i.e., faster than 0.4 m/s, only a few full weight-bearing steps were generated. Instead, there was a greater tendency to drag the paw during swing and to have dorsal paw placement during stance and subsequently, the hindlimbs frequently stumbled. In contrast, spinal cats that received daily locomotor training on the treadmill were able to maintain full plantar surface weight-bearing stepping at speeds up to 1.0 m/s. Unlike the non-trained spinal cats, trained spinal cats exhibited few step failures.

The reason for step failures in the non-trained spinal cats became evident in a subsequent study of the hindlimb coordination patterns during stepping (de Leon et al., 1999c). The frequent step failures were related to an abnormal timing of kinematic events in the two hindlimbs. During normal locomotion and in step-trained spinal cats, the ipsilateral hindlimb began the swing phase of the step cycle (i.e., toe-off) after the contralateral hindlimb had initiated the stance phase (i.e., after paw contact). In the non-trained spinal cats, however, the relative timing of ipsilateral toe-off and contralateral paw contact was altered. The two events of the step cycle occurred simultaneously or, in some cases, there was a reversal in the order, i.e., ipsilateral toe-off began before weight bearing occurred in the contralateral hindlimb, and as a result, stepping was disrupted and the hindlimbs stumbled. In addition, there was a gradual decrease in the duration of double support leading to failure to sustain stepping in spinal cats when neither leg was providing weight support (Fig. 3-3).

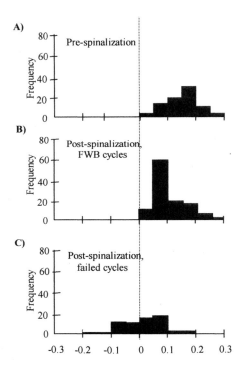

Figure 3-3. Step cycle properties associated with stepping success and failure. Histograms of the interval between ipsilateral toe-off and contralateral paw contact during pre-spinal (A), post-spinal full weight bearing (B), and post-spinal failed (C) step cycles. Pre-spinal and post-spinal (12 weeks after spinalization) data were collected from eight cats. The number of step cycles was 79, 104, and 68 for the pre-spinal, post-spinal full weight bearing and post-spinal failed cycles, respectively (8-15 cycles/cat). (Reproduced with permission; de Leon et al 1999, Progress in Brain Research 123: 341-348).

Our interpretation of these results is that the lumbar spinal circuitry can learn to step with exposure to repetitive step training. In the absence of any training, the patterns of motor output are generated in response to the stimuli associated with the movement of the treadmill. However, the probability that the correct patterns, i.e., synchronization of ipsilateral and contralateral events, will be generated in non-trained animals over a series of consecutive step cycles is low. The effect of training, therefore, is to repetitively activate the appropriate extensor and flexor networks in a specific temporal and spatial pattern so that the probability of that motor response occurring is greater than other potential responses. That the activity in neural networks of the lumbar spinal cord is to a large extent determined by the pattern of activity that the hindlimbs experience is further demonstrated by the effect of stand training in spinal cats. Spinal cats that were trained to perform full weight-bearing extension (maintaining posture) learned to stand for prolonged periods (de Leon et al., 1998b). However, when the ability to execute hindlimb stepping was tested, most stand-trained spinal cats could not generate even a few successful weight-supported steps. Thus, the spinal cord "learns" the specific motor task that it experiences during the training sessions. Furthermore, when the step training is stopped for 12 weeks, the stepping ability is as poor as it was prior to any training, i.e., the spinal cord appears to "forget" the specific motor task if it is not practiced (de Leon et al., 1999a).

3.2 Sensory stimulation

Another rehabilitative intervention is to present supra-threshold levels of sensory stimulation to facilitate a specific part of the step cycle that may be preventing the execution of the step. The most successful example of this has been the stimulation of the peroneal nerve at the end of the stance phase to induce a flexor reflex and facilitate the initiation of the swing phase. One approach in SCI patients has been to initiate the swing phase using a gravity sensor device to trigger the stimulation (Wieler et al., 1999). This device can be readily strapped around the leg below the knee and superficial to the peroneal nerve. The effectiveness of this device was examined in a multi-center trial involving 31 SCI subjects who had been injured for an average of 5.4 years. After using the device for an average of 1 year, the subjects improved their walking speed by 45%, with the largest gains occurring in the subjects that initially stepped the slowest. Importantly, among the subjects the acceptance of the device as one that could be used in the community was found to be good.

3.3   Spinal cord stimulation

Numerous experiments have demonstrated the ability to stimulate the dorsum of the upper lumbar spinal segments of cats (Edgerton and Grillner 1976; Avelev et al., 1997), rats (Iwahara et al., 1992), and humans (Dimitrijevic et al., 1983; Dimitrijevic and Dimitrijevic, 2002; Rosenfeld et al., 1995, Gerasimenko et al., 1996a,b; Shapkova et al., 1997). In the intact cat (under chloralose anesthesia), reciprocal interlimb activity in the hindlimb muscle group could be best obtained when the electrodes were placed over the L3-L4 segments (Avelev et al., 1997). It appears that generalized continuous stimulation can provide a level of activation to the spinal cord circuitry necessary to generate step-like movements in SCI subjects, if the afferents providing input to the lumbosacral segments are not damaged directly. There is one case reported in which a SCI subject (ASIA C, wheel chair-dependent) was implanted chronically with stimulating electrodes into the dorsal epidural space over the upper lumbar segments and trained to step on a treadmill while being stimulated (Herman et al., 2002). The stimulation strength was above the sensory threshold so that the subject could sense the stimulation, but below the motor threshold so that the leg muscles were not being stimulated directly. Overground stepping without stimulation was very slow and tiring. When partial body weight support was combined with electrical stimulation, there was immediate improvement in endurance and speed during overground walking. In any given training session, stimulation during full weight-bearing overground stepping reduced the energy cost of stepping by an average of 36%. The time required to walk 15 m overground improved from ~150 sec without stimulation to ~100 sec with stimulation over a span of several months of occasional training. After 1.5 months of a combination of stimulation and walking on a more regular basis (3 times/week) the time decreased to about 45 sec.

The encouraging results from these spinal cord stimulation experiments are that a generalized, nonspecific electrical stimulation of upper lumbar neurons may be sufficient to generate rhythmic stepping movements of the lower limbs. This approach potentially could eliminate the necessity to directly stimulate a specific axonal tract or a different set of motor pools during each phase of a step cycle to generate stepping or to sustain a standing posture. Considerable progress has been made using this approach in spinal cats (Mushahwar et al., 2000). The combined results from a wide range of experiments raise the possibility that when spinal cord stimulation is combined with ordered proprioceptive input to the cord, as occurs with alternating weight bearing, sufficient levels of activation and coordination can be attained more readily than if either method is applied alone. Dimitrijevic and colleagues suggest that stimulation at the same lumbar site, but at different frequencies, has distinctly different physiological effects (Pinter et al., 2000). Stimulation frequencies of 50-100 Hz at voltages ranging from 2-7 V were effective in controlling spasticity. Stimulating at 25-50 Hz at 7-10 V initiated and maintained rhythmical step-like hip and knee flexion in Asia A subjects (Dimitrijevic and Dimitrijevic, 2002). Additional evidence of a specific response at varying frequencies was shown when only extensor movements were produced when stimulating at < 15 Hz. Thus, the evidence to date seems to suggest that nonspecific stimulation of the upper lumbar segments with epidural electrodes combined with the sensory input that is generated by weight-bearing stepping has considerable potential as a therapeutic strategy to improve locomotor performance in SCI subjects.

3.4   Pharmacological Stimulation

The biochemical adaptations that occur within the spinal cord after an injury and in response to post-injury motor training may change markedly over time, and the pharmacological properties will reflect this (Rossignol et al., 2002). Some of these biochemical properties will be discussed in the context of therapeutic pharmacological manipulations. Some of the biochemical adaptations occur as a result of cell death, synaptic remodeling and sprouting. Together, these adaptations alter the capacity of the injured spinal cord to respond to any therapeutic agents used to enhance functional recovery. Therefore, it is crucial that we understand the nature of these ongoing adaptations, ranging from the level and sensitivity of membrane channels to the integration of inhibitory and excitatory input to multiple motor pools. Although this topic is deserving of a full review, only a few critical points related to its potential as a rehabilitative tool will be noted.

Numerous experiments ranging from patch clamping to systems physiology have shown that multiple neurotransmitter and neuromodulatory systems can either facilitate or degrade locomotor performance. Thus, it seems likely that there will be specific combinations of pharmacological agents that can achieve an optimal physiological state for the spinal circuitry to respond to the proprioceptive input associated with weight bearing in a manner that will facilitate stepping. Any approach to directly stimulate the sensory and motor circuitry of the spinal cord pharmacologically independent of sensory input seems infeasible. For example, it is impractical for one to ingest some pharmacological agent at the instant that one wants to step and then to ingest another agent when one wants to stop stepping or to sit or stand. Given our present understanding and technology, the pharmacological approach with the greatest potential is to manipulate the "state dependence" of the spinal circuitry so that the sensory input associated with weight-bearing stepping can be the controlling element. Consequently for any given functional state of the spinal cord after injury or after some degree of regeneration, there could be some optimal combination of partial weight support, electrical stimulation, and pharmacological treatment that could best optimize stepping after SCI.

Some of the pharmacological approaches that have been the most successful to date in facilitating locomotion in SCI cats have been the administration of glycinergic, e.g., strychnine, and GABAergic, e.g., bicuculline, antagonists which facilitate neuronal excitation by blocking the inhibitory effects of these neurotransmitters (de Leon et al., 1999b). Activation of serotonergic (5-HT) receptors (quipazine) and α-adrenergic receptors (clonidine) also facilitate locomotion (Robinson and Goldberger, 1986; Barbeau and Rossignol, 1990; 1991; Edgerton et al., 1997b; Chau et al., 1998a, b). In combination, the results of these experiments indicate that multiple neuromodulatory/neurotransmitter systems have the potential to enhance, or to diminish, the ability to step or stand.

Our recent studies on inhibitory neurotransmitter systems show that after spinalization, motor training can markedly change the physiological, biochemical, and pharmacological state of the spinal cord (Tillakaratne et al., 2000;2001;2002). Spinally transected cats and rats undergo changes in inhibitory signaling. After a complete spinal cord transection at ~T12 in adult cats, the levels of the GABA synthetic enzyme GAD_{67} increase in both the dorsal and ventral horns of L5 to L7 (Tillakaratne et al., 2000). Step training decreases these levels toward control (Tillakaratne et al., 2002). In addition, kinematic analyses show that this downward modulation is inversely correlated with stepping ability (Tillakaratne et al., 2002). Over a period of weeks, the specific changes in sensory information projecting to the spinal cord in response to weight-bearing, postural and cutaneous activity appear to affect GABA signaling by modulating GAD_{67} in spinal interneurons. Furthermore, we showed that the activity-dependent plasticity of GAD_{67} in the lumbar spinal cord in spinal cats varied depending on the specific motor task performed i.e., stepping or standing (Tillakaratne et al., 2002). Although hindlimb stepping and standing can be regained after a complete low-thoracic spinal cord transection in adult cats with appropriate post-traumatic training, a spinal cat that is trained to step cannot stand very well and vice versa, highlighting the task-specificity of motor learning in the spinal cord (de Leon et al., 1999b). This task-specificity was linked to the changes in GAD_{67} in the lumbosacral spinal cord. For example, there was increased GAD_{67} in the flexor motoneuron pools in stand-trained cats, whereas these levels in step-trained animals were comparable to those in control cats. Furthermore, the number of interneurons containing GAD_{67}

mRNA and the levels of GAD_{67} mRNA in laminae V and VI were higher in non-trained and stand-trained than in step-trained adult spinal cats (Tillakaratne et al., 2002).

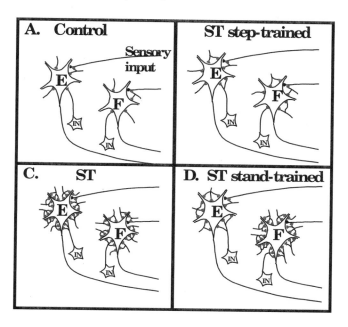

Figure 3-4. Schematic illustrating changes of GAD67 around extensor and flexor motor pools. The relative density of inhibitory synapses around flexor and extensor motoneurons for a control (A), spinal-transected, step-trained (B), spinal-transected, non-trained (C), and spinal-transected, stand-trained (D) animal is illustrated. The puncta on motoneurons are inhibitory inputs from interneurons (IN), and represent specific states of inhibitory signaling in the spinal cord. The motoneurons are shown to receive homonymous excitatory input and input from an Ia interneuron driven by its antagonist. The data shown in this study paper are consistent with the hypothesis that in the control animal, inhibitory inputs to both flexor (F) and extensor (E) neurons are modest (A), whereas spinal transection results in a marked increase in inhibitory inputs to both flexor and extensor motoneurons. Step training (B) results in a reversion to a more normal level of inhibitory input to both extensor and flexor motoneurons. In stand-trained animals (D), a relatively high level of inhibitory inputs is sustained to flexor motoneurons, whereas extensor motoneurons have slightly less inhibitory input compared with non-trained animals (C). Because of the chronic absence of weight-bearing and, therefore, extensor activity in spinal- transected, non-trained animals (B), we further hypothesize that the level of inhibitory input to the extensor motoneurons will exceed that observed for the flexor motoneurons. (Reproduced with permission; Tillakaratne et al 2002, J Neurosci 22: 3130-3143; Copyright 2002 by the Society for Neuroscience).

These findings suggest that the inability of stand-trained spinal animals to step is closely linked to an elevated level of inhibition of flexor motor pools. More recent work from our laboratory in rats that had undergone complete spinal cord transection at the mid-thoracic level confirms the motor pool-specific changes in GAD_{67} observed earlier in cats. Retrograde labeling of specific motor pools, i.e., the soleus, an ankle extensor, and the tibialis anterior (TA), an ankle flexor, with Fast Blue and Fluorogold showed that the changes in GAD_{67} are specific to certain motor pools depending on the motor task that was trained. As schematically shown in Figure 3-4, GAD_{67} is increased around both the soleus and TA motor pools after a complete spinal cord transection compared to controls. With 6 weeks of step training, the GAD_{67} levels around both motor pools return towards normal levels, whereas with unilateral stand training, GAD_{67} levels were reduced around the soleus motor pool of the trained side, but remained increased around the TA motoneurons. The implication of these data is that the type of locomotor training performed and the role of specific muscles in those tasks will impact the cellular status of localized regions and specific neurons of the spinal cord.

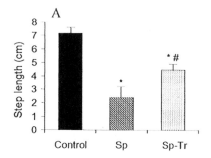
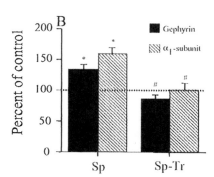

Figure 3-5. Step training of spinal rats normalizes glycine receptor concentration. A, mean step lengths of control, spinal cord transected (Sp) and step-trained (Sp-Tr) rats. Sp and Sp-Tr rats were spinal cord transected at 7 days of age. Sp-Tr rats were trained to step on a motorized treadmill at 0.09 ms-1 for 15 min per day, 5 days per week for 12 weeks. N=4 per group. B, mean amounts of gephyrin and the a1 subunit of the glycine receptor in the lumbar spinal cord of Sp and Sp-Tr rats. These data are normalized to a total of spinal cord protein and expressed as a percentage of control based on densitometer units from Western blots using monoclonal antibody 7A (Roche Biochemicals). *Significantly different from control; # significantly different from Sp at P≤0.05 according to a one-way analysis of variance and Fisher's least significant difference test. (Reproduced with permission; Edgerton et al 2001, J Physiol 533: 15-22)

Other related signaling molecules involved in the modulation of inhibitory neurotransmitters are altered as well. Changes in glycine and $GABA_A$ receptors in lumbar spinal cord sections were determined in step-trained and non-trained rats. Western blot analysis showed (Fig. 3-5) that the glycinergic system is sensitive to injury and motor training, i.e., the α_1-subunit of the glycine receptor and the receptor-associated protein gephyrin increased after spinal cord transection, and decreased in response to a subsequent 12-week period of step training (Edgerton et al., 2001a).

There is accumulating evidence that the modulation of neurotransmission after SCI also may have an effect on non-neuronal cell types. For example, compared to control rats, spinal transected rats show elevated immunostaining levels of the $GABA_A$ receptor γ-2 subunit in astroglial cells, i.e., cells positive for the cellular marker glial fibrillary acidic protein (GFAP) in the regions surrounding retrogradely-labeled soleus and TA motoneurons (Bravo et al., 2002). This elevated level returns towards nearly undetectable, control levels after 6 weeks of step training (Bravo et al., 2002). Since benzodiazepines, commonly used clinically as muscle relaxants, require the γ-2 subunit of the $GABA_A$ receptor for their effectiveness, further studies of γ-2 modulation in astrocytes as a means of modulating the excitability of spinal neurons may yield important information. An assessment of the effectiveness of benzodiazapines on sensory and motor function after SCI, theoretically, has considerable potential for the enhancement of motor function.

Other neurotransmitter systems are altered by SCI as well. Autoradiographic receptor binding studies in adult cats show that noradrenergic (NA) and 5-HT receptor densities change in selected laminae of lumbar segments after a complete spinal cord transection at T13. Shortly after transection the labeling of spinal α1- and α2-NA and 5-HT1A receptors were elevated, but returned to control values after one month (Giroux et al., 1999). However, when spinal rats received a fetal spinal cord transplant (E14), and were trained to step on a treadmill, the 5-HT agonists quipazine and DOI ($\pm$ -1-[2,5]-dimethoxy-4-iodophenyl-2-aminopropane) facilitated locomotion, suggesting a role for 5-HT in enhancing locomotor capacity (Kim et al., 1999). In adult spinal rats, the spinal cord caudal to the transection site was totally devoid of NA-immunoreactive fibers (Roudet et al., 1995). Furthermore, the densities of both α-1 and α-2 receptors showed denervation-induced increases in different gray matter regions (Roudet et al., 1995; Roudet et al., 1996). Five weeks after implantation of embryonic NA neurons from the locus coeruleus, NA-immunoreactive fibers were observed in the lumbar spinal cord whereas α-1 receptor density decreased towards control levels in the reinnervated regions (Roudet et al., 1996). α-2 receptor reversal was not as robust as that of the α-1 receptors. To what degree the state of locomotor training affected the level and the rate of change in NA or 5-HT receptor densities has not been studied in the spinal cord of transected rats or cats.

Pharmacological studies have provided complementary insight to some of the molecules that may be involved in the recovery of locomotion after a complete spinal cord transection. Pharmacologically applied agents associated with NA, 5-HT, GABAergic, glycinergic and glutamatergic neurotransmitters/modulators induce changes in locomotion and/or standing of spinal animals and humans (Rossignol et al., 2002). For example, clonidine, a NA α-2 agonist, facilitates weight-bearing stepping in some adult, completely spinalized cats (Chau et al., 1998a; Giroux et al., 2001), as does the NA α-1 agonist methoxyamine (Chau et al., 1998b). Interestingly, the α-1 antagonist, prazosin inhibits the locomotor capacity of spinal transected cats that can step without pharmacological assistance, whereas the α-2 antagonist yohimbine has no negative impact on the locomotor ability of these cats (Giroux et al., 2001). Furthermore, clonidine administration is prohibitive to stepping in cats with an incomplete SCI, possibly due to its action at presynaptic synapses, whereas it has no effect on the stepping ability in animals with a complete spinal cord transection, where it has only a post-synaptic action (Giroux et al., 1999). These findings highlight the importance of understanding the mechanisms of action of the molecules involved and their site of action when developing pharmacological therapeutic interventions.

Pharmacological modulation of the glycinergic system also affects locomotion. When given small doses of strychnine, a glycine receptor antagonist, adult spinal cats that have never had any motor training can perform some weight-bearing alternating stepping, suggesting a role for glycine in the cellular basis of locomotion (de Leon et al., 1999b). The GABA$_B$ receptor agonist baclofen is commonly used as an anti-spastic medication in SCI patients, but it also reduces their capacity for locomotion (Norman et al., 1998). We do not know if or how the GABA$_B$ receptors are changed by SCI or by motor training. Such studies, however, may provide more definitive information about the usefulness of these drugs after a SCI.

Ultimately, the cellular status of the spinal cord will be reflective of its response to pharmacological treatment. Since we know that cellular changes vary with the severity of and time after an injury, as well as with training, a drug that may be appropriate for an acute SCI patient may not be appropriate for a more chronic patient. For example, when clonidine is given daily to spinal cats in concert with motor training, the onset of weight-bearing stepping ability is about 3 weeks faster than in cats receiving motor training alone, suggesting a synergistic effect of this combined therapy (Rossignol et al., 2002). However, spinal cats that were trained to step and had good stepping ability showed no locomotor changes after the administration of strychnine, whereas animals that stepped poorly, such as cats trained to stand, performed quite well after being given strychnine (de Leon et al., 1999b). Similar results have been observed by Yi and colleagues (Yi et al., 1998). Rats were neonatally spinally transected, and a subset received a transplant of E14 spinal cord tissue into the lesion site. Following several weeks of locomotor training, they were given either quipazine or DOI (5-HT agonists) and their locomotor ability was tested. Only poorly stepping transplant rats showed improvement after 5-HT agonist administration, while those that could step well before drug administration showed no further improvement (Yi et al., 1998). Thus, the efficacy of pharmacological therapy on the recovery of locomotor function appears to depend on the cellular status of the spinal cord at the time of the pharmacological intervention.

3.5 Growth factors

Interest in the potential of growth factors to facilitate recovery of motor function continues as we learn more about their effects and their biological bases. Perhaps the greatest interest in growth factors, particularly those referred to as nerve growth factors or neurotrophins, has been in their possible role in stimulating axons to grow across the lesion of an injured spinal cord (Tuszynski et al., 1994). Although some success has been reported in demonstrating elongations and projections of descending and ascending axons, the net functional effects shown by both sensory and motor tests have not been impressive as a whole. A second mode of action of one growth factor, brain-derived neurotrophic factor (BDNF) has been a physiological one. For example, stepping behavior has been induced in spinal rats given BDNF (Jakeman et al., 1998). It is not clear whether BDNF acted directly on those neurons that induce CPG or whether it played a facilitating role, as it appears that strychnine and bicuculline do. It appears that these agents enable the spinal circuitry to process the proprioceptive input associated with load-bearing, allowing stepping to occur. A third aspect of nerve growth factors of interest from a rehabilitative perspective is the modulation of their expression in response to exercise. A series of studies show that BDNF mRNA and protein in the hippocampus (Kesslak et al., 1998; Molteni et al., 2002), spinal cord (Gomez-Pinilla et al., 2002) and skeletal muscle (Gomez-Pinilla et al., 2001) are highly responsive to the amount of wheel running in rats. These observations raise the question as to what role BDNF, as well as other growth factors such as neurotrophin-3 (NT-3), may play in the adaptations associated with step training after SCI.

3.6   Cell implants

Multiple cellular implant approaches are being used to promote regeneration in the spinal cord after SCI (Bartolomei and Greer, 2000; also see Progress in Brain Research Vol. 137). In this chapter we will focus on and highlight some studies that have assessed the level of motor recovery using four transplant procedures after a complete spinal cord transection: 1) Schwann cells; 2) peripheral nerves; 3) fetal tissue; and 4) olfactory ensheathing cells (OECs).

3.6.1   Schwann cell implants

Schwann cells normally ensheath and myelinate axons and promote regeneration in both the PNS and CNS (Dezawa, 2002). Therefore, they have been used extensively in attempts to promote regrowth of descending or ascending axons across a complete spinal cord transection. Guest et al. (1997) used copolymer guidance channels filled with human Schwann cells to form a bridge, plus the administration of methylprednisolone to decrease secondary infection, in completely transected (T8) athymic nude female rats. Based on retrograde labeling ~6 weeks after surgery, no fibers from descending pathways (5-HT positive fibers) were observed in the spinal cord below the lesion indicating that there was no regeneration of the corticospinal tract (a major motor pathway). Also, no ascending fibers were observed to reach any brainstem area. Thus, only propriospinal and sensory neurons reentered the host spinal cord, with only about 1% of these fibers crossing the lesion site. BBB scores ranged between 7 and 8 in the transplant rats (Basso et al., 1995). No data were provided on non-transplant rats, although other studies have reported BBB scores of ~1-2 in non-treated spinal rats (Lee et al., 2002). There also was no evidence of sensory awareness to tactile stimuli. Similar results were observed when spinal cord cells distal to the Schwann cell bridge were transduced with adeno-associated viral vectors encoding BDNF and NT-3 in female adult Fischer rats (Blits et al., 2003). After 12 weeks, BBB scores were slightly, but significantly, higher in the neurotrophin-treated spinal rats (6.9) than in green fluorescent protein-treated (5.8) or saline-treated (5.6) spinal transplant rats. From a functional perspective, Pinzon et al. (2001) implanted Schwann cells mixed in Matrigel in adult female Fischer rats and after 3 months were able to record measurable evoked potentials rostral to the transection site following electrical stimulation of regenerated axons within the graft. This was observed in 2/9 rats implanted with Schwann cells plus Matrigel and none of the rats implanted only with Matrigel. Thus, functional connectivity between neural structures above and below the site was observed in some implanted rats.

3.6.2   Peripheral nerve implants

Peripheral nerve implants include Schwann cells and also provide a biological conduit for axonal growth. Such implants have been used in combination with a number of growth factors, matrices, etc. to promote regeneration in the CNS. Cheng and co-workers reported partial motor function recovery in the hindlimbs of compete spinal cord transected (T8) adult rats using a peripheral nerve bridge implant system (Cheng et al., 1995; 1996; 1997). In this model, a close apposition of the cut ends of the spinal cord is maintained via a shortening and fixation of the vertebral column at the level of the spinal cord transection and the application of fibrin glue at the transection site (Cheng and Olson, 1995; Cheng et al., 1997). In subsequent experiments, multiple intercosal nerves were implanted to bridge the remaining gap between the cut ends of the spinal cord to reroute the regenerating pathways from non-permissive white to permissive gray matter (Cheng et al., 1996;1997). In addition, acidic fibroblast growth factor (aFGF), a substance known to enhance nerve fiber development, was mixed in the fibrin glue. Rats were maintained from 1 to 12 months. Retrograde and anterograde labeling indicated fiber tract regeneration, particularly for the corticospinal tract. Functional assessments included an Open Field Walking Scale (Blight and Young, 1989), a Combined Behavioral Score (BBB scale) (Basso et al., 1995) and an in-house gait analysis procedure (Cheng et al., 1997). Compared to non-treated spinal rats, the spinal rats receiving white matter to gray matter bridges and aFGF-containing fibrin glue showed significant improvement in their motor capabilities: 1) appearance of functional posture, i.e., flexion at the hips, knees and ankles; 2) some alternating hindlimb movements and partial weight support with a trend for improved performance during repeated trials; 4) some coordination between the forelimbs and hindlimbs; 3) improved contact placing; and 4) some gait patterns similar to those observed in non-spinalized rats. Lin and co-workers (Lee et al., 2002) used the same preparation (Cheng et al., 1996) and tested the spontaneous motor recovery in completely transected (T8) adult female Sprague-Dawley rats implanted with autografts of intercostals nerves, fibrin mixture containing aFGF, or both. The rats were maintained for 6 months with no exercise intervention. BBB scores were highest in the group receiving both treatments, reaching a mean score of 6.75 after 24 weeks, whereas the other groups had scores of 1 or below. Contact placing scores were also higher in the group receiving both treatments than in the other transected groups. Motor evoked potential amplitudes (stimulation of the motor cortex, recording from the quadriceps, adductor and gastrocnemius muscles) and somatosensory evoked potentials (stimulation of the sciatic nerve, recording from somatosensory cortex) amplitudes were higher in the group receiving both treatments than in the other groups. These findings indicate that both some sensory and motor (presumably corticospinal) pathways were formed. Retransection eliminated both responses and this was interpreted as evidence that the recovery of motor control was due to reconnection of long descending tracts. Neurofilament-positive fibers (reflecting regenerating axons) extended through the graft (proximal to distal). Motor evoked potential amplitudes and BBB scores were highly correlated ($r = 0.85$) only in the group receiving both treatments. These results are generally consistent with those of (Cheng et al., 1996), and indicate some positive results for motor recovery using this type of implant.

3.6.3   Fetal tissue implants

The rationale for using fetal cell transplants after SCI includes the following: 1) to restore damaged neurons that form local intersegmental synaptic circuits; 2) to serve as a bridge for severed axons to cross the lesion site in either direction; 3) to serve as a relay for supraspinal input to synapse with donor graft neurons, which then can synapse with distal target neurons; and 4) to serve as a source of trophic support to rescue host neurons from cell death (Bregman, 1998; Howland et al., 1995).

Coumans et al. (2001) transected the spinal cords of 6 to 8 week-old rats at T6-T8 and whole pieces of embryonic day 14 fetal spinal cord tissue were implanted immediately or 2 or 4 weeks after the surgery. In addition, at the time of the fetal implants, osmotic minipumps delivering either human recombinant BDNF or NT-3 were inserted subcutaneously in some groups and maintained for 2 weeks. Rats that had a delayed fetal implant plus administration of either neurotrophic factor had a higher density of 5-HT and NA axons within the transplant site and in the spinal cord regions distal to the transection site than in rats having only a transection or having an immediate fetal transplant, with or without neurotrophin treatment. Based on retrograde labeling, the transection plus neurotrophin-treated rats showed connectivity between the spinal cord below the transection site and the red nucleus, locus coeruleus, raphe nucleus, lateral vestibular nucleus, reticular formation and layer V of the hindlimb sensorimotor cortex. In addition, anterograde labeling from the motor cortex revealed connectivity of the corticospinal tract as far as the lumbar enlargement. Behavioral analyses showed that the delayed-implanted and neurotrophin-treated group performed closer to normal than any other experimental group. Three to four weeks after transection, these rats exhibited partial weight support and extended periods of forelimb-hindlimb coordination during runway walking (with postural support from the sides), reciprocal and bilateral stepping during assisted (adjustable Velcro harness fitted between the forelimbs and external postural support for the hindquarters provided as needed to provide postural stability) quadrupedal locomotion on the treadmill, and coordinated forelimb-hindlimb stepping up stairs. None of these abilities were observed in the other experimental groups. When the spinal rats were trained on the behavioral tasks for one week (week 5 after surgery), the delayed-implanted and neurotrophin-treated group showed the most improvement, with about 50% of these rats showing consistent or inconsistent weight-supported plantar stepping capabilities and a significantly higher number of weight-supported steps during stair climbing than any other group. Retransection of the cord at 4 months post-transection abolished the recovery of these motor tasks and no improvement was observed over the next 5 weeks. Thus, at least some of the recovery of motor function seemed to be related to the anatomical reconnections in these spinal rats, although there was not a close relationship between the level of functional improvement and the extent of supraspinal fiber growth.

The two major conclusions from this study were: 1) the presence of neurotrophins is required for supraspinal axons to grow across the lesion; and 2) the procedures for the delayed implant, i.e., clearing away the glial scar at the injury site prior to the implantation of the fetal tissue, may serve as a "conditioning lesion" and enhance the capability of regrowth across the transection site. It appears that immediate fetal transplant plus neurotrophin therapy enhances propriospinal input, whereas a similar delayed therapy enhances both propriospinal and supraspinal input across the transplant, with a concomitant improvement in locomotor function.

Implants using other types of fetal neural tissue have shown promise for restoring some motor function. For example, Giménez y Ribotta et al. (1998, 2002; Ribotta et al., 2000) spinally transected adult rats at T9 or T11. One week later, they injected a suspension containing embryonic raphe cells into the spinal cord at the same levels. Two months later the T11 transplanted-spinal rats could stand supporting their body weight for short periods when placed in a convenient posture and could walk on a treadmill with tail pinching. The kinematics and EMG patterns observed during the treadmill locomotion were remarkable similar to that observed in control rats. Non-transplanted spinal rats were unresponsive and T9-transplanted spinal rats were much less responsive to the same stimuli. The spinal cords of non-transplanted spinal rats were devoid of 5-HT terminals below the level of the lesion, whereas both transplanted groups showed extensive immunoreactivity for 5-HT. A significant difference, however, was that 5-

HT fibers were detectable at the L1-L2 level in the T11, but not T9, transplanted rats. Retransection of the spinal cord at the same level had no effect on the motor performance, clearly showing that the improved function was due to a direct effect of the transplant on the lumbar spinal cord circuitry (CPG) and not to any supraspinal influences. These results indicate that 5-HT reinnervation of the upper lumbar regions is a key element for the generation of locomotor rhythmic activity in vivo.

Similar results were observed by Slawinska et al. (2000) when they transected (T9-T10) the spinal cord of young adult Wistar rats, waited 1 month and then implanted a piece of embryonic tissue (raphe nucei from E14-15 embryos) at the transection site. After two months, the upper body of the rat was held on a trolley that was moved along a horizontal path while the tail was pinched continuously. The hindlimbs of rats with a transplant showed alternate stepping movements with proper plantar placement during stance and proper dorsiflexion during swing. In addition, the EMG burst waveforms in the soleus (primary plantarflexor) and TA (primary dorsiflexor) muscles were similar to normal and there was appropriate alternation in the EMG bursts of these antagonists. Non-transplanted rats did not show these appropriate kinematic or EMG responses. Immunohistochemical analyses revealed that 5-HT positive axons were present in the graft region and in the spinal cord distal to the lesion in the implanted, but not the non-implanted rats. Thus, the effects of an embryonic raphe implant on both the anatomy of the spinal cord and on motor recovery of completely transected adult rats were similar whether the implant was received one week (Giménez y Ribotta et al., 1998, 2002; Ribotta et al. 2000) or one month (Slawinska et al., 2000) post-transection.

These results using implants of fetal raphe tissue (Giménez y Ribotta et al., 1998, 2002; Ribotta et al., 2000; Slawinska et al., 2000) differ in two important ways from the implants using fetal spinal cord tissue (Coumans et al., 2001). First, Coumans et al. (2001) attributed the improvement in performance to the growth of axons across the lesion site, whereas Giménez y Ribotta et al. (1998, 2002; Ribotta et al., 2000) attributed all of the behavioral improvement to adaptations intrinsic to the lumbosacral spinal cord. Second, Coumans et al. (2001) attributed most of their success to the delayed strategy of the implant, while the post-injury timing made no difference to the outcome in the studies by Giménez y Ribotta et al. (1998, 2002; Ribotta et al., 2000) and Slawinska et al. (2001).

More extensive study of the effects of fetal tissue transplants has been performed in neonatal animals. Reasons for this include the fact that, in general, the plasticity of the nervous system is greater and that spontaneous recovery is more robust at a younger than an older age. Nakamura and colleagues (Hase et al., 2002a, b); Asada et al. 1998) transected rats pups at post-natal day 2 (P2) at T10-11 and implanted with embryonic spinal cord (E14, oriented in normal position) or peripheral nerve grafts (sciatic nerves from 2 week-old rats that were axotomized 48 hours prior to implant) at the transection site. After 6 weeks, some spinal cord implanted rats showed locomotor capabilities that were "virtually normal" with a delay of 5 days compared to normal rats. Their scores ranged from 6-20 (average score, ~15) on the BBB scale, whereas the non-grafted rats ranged from 0-6 (average score, ~3) and for peripheral nerve implant rats it was ~10. Since this was during development, there was an increase in locomotor performance in all groups (including controls) with the control peaking at about 17 days and both graft groups at about 22 days (5-day delay). Retrograde labeling showed a wide range of labeled neurons in the raphe nuclei, vestibular nuclei, red nucleus and sensorimotor cortex in the rats receiving the embryonic graft but not in the non-graft rats. Locomotor ability was highly correlated with the number of labeled cells in the raphe nuclei and sensorimotor cortex, but not in the vestibular or red nuclei. Thus, it appears that the transplant affected specific motor tracts.

Iwashita et al. (1994) completely transected 1-2 day old rats at T10-11 and implanted either embryonic spinal cord tissue (E14-E17, T10-T11 segments) in a normal or inverted dorso-ventral or rostro-caudal orientation, or sciatic nerve at the transection site. Rats were maintained for 1 to 4 months. Retrograde labeling from the lumbar region of the spinal cord showed staining in the upper brain structures. Anterograde labeling in the sensorimotor cortex labeled the corticospinal tract and both the spinocerebellar (cerebellum) and spinothalamic (thalamus) tracts were labeled in the lumbar spinal cord. The rats with normally oriented grafts showed good motor performance with hindlimb-forelimb coordination during walking and climbing a grid. When thrown in the air, these rats showed a righting reflex and landed on all four extremities. The other groups did not show this level of motor coordination, suggesting that some guidance cues were related to the orientation of the implant tissue.

Murray and coworkers (Kim et al., 1999; Miya et al., 1997; Tessler et al., 1997) used a model of complete transection at T8-T9 and implantation of fetal spinal cord tissue (E14) in rat pups within 48 hours of birth. Miya et al. (1997) trained (3 to 5 days/week) the rats for 5 to 10 weeks after weaning. Training consisted of walking on a treadmill (bipedal and quadrupedal stepping, 15 minutes, 4 days/week), crossing wide and narrow runways (4 crossings/day, 3 times/week), descending stairs (5 trials, 3 times/week), and crossing a horizontal ladder for a water reward (4 crossings/day, 3 times/week). Anatomical analyses included: 5-HT antibodies; myelin-associated protein (MAP-2, to identify neurons within the transplant); calcitonin gene related peptide (CGRP, to identify dorsal root axons); GFAP (to identify astrocytes); and Ox-42 (to stain activated monocyte-derived cells). Bipedal locomotion was similar in all 3 groups studied, i.e., control, spinal, spinal-transplant. During quadrupedal treadmill locomotion, the spinal-transplant rats performed better than the spinal rats, i.e., showed more weight-supported steps, although both groups showed a wide variability in performance. All rats crossed the runway without hesitation. Spinal rats basically showed no weight-supported hindlimb steps when crossing the wide runway, and some nonweight-bearing scissor-like stepping motions in the hindlimbs. Spinal-transplants rats crossed the runway with poor quadrupedal stepping that was described as "intermittent control of hindlimb placing and stepping" and included dorsal stepping, and exaggerated hindlimb abduction. On the narrow runway, the spinal rats performed similarly kinematically to that on the wide runway, but also often fell and "froze" rather than crossing the runway. The spinal-transplant rats actually performed better on the narrow than wide runway, i.e., had better intralimb and interlimb coordination, no falls, no freezing, and little dorsal stepping. Thus, the "challenge" of walking on the narrow beam improved the kinematics and coordination of stepping in the spinal-transplant rats. On the stair descent task, the spinal rats showed few or no hindlimb weight-supported steps (dragging the hindlimbs down) and had many falls. Some spinal-transplant rats used hindlimb weight-supported steps to descend the stairs, i.e., getting both the hindlimbs and forelimbs on the same step each time. Spinal rats crossed the ladder by dragging their hindlimbs. Spinal-transplant rats performed better than the spinal rats, but still were poor in performing this task. This was the most difficult motor test and the authors suggested that the poor performance was due to "a poor level of motor precision and impaired ability to coordinate sensory information between the fore and hindlimbs". Retransection as adults resulted in little change in the performance (during 5 weeks of recovery) of the spinal rats, but resulted in a decrease in performance of the spinal-transplant rats. In rats receiving the implant, MAP2 staining at the lesion site was consistent with survival of the implanted cells. Although 5-HT staining was observed caudal to the transplant in some rats, neither the density nor the distribution of the 5-HT staining seemed to correlate with motor performance. DRG axons sometimes entered the transplant (based on CGRP staining).

The authors suggested that spinal rats could locomote bipedally via "reflex hindlimb stepping", i.e., probably via proproceptive input to the CPG. The spinal rats could perform limited weight-supported steps during quadrupedal stepping and other motor tasks tested, i.e., little interlimb coordination or postural adjustment capability. Spinal-transplant rats on the other hand, performed much better on most tasks and showed weight-supported steps and postural adjustment capabilities, perhaps via the transplant allowing descending pathways to make connections with the spinal circuits below the transection. However, the limited extent of the connections may be reflected in the poor performance in the ladder-crossing task. The presence of 5-HT staining caudal to the lesion would suggest reconnectivity as a possible mechanism for recovery, but there was a poor correlation between 5-HT staining and motor performance. The authors suggested that the transplanted tissue could have had an indirect or trophic effect on the axotomized host neurons and have kept these cells alive.

Kim et al. (1999) examined the effects of the 5-HT agonists quipazine and DOI on weight-supported hindlimb locomotion in spinal and transplanted rats (same model as Miya et al., 1997). Rats were trained on a treadmill at 3 speeds (2, 5 and 10 cm/sec) one 9-minute session/day, 5 days/week for 4 to 6 weeks. Some rats were injected with quipazine (i.p., 3 doses) or DOI (i.p., 2 doses) with 2 days between each testing (and injection) day. Reuptake inhibitors also were administered to another series of rats (sertraline hydrochloride plus D-fenfluramine hydrochloride). As observed by Miya et al. (1997), the transplanted rats showed more weight-bearing stepping than spinal rats. Both quipazine and DOI increased the number of weight-bearing steps, the number of consecutive weight-bearing steps (no change in total number of steps), increased trunk elevation and hindlimb support during stepping. But this effect was observed primarily in the transplant rats that had moderate or poor walking capabilities. The selective serotonin reuptake inhibitors had no effects, suggesting that endogenously available serotonin does not contribute to the motor function provided by the transplants or agonists. As seen by Miya et al. (1997), 5-HT staining was evident caudal to the transection in the transplanted rats, but there was a poor relationship between motor performance, responsiveness to the drugs and the extent of 5-HT staining. No 5-HT axons extended to the lumbar region.

Kim et al. (2001) subsequently injected a single daily dose of m-CPP (which has a high affinity for 5-HT$_{2C}$ receptors) 5 minutes before treadmill testing each day for 14 days in one set of rats or CMI (a 5-HT/NA reuptake inhibitor) for 21 days in another group of rats. All rats were trained on a treadmill starting after weaning for 5-minute sessions, 5 days/week for 3 to 5 weeks. M-CPP treatment dramatically increased weight-supported stepping in both spinal and transplant rats. Combined these studies suggest that stimulating spinal 5-HT$_{2C}$ receptors is beneficial for motor function in spinal animals. CMI had no effect on weight support, and this finding was consistent with the earlier work suggesting that endogenous 5-HT produced by any 5-HT axons that regenerated through the fetal grafts was inadequate to restore function.

Similar results using fetal transplants have been observed in kittens. Howland et al. (1995) completely transected kittens at a low thoracic level within 2 days of birth. Some kittens received fetal spinal cord (low thoracic-high lumbar) transplants (E21-E26). Spinal kittens were trained (bipedal and quadrupedal treadmill and overground locomotion) 4 to 6 times/week for at least 5 months. All spinal kittens with transplants developed full weight-supporting overground locomotion by 6 weeks, when non-transplanted kittens did not even show partial support. In addition, balance and coordination between the forelimbs and hindlimbs was attained in kittens with transplants by 20 weeks, whereas spinal kittens without transplants did not show this level of motor recovery. Detailed kinematic analyses during quadrupedal locomotion on a treadmill showed that the forelimbs and hindlimbs of the non-transplant kittens were poorly coordinated. The non-transplant kittens also did not have appropriate knee flexion during swing and were unable to maintain pace with the treadmill as the transplant kittens.

Spinal cord immunohistochemical staining showed that the transplants had a wide range of neuronal density, but that the density was not related to locomotor ability. Both segmental (CGRP reactive, either motoneurons or cervical dorsal horn neurons) and descending (5-HT and NA, brain stem neurons) host axons grew into the transplant. These authors suggested that propriospinal neurons likely contributed to the improved forelimb-hindlimb coordination, that NA axons from the brainstem may have affected the CPG, and that 5-HT axons may have increased muscle tone and greater weight support in the transplant vs. non-transplant animals. However, there was no direct test to show that the improved function could be attributed to NA neurons or any other neuron type from the brainstem, nor whether any descending axons contributed to the improved motor performance.

3.6.4   OEC implants

OECs have been studied extensively in the past few years. Neurons in the olfactory mucosa are unique in that they are the only neurons that continue to divide throughout life. This growth is supported by specialized glial cells from the mucosa called OECs (Ramon-Cueto and Avila, 1998). In turn these OECs are the only glial cells that can cross the PNS-CNS boundary and, therefore, have the potential to serve as a "guide" for regenerating axons. Ramon-Cueto et al. (1998) implanted Schwann cell-seeded guidance channels that crossed the lesion site and then injected suspensions of OECs into both the proximal and distal stumps of the transected spinal cord. After 6 weeks, there was extensive regeneration of raphespinal and propriospinal axons. The transplanted OEC cells were evident throughout the spinal axis and consistently accompanied the regenerating axons well beyond the host-graft interface and glial scar. Subsequently, Ramon-Cueto et al. (2000)

completely transected (T8-T9) the spinal cord of adult female rats and injected OECs at 4 sites at the midline of both spinal cord stumps. The rats were trained to climb an inclined grid (4 incline levels) to reach a platform for a food reward. Hindlimb movement, plantar placement of the paw, and body weight support were required to reach the platform, i.e., the hindlimbs had to contact the rungs and push upward to reach the platform. The rats were maintained for ~7 months and were tested weekly for voluntary motor function during months 3 to 7. The OEC-injected spinal rats showed a variable amount of recovery in the climbing task, but all rats in this group showed some improvement over the course of the experiment, and were able to reach the platform at least at one incline level. None of the spinal non-injected rats could accomplish this task. All OEC-injected rats also could respond to light touch (contact placing) and joint bending (proprioceptive) stimuli. There was a strong positive correlation ($r^2 = 0.95$) between climbing ability and the response to sensory stimulation. Based on anterograde labeling, both NA and 5-HT axons were observed in the distal stumps of OEC-treated but not non-treated rats, indicating some regeneration of the corticospinal tract. The distance traveled by the axons into the distal stump and the level of motor recovery at 7 months also were highly correlated ($r^2 = 0.95$). These are some of the most dramatic improvements that have been observed in functional recovery after a complete spinal cord transection using a regeneration-promoting intervention.

Lu et al. (2001) transected (vertebral level T10) adult female Sprague-Dawley rats and either OECs (50% pure, compared to 98% pure in Ramon-Cueto et al. (2000) cultured from olfactory lamina propria were injected into the midline region of the two stumps of the transected cord or pieces of lamina propria from the olfactory mucosa were implanted at the transection site. This method is different from that of Ramon-Cueto's (2000) work in that the OECs are derived from the olfactory mucosa of the nasal septum of adult rats rather than from the olfactory bulb from embryos. Beginning at 4 weeks after surgery, BBB scores were higher in both transplant groups compared to medium-injected rats. After 10 weeks, the mean BBB score was ~2 in controls and between 5 and 6 in the transplant groups. Three weeks after retransection, the BBB scores of the transplant groups were ~2, indicating that the improvement in motor performance was dependent on regrowth of axons through the lesion site. Retrograde labeling from the distal spinal cord appeared in the brainstem (raphe and gigantocellularis neurons) and anterograde labeling and immunohistochemical analyses demonstrated 5-HT fibers within the transection site and in the spinal cord below the lesion. In a subsequent study, Lu et al. (2002) reported similar improvements in motor performance and regeneration across the transection site after implantation of olfactory lamina propria containing OECs 4 weeks after a complete transection of the cord in adult rats. These data are significant in that they provide evidence that these cells can promote functional recovery even when transplantation is delayed after SCI.

3.6.5   Perspective on recovery of motor performance after cell implants

The above short overview indicates that a variety of cell implants have some effect on motor recovery after a complete spinal cord transection in adult or neonatal animals. The extent of recovery appears to be more dramatic in neonatal than adult animals, most likely reflecting a generally higher plasticity in developing than mature systems. Although some axonal regeneration has been observed via both anterograde and retrograde labeling techniques, the amount of regeneration generally is limited, particularly from supraspinal sources. A significant point is that the extent of motor recovery reported in many of these studies, especially in animals transected as adults, also is not overly impressive. In addition, it is difficult to compare the improvements in motor function across studies because of the varied and subjective nature of the behavioral tests used.

It is also becoming more apparent that a combination of therapies may be required to get complete or near complete recovery. For example, the more robust results reported after the OEC implantations in completely transected rats might be improved further if combined with an exercise training paradigm. This is particularly important since regular bouts of exercise 1) increase the availability of specific neurotrophins in the spinal cord that may enhance regeneration and/or stimulate CPGs (Gomez-Pinilla et al., 2001; 2002) and 2) can directly impact CPGs and may make the spinal circuitry below the lesion more receptive to axonal elongation from the implant. It is conceivable, for example, that new connections indeed could be formed as a result of the newly implanted cells with or without the supplementation of growth factors, but the spinal circuitry could be in a non-trained state such that the new connections could not manifest their influence on the locomotor potential of the spinal animals. Therefore, a prudent approach would be to execute a repair strategy in an animal that has a physiologically "fit" spinal cord and skeletal muscular system.

4.   OVERALL SUMMARY

Significant levels of postural and locomotor control can be recovered following a SCI. This can occur because many of the details of the neural control for these motor tasks are embedded within the spinal cord itself.

Even though the spinal circuitry normally depends on and interacts closely with supraspinal control, relatively complex and highly adaptive tasks can be performed without supraspinal-spinal connectivity. The control of these complicated motor tasks can occur because of the ability of the spinal cord to process and interpret peripheral sensory input associated with weight bearing that occurs during standing and stepping. These peripheral inputs project to the spinal circuitry, some of which may be responsible for activating the CPGs, and thus help to define the physiological state of the spinal cord at any given instant.

Functional motor recovery is also possible because the spinal cord has a robust ability to adapt to any imposed activity pattern following SCI. The nature of the adaptation appears to be defined by the specific activity patterns of the remaining functional circuitry that occurs over a period of weeks to months. In spite of the functional plasticity that can occur within the spinal circuitry and in the sensory input to the circuitry, there remains a severe deficit in the ability to execute independent stepping and standing after a SCI. To solve this problem, a large number of intervention strategies are being tested. Although there have been some positive results, e.g., using cell implants, neurotrophic/growth factors, pharmacological agents, and/or motor training, the level of improvement for independent and functional control of posture and locomotion has been fairly modest. It is also important to note that, in most cases, the postural and locomotor tests that have been used to determine the effectiveness of an intervention are too imprecise. It seems obvious that a combination of these therapeutic strategies will be necessary to achieve the ultimate goal, i.e., attaining a critical level of success in forming functional connections between supraspinal and spinal neurons to allow the most severely injured SCI subjects to achieve a reasonable level of recovery of interlimb coordination and posture and locomotor ability.

REFERENCES

Asada Y, Kawaguchi S, Hayashi H, Nakamura T (1998) Neural repair of the injured spinal cord by grafting: comparison between peripheral nerve segments and embryonic homologous structures as a conduit of CNS axons. Neurosci Res 31: 241-249

Avelev V, Anissimova N, Khoroshikh E, Gerasimenko Y (1997) Activation of central pattern generator in cat by epidural spinal cord stimulation. In: Gurfinkel VS, Levik Yu S (eds) Proceedings of the International Symposium on Brain and Movement St. Petersburg, Moscow, pp.26.

Baev, KV (1998) Biological Neural Networks: Hierachical Concept of Brain Function, Birkhauser, Boston, pp. 1-273

Barbeau H, Rossignol S (1987) Recovery of locomotion after chronic spinalization in the adult cat. Brain Res 412: 84-95

Barbeau H, Rossignol S (1990) The effects of serotonergic drugs on the locomotor pattern and on cutaneous reflexes of the adult chronic spinal cat. Brain Res 514: 55-67

Barbeau H, Rossignol S (1991) Initiation and modulation of the locomotor pattern in the adult chronic spinal cat by noradrenergic, serotonergic and dopaminergic drugs. Brain Res 546: 250-260

Bartolomei JC, Greer CA (2000) Olfactory ensheathing cells: bridging the gap in spinal cord injury. Neurosurgery 47: 1057-1069

Basso DM, Beattie MS, Bresnahan JC (1995) A sensitive and reliable locomotor rating scale for open field testing in rats. J Neurotrauma 12: 1-21

Belanger M, Drew T, Provencher J, Rossignol S (1996) A comparison of treadmill locomotion in adult cats before and after spinal transection. J Neurophysiol 76: 471-491

Blight AR, Young W (1989) Central axons in injured cat spinal cord recover electrophysiological function following remyelination by Schwann cells. J Neurol Sci 91: 15-34

Blits B, Oudega M, Boer GJ, Bartlett Bunge M, Verhaagen J (2003) Adeno-associated viral vector-mediated neurotrophin gene transfer in the injured adult rat spinal cord improves hind-limb function. Neuroscience 118: 271-281

Bravo AB, Bigbee AJ, Roy RR, Edgerton VR, Tobin AJ, Tillakaratne NJK (2002) Gamma2 subunit of the GABAA receptor is increased in lumbar astrocytes in neonatally spinal cord transected rats. Soc Neurosci Abstr 32: 853.5

Bregman BS (1998) Regeneration in the spinal cord. Curr Opin Neurobiol 8: 800-807

Buerger AA, Fennessy A (1971) Long-term alteration of leg position due to shock avoidance by spinal rats. Exp Neurol 30: 195-211

Chau C, Barbeau H, Rossignol S (1998a) Early locomotor training with clonidine in spinal cats. J Neurophysiol 79: 392-409

Chau C, Barbeau H, Rossignol S (1998b) Effects of intrathecal alpha1- and alpha2-noradrenergic agonists and norepinephrine on locomotion in chronic spinal cats. J Neurophysiol 79: 2941-2963

Cheng H, Olson L (1995) A new surgical technique that allows proximodistal regeneration of 5-HT fibers after complete transection of the rat spinal cord. Exp Neurol 136: 149-161

Cheng H, Cao Y, Olson L (1996) Spinal cord repair in adult paraplegic rats: partial restoration of hind limb function. Science 273: 510-513

Cheng H, Almstrom S, Gimenez-Llort L, Chang R, Ove Ogren S, Hoffer B, Olson L (1997) Gait analysis of adult paraplegic rats after spinal cord repair. Exp Neurol 148: 544-557

Coumans JV, Lin TT, Dai HN, MacArthur L, McAtee M, Nash C, Bregman BS (2001) Axonal regeneration and functional recovery after complete spinal cord transection in rats by delayed treatment with transplants and neurotrophins. J Neurosci 21: 9334-9344

de Guzman CP, Roy RR, Hodgson JA, Edgerton VR (1991) Coordination of motor pools controlling the ankle musculature in adult spinal cats during treadmill walking. Brain Res 555: 202-214

de Leon RD, Hodgson JA, Roy RR, Edgerton VR (1998a) Locomotor capacity attributable to step training versus spontaneous recovery after spinalization in adult cats. J Neurophysiol 79: 1329-1340

de Leon RD, Hodgson JA, Roy RR, Edgerton VR (1998b) Full weight-bearing hindlimb standing following stand training in the adult spinal cat. J Neurophysiol 80: 83-91

de Leon RD, Hodgson JA, Roy RR, Edgerton VR (1999a) Retention of hindlimb stepping ability in adult spinal cats after the cessation of step training. J Neurophysiol 81: 85-94 de Leon RD, Tamaki H, Hodgson JA, Roy RR, Edgerton VR (1999b) Hindlimb locomotor and postural training modulates glycinergic inhibition in the spinal cord of the adult spinal cat. J Neurophysiol 82: 359-369

de Leon RD, London NJ, Roy RR, Edgerton VR (1999c) Failure analysis of stepping in adult spinal cats. Prog Brain Res 123: 341-348

Dezawa M (2002) Central and peripheral nerve regeneration by transplantation of Schwann cells and transdifferentiated bone marrow stromal cells. Anat Sci Int 77: 12-25

Dietz V, Wirz M, Curt A, Colombo G (1998) Locomotor pattern in paraplegic patients: training effects and recovery of spinal cord function. Spinal Cord 36: 380-390

Dimitrijevic MM, Dimitrijevic MR (2002) Clinical elements for the neuromuscular stimulation and functional electrical stimulation protocols in the practice of neurorehabilitation. Artif Organs 26: 256-259

Dimitrijevic MR, Faganel J, Sherwood AM (1983) Spinal cord stimulation as a tool for physiological research. Appl Neurophysiol 46: 245-253

Durkovic RG (1975) Classical conditioning, sensitization and habituation in the spinal cat. Physiol Behav 14: 297-304

Duysens J, Pearson KG (1980) Inhibition of flexor burst generation by loading ankle extensor muscles in walking cats. Brain Res 187: 321-332

Edgerton, V.R., Grillner, S., Sjostrom, A. and Zangger, P. Central generation of locomotion in vertebrates. IN: *Neural Control of Locomotion.* R.N. Herman, S. Grillner, P.S.G. Stein and D.G. Stuart (Eds.) Plenum Publishing Corporation, New York, 1976, pp. 439-464.

Edgerton VR, de Guzman CP, Gregor RJ, Roy RR, Hodgson JA, Lovely RG (1991) Trainability of the spinal cord to generate hindlimb stepping patterns in adult spinalized cats. In: Shimamura MM, Grillner S, Edgerton VR (eds) Neurophysiological Bases of Human Locomotion, Japan Scientific Societies Press, Tokyo, pp. 411-423

Edgerton VR, Roy RR, de Leon R, Tillakaratne N, Hodgson JA (1997a) Does motor learning occur in the spinal cord? Neuroscientist 3: 287-294

Edgerton VR, de Leon RD, Tillakaratne N, Recktenwald MR, Hodgson JA, Roy RR (1997b) Use-dependent plasticity in spinal stepping and standing. Adv Neurol 72: 233-247

Edgerton VR, de Leon RD, Harkema SJ, Hodgson JA, London N, Reinkensmeyer DJ, Roy RR, Talmadge RJ, Tillakaratne NJ, Timoszyk W, Tobin A (2001a) Retraining the injured spinal cord. J Physiol 533: 15-22

Edgerton VR, Roy RR, de Leon RD (2001b) Neural darwinism in the mammalian spinal cord. In: Patterson MM, Grau JW (eds) Spinal Cord Plasticity; Alterations in Reflex Function, Kluwer Academic Publishers, pp. 185-206

Forssberg H, Grillner S, Halbertsma J (1980a) The locomotion of the low spinal cat. I. Coordination within a hindlimb. Acta Physiol Scand 108: 269-281

Forssberg H, Grillner S, Halbertsma J, Rossignol S (1980b) The locomotion of the low spinal cat. II. Interlimb coordination. Acta Physiol Scand 108: 283-295

Forssberg H, Grillner S, Rossignol S (1975) Phase dependent reflex reversal during walking in chronic spinal cats. Brain Res 85: 103-107

Gerasimenko YP, Makarovsky AN (1996a) Neurophysiological evaluation of the effects of spinal cord stimulation in spinal patients.In: Stuart DG, Gatchev GN, Gurfunkel VS, Wiesendangerr M (eds) Motor Control VII, Motor Control Press, Tucson, pp. 153–7

Gerasimenko Y, McKay WB, Pollo FE, Dimitrijevic MR (1996b) Stepping movements in paraplegic patients induced by epidural spinal cord stimulation. Soc Neurosci Abstr 22:1372

Gimenéz y Ribotta M, Orsal D, Feraboli-Lohnherr D, Privat A (1998) Recovery of locomotion following transplantation of monoaminergic neurons in the spinal cord of paraplegic rats. Ann N Y Acad Sci 860: 393-341

Gimenéz y Ribotta M, Gavaria M, Menet V, Privat A (2002) Strategies for regeneration and **repair** in spinal cord traumatic injury. In: McKerracher L, Doucet G, Rossignol S (eds) **Progre**ss in Brain Research 137:191-212 Elsevier

Giroux N, Rossignol S, Reader TA (1999) Autoradiographic study of alpha1- and alpha2-noradrenergic and serotonin1A receptors in the spinal cord of normal and chronically transected cats. J Comp Neurol 406: 402-414

Giroux N, Reader TA, Rossignol S (2001) Comparison of the effect of intrathecal administration of clonidine and yohimbine on the locomotion of intact and spinal cats. J Neurophysiol 85: 2516-2536

Gomez-Pinilla F, Ying Z, Opazo P, Roy RR, Edgerton VR (2001) Differential regulation by exercise of BDNF and NT-3 in rat spinal cord and skeletal muscle. Eur J Neurosci 13: 1078-1084

Gomez-Pinilla F, Ying, Z, Roy RR, Molteni R, Edgerton VR (2002) Voluntary exercise induces a BDNF-mediated mechanism that promotes neuroplasticity. J Neurophysiol 88: 2187-2195

Grau JW, Joynes RL (2001) Pavlovian and instrumental conditioning within the spinal cord: Methodological issues. In: Patterson MM, Grau JW (eds) Spinal Cord Plasticity; Alterations in Reflex Function, Kluwer Academic Publishers, pp. 13-54

Grillner S (1985) Neurobiological bases of rhythmic motor acts in vertebrates. Science 228: **143-149**

Grillner S, Ekeberg, El Manira A, Lansner A, Parker D, Tegner J, Wallen P (1998) Intrinsic **function of a neuronal** network - a vertebrate central pattern generator. Brain Res Rev 26: 184-197

Guest JD, Rao A, Olson L, Bunge MB, Bunge RP (1997) The ability of human Schwann cell grafts to promote regeneration in the transected nude rat spinal cord. Exp Neurol 148: 502-522

Harkema SJ, Hurley SL, Patel UK, Requejo PS, Dobkin BH, Edgerton VR (1997) Human lumbosacral spinal cord interprets loading during stepping. J Neurophysiol 77: 797-811

Hase T, Kawaguchi S, Hayashi H, Nishio T, Asada Y, Nakamura T (2002a) Locomotor performance of the rat after neonatal repairing of spinal cord injuries: quantitative assessment and electromyographic study. J Neurotrauma 19: 267-277

Hase T, Kawaguchi S, Hayashi H, Nishio T, Mizoguchi A, Nakamura T (2002b) Spinal cord **repair in neonatal** rats: a correlation between axonal regeneration and functional recovery. **Eur J Neurosci 15: 969-974**

Herman R, He J, D'Luzansky S, Willis W, Dilli S (2002) Spinal cord stimulation facilitates functional walking in a chronic, incomplete spinal cord injured. Spinal Cord 40: 65-68

Howland DR, Bregman BS, Tessler A, Goldberger ME (1995) Transplants enhance locomotion in neonatal kittens whose spinal cords are transected: a behavioral and anatomical study. Exp Neurol 135: 123-145

Iwahara T, Atsuta Y, Garcia-Rill E, Skinner RD (1992) Spinal cord stimulation-induced locomotion in the adult cat. Brain Res Bull 28: 99-105

Iwashita Y, Kawaguchi S, Murata M (1994) Restoration of function by replacement of spinal cord segments in the rat. Nature 367: 167-170

Jakeman LB, Wei P, Guan Z, Stokes BT (1998) Brain-derived neurotrophic factor stimulates hindlimb stepping and sprouting of cholinergic fibers after spinal cord injury. Exp Neurol 154: 170-184

Kesslak JP, So V, Choi J, Cotman CW, Gomez-Pinilla F (1998) Learning upregulates brain-derived neurotrophic factor messenger ribonucleic acid: a mechanism to facilitate encoding and circuit maintenance? Behav Neurosci 112: 1012-1019

Kim D, Adipudi V, Shibayama M, Giszter S, Tessler A, Murray M, Simansky KJ (1999) Direct agonists for serotonin receptors enhance locomotor function in rats that received neural transplants after neonatal spinal transection. J Neurosci 19: 6213-6224

Kim D, Murray M, Simansky KJ (2001) The serotonergic 5HT (2C) agonist m-chlorophenylpiperazine increases weight-supported locomotion without development of tolerance in rats with spinal transections. Exp Neurol 169: 496-500

Lee YS, Hsiao I, Lin V (2002) Peripheral nerve grafts and aFGF restore partial hindlimb function in adult paraplegic rats. J Neurotrauma 19: 1203-1216

Lovely RG, Gregor RJ, Roy RR, Edgerton VR (1986) Effects of training on the recovery of full-weight-bearing stepping in the adult spinal cat. Exp Neurol 92: 421-435

Lovely RG, Gregor RJ, Roy RR, Edgerton VR (1990) Weight-bearing hindlimb stepping in treadmill-exercised adult spinal cats. Brain Res 514: 206-218

Lu J, Feron F, Ho SM, Mackay-Sim A, Waite PM (2001) Transplantation of nasal olfactory tissue promotes partial recovery in paraplegic adult rats. Brain Res 889: 344-357

Lu J, Feron F, Mackay-Sim A, Waite PM (2002) Olfactory ensheathing cells promote locomotor recovery after delayed transplantation into transected spinal cord. Brain 125: 14-21

McKerracher L, Doucet G, Rossignol S (eds) (2002) In: Spinal Cord Trauma: Regeneration, Neural Repair and Functional Recovery. Progress in Brain Research 137: 1-470 Elsevier

Miya D, Giszter S, Mori F, Adipudi V, Tessler A, Murray M (1997) Fetal transplants alter the development of function after spinal cord transection in newborn rats. J Neurosci 17: 4856-4872

Molteni R, Ying Z, Gomez-Pinilla F (2002) Differential effects of acute and chronic exercise on plasticity-related genes in the rat hippocampus revealed by microarray. Eur J Neurosci 16: 1107-1116

Mushahwar VK, Collins DF, Prochazka A (2000) Spinal cord microstimulation generates functional limb movements in chronically implanted cats. Exp Neurol 163: 422-429

Nesmeyanova TN (1977) Experimental Studies in the Regeneration of Spinal Neurons, VH Winston & Sons, Washington, D.C., pp. 61-64

Norman KE, Pepin A, Barbeau H (1998) Effects of drugs on walking after spinal cord injury. Spinal Cord 36: 699-715

Pinter MM, Gerstenbrand F, Dimitrijevic MR (2000) Epidural electrical stimulation of posterior structures of the human lumbosacral cord: 3. Control of spasticity. Spinal Cord 38: 524-531

Pinzon A, Calancie B, Oudega M, Noga BR (2001) Conduction of impulses by axons regenerated in a Schwann cell graft in the transected adult rat thoracic spinal cord. J Neurosci Res 64: 533-541

Prochazka A, Gorassini M (1998) Ensemble firing of muscle afferents recorded during normal locomotion in cats. J Physiol 507: 293-304

Ramon-Cueto A, Avila J (1998) Olfactory ensheathing glia: properties and function. Brain Res Bull 46: 175-187

Ramon-Cueto A, Plant GW, Avila J, Bunge MB (1998) Long-distance axonal regeneration in the transected adult rat spinal cord is promoted by olfactory ensheathing glia transplants. J Neurosci 18: 3803-3815

Ramon-Cueto A, Cordero MI, Santos-Benito FF, Avila J (2000) Functional recovery of paraplegic rats and motor axon regeneration in their spinal cords by olfactory ensheathing glia. Neuron 25: 425-435

Ribotta MG, Provencher J, Feraboli-Lohnherr D, Rossignol S, Privat A, Orsal D (2000) Activation of locomotion in adult chronic spinal rats is achieved by transplantation of embryonic raphe cells reinnervating a precise lumbar level. J Neurosci 20: 5144-5152

Robinson GA, Goldberger ME (1986) The development and recovery of motor function in spinal cats. II. Pharmacological enhancement of recovery. Exp Brain Res 62: 387-400

Rosenfeld JE, Sherwood AM, Halter JA, Dimitrijevic MR (1995) Evidence of a pattern generator in paralyzed subject with spinal cord stimulation. Soc Neurosci Abstr 21: 688

Rossignol S, Chau C, Giroux N, Brustein E, Bouyer L, Marcoux J, Langlet C, Barthelemy D, Provencher J, Leblond H, Barbeau H, Reader TA (2002) The cat model of spinal injury. Prog Brain Res 137: 151-168

Roudet C, Gimenez y Ribotta M, Privat A, Feuerstein C, Savasta M (1995) Intraspinal noradrenergic-rich implants reverse the increase of alpha 1 adrenoceptors densities caused by complete spinal cord transection or selective chemical denervation: a quantitative autoradiographic study. Brain Res 677: 1-12

Roudet C, Gimenez Ribotta M, Privat A, Feuerstein C, Savasta M (1996) Regional study of spinal alpha 2-adrenoceptor densities after intraspinal noradrenergic-rich implants on adult rats bearing complete spinal cord transection or selective chemical noradrenergic denervation. Neurosci Lett 208: 89-92

Roy RR, Sacks RD, Baldwin KM, Short M, Edgerton VR (1984) Interrelationships of contraction time, Vmax and myosin ATPase after spinal transection. J Appl Physiol 56: 1594-1601

Roy RR, Acosta L (1986) Fiber type and fiber size changes in selected thigh muscles six months after low thoracic spinal cord transection in adult cats: exercise effects. Exp Neurol 92: 675-685

Roy RR, Baldwin KM, Edgerton VR (1991) The plasticity of skeletal muscle: effects of neuromuscular activity. Exerc Sports Sci Rev 19: 269-312

Roy RR, Talmadge RJ, Hodgson JA, Zhong H, Baldwin KM, Edgerton VR (1998) Training effects on soleus of cats spinal cord transected (T12-13) as adults. Muscle Nerve 21: 63-71

Roy RR, Talmadge RJ, Hodgson JA, Oishi Y, Baldwin KM, Edgerton VR (1999) Differential response of fast hindlimb extensor and flexor muscles to exercise in cats spinalized as adults. Muscle & Nerve 22: 230-241

Shapkova HY, Shapkov YT, Mushkin AY (1997) Locomotor activity in paraplegic children induced by spinal cord stimulation. In: Gurfinkel VS, Levik Yu S (eds) Proceedings of the International Symposium on Brain and Movement, St. Petersburg, Moscow. Pp. 170-1

Shurrager PS (1955) Walking in spinal kittens and puppies. In: WF Windle (ed), Regeneration in the Central Nervous System, Thomas, Springfield, Illinois, pp. 208-218

Slawinska U, Majczynski H, Djavadian R (2000) Recovery of hindlimb motor functions after spinal cord transection is enhanced by grafts of the embryonic raphe nuclei. Exp Brain Res 132: 27-38

Tessler A, Fischer I, Giszter S, Himes BT, Miya D, Mori F, Murray M (1997) Embryonic spinal cord transplants enhance locomotor performance in spinalized newborn rats. Adv Neurol 72: 291-303

Tillakaratne NJ, Mouria M, Ziv NB, Roy RR, Edgerton VR, Tobin AJ (2000) Increased expression of glutamate decarboxylase (GAD(67)) in feline lumbar spinal cord after complete thoracic spinal cord transection. J Neurosci Res 60: 219-230

Tillakaratne NJK, de Leon RD, Bigbee AJ, Sebata HS, Hoang TX, Tran CL, Vaghasia N, Shah R, Roy RR, Edgerton VR, Tobin AJ (2001) Hindlimb activity triggers muscle-specific changes of GAD_{67} in spinal transected rats. Soc. Neurosci. Abstr. 31: 297.17

Tillakaratne NJ, de Leon RD, Hoang TX, Roy RR, Edgerton VR, Tobin AJ (2002) Use-dependent modulation of inhibitory capacity in the feline lumbar spinal cord. J Neurosci 22: 3130-3143

Tuszynski MH, Peterson DA, Ray J, Baird A, Nakahara Y, Gage FH (1994) Fibroblasts genetically modified to produce nerve growth factor induce robust neuritic ingrowth after grafting to the spinal cord. Exp Neurol 126: 1-14

Wernig A, Müller S (1991) Improvement of walking in spinal cord injured persons after treadmill training. In: A. Wenig (ed) Plasticity of Motoneuronal Connections, Elsevier, pp. 475-485

Wieler M, Stein RB, Ladouceur M, Whittaker M, Smith AW, Naaman S, Barbeau H, Bugaresti J, Aimone E (1999) Multicenter evaluation of electrical stimulation systems for walking. Arch Phys Med Rehabil 80: 495-500

Yi DK, Adipudi V, Shibayama M, Anderson KN, Simansky KJ, Murray M (1998) Transplant-mediated locomotion is improved by selective serotonergic agonists. Ann N Y Acad Sci 860: 524-527

Chapter 4

SPINAL NETWORKS INVOLVED IN INTERLIMB CO-ORDINATION AND REFLEX REGULATION OF LOCOMOTION

Volker Dietz
University Hospital Balgrist, Zürich, Switserland

Abstract: There is increasing evidence that neuronal networks within the spinal cord are involved in human gait, similar to those known from the cat. Does it indicate that the control of human gait is still based on that of quadrupedal locomotion? Tackling this question is of basic and practical relevance. With the evolution of upright stance and gait, a greater influence of the direct cortico-motoneuronal system paralleled advanced hand function which might have replaced the phylogenetically older control of arm movements. Nevertheless, there is some recent evidence that besides the direct connections, corticospinal excitation of upper limb motoneurons can also be mediated indirectly via propriospinal neurons in the cervical cord, for example during locomotion. Such alternative pathways allow a task-dependent neuronal linkage of cervical and thoraco-lumbar propriospinal circuits controlling leg and arm movements during human locomotor activities. Consequently, interlimb co-ordination and reflex regulation during human locomotion appear to be organised along similar lines to that in the cat. The persistence of such a mode of movement control has consequences for rehabilitation.

Key words: Human locomotion; Interlimb co-ordination; Cortico-motoneuronal control; Propriospinal neuronal circuits; Skilled hand movements; Cyclical movements; Quadrupedal locomotion; Evolution; Reflex regulation.

1. INTRODUCTION

The co-ordination of fore- and hind-limb rhythmic activities represents one of the main features characterising quadruped locomotion (Gans et al.

1997, see also Clarac et al., Chapter 2). Specialised neural circuits located in the caudal spinal cord (the so-called central pattern generator (CPG) for locomotion) organise hindlimb locomotor activity, while others located in the rostral spinal cord control forelimb movements (for reviews see Grillner 1981; Duysens and Van de Crommert 1998; Duysens et al., Chapter 1). The co-ordination of both circuits is mediated by propriospinal neurons (see Box 4-1) with long axons that couple the cervical and lumbar enlargements of the spinal cord (Cazelets and Betrand 2000).

In many respects, bipedal and quadrupedal locomotion share common spinal neuronal control mechanisms. As in quadrupeds, long projecting propriospinal neurons couple the cervical and lumbar enlargements in humans (Nathan et al. 1996). Furthermore, the co-ordination of limb movements during walking is similar in human infant (Yang et al 1998; Pang and Yang 2000), adults (for reviews see Dietz 1992; 1997), and quadrupeds (for reviews see Grillner 1981, 1986). Even during normal gait, swinging of the arms is regarded as being an integral part of progression, as it serves to counteract torsional movements of the trunk (Elftman 1939). Nevertheless, there are also distinct differences because the upper limbs in primates have become specialised to perform skilled hand movements. This evolution of upright stance and gait, in association with a differentiation of hand movements represents a basic requirement for human cultural development (Herder 1785).

This review deals with the networks involved in human locomotor activities: (1) the interlimb co-ordination between the legs and between the arms, (2) the associated reflex regulation and, (3) the similarities between bipedal and quadrupedal locomotion.

2.   INTERLIMB CO-ORDINATION OF THE LEGS

The regulation of human walking requires a close co-ordination of muscle activation between the two legs. During stance and gait, both legs act in a cooperative manner, each limb affects the strength of muscle activation and the time-space behavior of the other. There exists some evidence that this interlimb co-ordination is mediated by spinal interneuronal circuits, which themselves are under supra-spinal control (Dietz 1992).

Box 4-1 Neural mechanisms of locomotion

Propriospinal neuronal system: Circuits of interneurons within the spinal cord. A convergence of afferent input from peripheral receptors within the limbs and trunk and of descending signals from supraspinal centres takes place on these interneuronal circuits for the generation of appropriate motor responses.

Direct/indirect cortico-motoneuronal system: Skilled hand and finger movements are controlled by direct cortico-motoneuronal connections. However, most probably also indirect cortico-motoneuronal connections (i.e. mediated by propriospinal interneurons) persisted in humans and are predominantly involved in the control of cyclic movements of proximal arm muscles (swimming) crawling, swinging of arms during gait.

Flexor reflex: Complex withdrawing reaction of leg muscles following a noxious stimulus applied to a leg, associated with a flexion movement of the ipsilateral and an extension movement of the contralateral leg.

2.1   Perturbations of stance

Information about the organization and functional significance of the interlimb co-ordination of leg muscle activity in the control of bipedal stance and gait can be obtained when displacements are applied during stance on a treadmill with split belts (Dietz et al. 1989). Unidirectional bilateral perturbations are followed by larger EMG responses of synergistic muscle groups in both legs than during unilateral displacements. The EMG amplitude in one leg is about equal to the sum of the EMG amplitude of both the displaced and the nondisplaced leg obtained during unilateral displacement (Dietz et al. 1989). This is explained by the fact, that a unilateral displacement is always followed by a compensatory response in the displaced and the nondisplaced leg, appearing on both sides with about the same latency and amplitude. The inverse result is obtained when the legs are simultaneously displaced in opposite directions. In this case, the EMG responses in both legs are significantly smaller than those obtained during unilateral displacements. Consequently, during bilateral displacements, the EMG activity is linearly summed or subtracted, depending on whether the legs are displaced in the same or opposite directions (Dietz et al. 1989).

From a functional point of view, this interlimb co-ordination is necessary to keep the body's center of gravity over the feet. The bilateral muscle activation during unilateral displacements produces rapid automatic co-

contraction also of the nondisplaced leg. This mechanism provides a more stable base to compensate for the displacement. During bilateral displacements, the destabilisation of the body is greater. In this case the mutual enhancement of the response amplitude allows more effective contractile force to compensate for the perturbation. A displacement of the legs in opposite directions causes the body's center of gravity to fall between the legs. In this case a reduced level of compensatory responses is needed to gain body equilibrium.

2.2   Perturbations of locomotion

Perturbations of gait, for example a short obstruction of the swing phase, evoke a bilateral response pattern (Dietz et al. 1986). This interlimb co-ordination is similarly organised in both cats and humans. According to the short latencies of the electromyographic (EMG) responses, it is thought to be mediated at a spinal level (Dietz et al. 1986), as in the cat (Gorassini et al. 1994).

During stepping on a split-belt treadmill with the belts running at different speeds the legs act in a co-operative manner in infants (Thelen et al. 1987) and adults (Prokop et al. 1995; Erni and Dietz 2001; Dietz et al. 1994), each limb affecting the time-space behaviour of the other. In line with studies on the spinal cat (Grillner 1981), this indicates that the spinal cord contains networks responsible for each limb, which can be interconnected in a flexible manner. In general, initiation of the swing phase on one side is contingent on the contralateral limb being in the stance phase in adult humans (for review see Dietz 1992; 1997) and already in infants (Yang et al. 1998; Pang and Yang 2000; 2001). A bilateral co-ordination of leg muscle activation is already present in early infancy, i.e. well before the onset of independent walking (Pang and Yang, 2001).

Such a pattern of interlimb co-ordination as described above has also been reported for a variety of preparations in the cat (Gorassini et al. 1994; Hiebert et al. 1996; Schomburg et al. 1998) and agrees with the 'half-centre' model proposed for the organisation of the neuronal circuits within the spinal cord that can generate locomotor movements (CPG) (Brown 1914; Lundberg 1980). In this model the neuronal circuits that co-ordinate the leg flexor activity of both sides during the swing phase of locomotion (i.e. the flexor half-centres) mutually inhibit each other. In contrast, the extensor half-centres on each side have no mutual inhibitory connections, agreeing with the co-existence of the stance phase on the two sides.

An interlimb neuronal mechanism that coordinates the activity between muscles of both legs was also described for pedalling movements (Ting et al., 2000). In these experiments, an influence of contralateral extensor phase afferent input on the ipsilateral flexion movements indicated a bilateral coupling of gain control mechanisms. Obviously there exists an innate capacity of the neuronal circuitry that controls walking.

2.3 Factors influencing patterned leg muscle activation

Recently, it could be demonstrated that unilateral stepping movements can lead to a patterned activation of the contralateral static, but rhythmically loaded leg (Dietz et al. 2002). This influence of locomotor activity on the contralateral static leg is obviously based on the interlimb co-ordination discussed above. However, this patterned activity in the non-moving leg was mostly restricted to the leg flexor muscles. The preserved activation of leg flexor but strongly reduced EMG activity in the leg extensor muscles observed in these experiments was suggested to be due to the well established differential control of these muscles with a central dominance in the control of leg flexor activity (for a review see Dietz 1992; 2002 b). During normal locomotion, the leg extensor activity is modulated continuously by proprioceptive feedback during the stance phase. Therefore, in the condition reported above, with a lack of roll off the body over the standing leg, this EMG activity is expected to be diminished.

In contrast, no significant EMG activity was present in the muscles of the non-moving leg in a group of patients with complete paraplegia (Dietz et al., 2002). This indicates, that the spinal co-ordination of bilateral leg muscle activation depends on a facilitation by supraspinal centres. Indeed, a cerebellar contribution via reticulo-spinal neurons has been suggested in both cats (Ito, 1984) and humans (Bonnet et al., 1976), and recently, evidence was presented for a cortical (supplementary motor area) control of interlimb co-ordination (Debaere et al., 2001, see also Chapter 5).

2.4 Reflex modulation

The spinal neuronal control of walking and its similarities with quadrupedal locomotion is also reflected in the modulation of cutaneous reflexes evoked in the lower limbs. A task-dependency of cutaneous reflexes has been shown in standing versus running (Duysens et al. 1993; Tax et al. 1995) and cycling versus static contraction (Zehr et al. 2001 a). Cutaneous reflexes in leg muscles are sensitive to the specific motor task that is performed (Bastiaanse et al. 2000; for review see Zehr and Stein 1999). Furthermore, a nerve specificity of cutaneous reflex modulation exists, which seems to be functionally important (Van Wezel et al. 1997; Zehr et al. 1997; 1998). Certain features of this reflex modulation have been suggested to be determined by the CPG (Duysens and Van de Crommert 1998; Zehr et al. 2001 a,b; Komijama et al. 2000). This would be consistent with observations made in the cat (Brooke et al. 1997; Drew and Rossignol 1987).

3.   UPPER INTERLIMB CO-ORDINATION DURING CYCLICAL MOVEMENTS

An interlimb co-ordination also exists between the arms during a great variety of tasks. There exists evidence for a coupling between the upper limbs, even during bimanual discrete tasks, i.e. when they perform different movements simultaneously (Swinnen et al. 1991, see also Wenderoth et al., Chapter 8). During rhythmic movements the co-ordination of reflex activities seems to be similarly organised as that of the legs. For example, the modulation of cutaneous reflexes during rhythmic cyclical arm movements (Zehr and Chua 2000; Zehr and Kido 2001) corresponds with that of the legs (Zehr et al. 1997), in respect to their task-dependency, nerve specificity and phase-dependency. This implicates similar control mechanisms for the reflex modulation in upper and lower limbs (Zehr and Kido 2001; Zehr et al. 2001 a).

These observations are again in accordance with studies in the cat, where the organisation and pattern of cutaneous reflex modulation was quite similar when comparing the fore- and hind-limbs (Drew and Rossignol 1987).

4.   CO-ORDINATION BETWEEN ARM AND LEG MOVEMENTS

Some evidence has been established that interlimb co-ordination is similarly organised in lower and upper limbs during cyclic movements in both humans and the cat. This indicates that the neuronal co-ordination and patterns of reflex modulation are conserved within the human lumbar and cervical spinal cord. This may serve as a basis for the remaining piece of quadrupedal limb co-ordination during human bipedal walking, the similarity of co-ordination between upper and lower limb muscles during locomotor activities in humans and that of fore- and hind-limbs in quadrupeds (cf. Jones 2000; for review see Dietz 2002 a).

4.1   Evidence for a neuronal coupling

Recent experiments have indicated a neuronal coupling of upper and lower limb muscles during various human locomotor activities (Wannier et al. 2001). A linkage between the cervical and lumbar enlargement of the spinal cord by propriospinal neuronal circuits with long axons can also be suggested on the basis of H-reflex studies (Delwaide and Crenna 1984; Baldissera et al. 1998). Using this technique, the monosynaptic spinal reflex excitability is tested by electrical stimulation of group Ia afferents. For example, during rhythmical movements of one foot, a cyclic H-reflex modulation was observed in the upper limbs (Baldissera et al. 1998). According to recent studies using functional magnetic resonance imaging (Debaere et al. 2001), the supplementary motor area might be involved in the supraspinal control of this coupling between upper and lower limb movements.

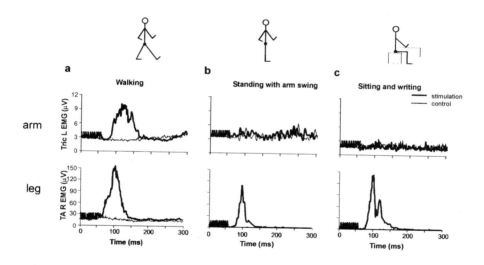

Figure 4-1. Interlimb reflexes: Comparison between walking, standing and sitting. Rectified and averaged tibialis anterior (TA) and left triceps brachii (Tric) EMG responses to a train of electrical stimuli applied to the right distal tibial nerve during walking (a, n=100), standing with voluntary arm swing (b, n=20) and sitting while writing (c, n=20). The data was obtained from an individual subject. Control EMG recordings without stimulation (thin lines) are displayed for comparison. During walking right tibial nerve stimulation is followed by leg and arm muscle EMG responses. However, arm muscle responses are absent when stimuli are applied during standing and sitting. This suggests that proximal arm muscle responses are associated with the swinging of the arm during walking as a residual function of quadrupedal locomotion. - Note the different calibration of the leg and arm muscle EMG . On the left side of each graph stimulus artifacts are present (adapted from Dietz et al. 2001)

4.2   Task-dependency of neuronal coupling

When mechanical impulses are applied to one leg during sitting, stance or walking (mid or end of the stance phase), only during walking distinct bilateral arm EMG responses appear. These are larger in deltoideus and triceps than in biceps brachii and are modulated within the stance phase (Dietz et al. 2001).

Similar interlimb reflex responses in arm muscles are also obtained following electrical stimulation (above motor threshold) of the distal tibial nerve (mixed nerve that innervates plantar foot muscles and the skin of foot sole) during walking (Dietz et al. 2001). In contrast, arm muscle responses are absent when stimuli are applied during standing with voluntary arm swing or sitting while writing, i.e. with a comparable background EMG activity (Fig. 4-1). These observations indicate a flexible task-dependent neuronal coupling between upper and lower limbs. The pathway coupling upper and lower limb movements obviously becomes gated by the activity of the CPG during walking. It has been concluded that a stimulus applied to a leg can exert a direct influence not only on the compensatory leg muscle EMG activity, but also on the neuronal control of upper arm muscles of both sides in a task-dependent manner (Dietz et al. 2001). The range of movements across which such a task-dependent neuronal coupling of upper and lower limb movements occurs, has yet to be established.The stronger impact of leg flexors in interlimb co-ordination is in line with the increasing evidence that leg flexor and extensor muscles are differentially controlled in both animals (Cheng et al. 1998) and humans (Brouwer and Ashby 1992; Schubert et al. 1997; Nielsen et al. 1993; for review see Dietz 1992; 2002 a).

4.3   Interlimb coupling within the isolated human spinal cord

Further evidence for a neuronal coupling between upper and lower limbs has been provided by studies in patients with cervical spinal cord injury. So-called "interlimb reflex responses" can be evoked in distal muscles of upper limbs with short-latency by electrical stimulation of the tibial nerve at the ankle (Calancie et al. 1996). These reflex connections might be due to a loss of supraspinal inhibition or, alternatively, to a sprouting of ascending propriospinal systems occupying synaptic locations vacated by degenerating descending connections. The interlimb reflexes can appear months after a spinal cord injury and are suggested to be due to a synaptic plasticity becoming effective after a spinal cord injury (Calancie et al. 2002). Functional and anatomical evidence indicates that plasticity of neuronal circuits exists below the level of spinal cord lesion (for review see Raineteau and Schwab 2001), that can be potentated by training and experimental manipulations (Dietz et al. 1995; Dietz 2001).

Furthermore, it has been shown that the higher the level of spinal cord lesion, the more "normal" is the locomotor pattern induced in patients with complete para-/tetraplegia (Dietz et al.1999). This observation indicates that neuronal circuits underlying locomotor "pattern generation" in humans are not restricted to any specific level of the spinal cord, but that an intricate neuronal network contributing to bipedal locomotion extends from thoraco-lumbar to cervical levels (Dietz et al. 1999). Such an organisation is in line with that of quadrupedal locomotion requiring the co-ordination between neuronal circuits at cervical and lumbo-sacral levels, with the brainstem being the next higher co-ordinating centre (Grillner et al. 1995).

However, it has to be pointed out that in primates, more than in cat or rat, cortico-spinal drive is of greater importance to generate and maintain locomotor movements (cf. Vilensky 1987). Therefore, at present we only can take into account that rhythmic neuronal circuits, such as the CPG, can be extremely flexible in response of varying behavioural conditions (Marder 2000).

4.4   Functional evidence

Recently it has been shown that arm and leg muscle activity is well co-ordinated during walking, creeping on all four limbs or swimming (Wannier et al. 2001). In such conditions, arm and leg movements are locked with a fixed frequency relationship. Even if the movements of the legs are slowed down by flippers or if the mechanical interactions between the limbs are minimised (swimming and crawling in the air), this co-ordination remains preserved (Wannier et al. 2001). This indicates a coupling of neuronal circuits controlling arm and leg movements that is again under supraspinal control. The frequency relationship characterising this co-ordination corresponds to that observed in well-defined biological systems consisting of coupled oscillators (Bartos et al. 1999).

5.   DIFFERENTIAL CONTROL OF UPPER LIMB MOVEMENTS

The task-dependency of the neuronal coupling between upper and lower limbs might be associated with a differential neuronal control of upper limbs during skilled hand movements or, alternatively, during locomotor activities.

V. Dietz

Direct cortico-motoneuronal connections (see Box 4-1) to hand muscles are considered to determine the degree of dexterity in humans and non-human primates (Nakajiama et al. 2000; Lemon 1999). It has been suggested that these phylogenetically new components are integrated into pre-existing neuronal circuits (Georgopoulos and Grillner 1989). Nevertheless, evidence obtained by different experimental approaches indicates, that also indirect cortico-motoneuronal connections via propriospinal neuronal circuits, corresponding to those described for cat (for review see Lundberg 1999) persist (Fig. 4-2). They most likely remain involved in the control of arm movement (Nicolas et al. 2001; Marchand-Pauvert et al. 1999; for review see Pierrot-Deseilligny 1996).

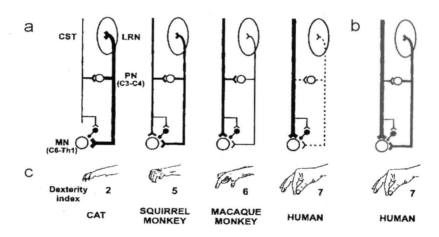

Figure 4-2. Contribution of propriospinal and cortico-motoneuronal excitation of upper limb motoneurons in cat, squirrel monkey, macaque monkey, and human. (a) Schematic diagram of the presumed propriospinal and corticospinal inputs to upper limb motoneurons in different species (Dietz 2002 a; adapted from Nakajiama et al. 2000). According to this speculative view, the propriospinal neuronal system becomes progressively weaker from squirrel to macaque monkey and this change is accompanied by a progressive increase in the strength of the cortico-motoneuronal connection, which is stronger again in humans. It has been suggested that by extrapolation, these changes predict that the propriospinal neuronal system is unlikely to play a major role in transmitting corticospinal excitation in humans (dashed lines). – (b) According to the evidence established in this review, the propriospinal neuronal system persists in humans for locomotor-like movements despite the strong cortico-motoneuronal connections (CST corticospinal tract; LRN lateral reticular nucleus; PN propriospinal neurons; MN motoneurons). (b) Index of dexterity for 4 species (Heffner and Masterton 1975).

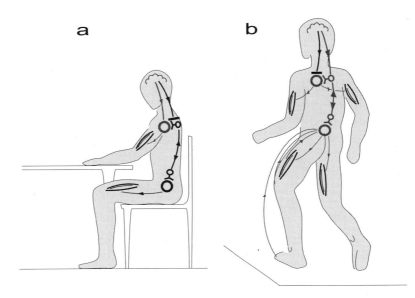

Figure 4-3. Schematic illustration of movement control during different motor tasks. According to the research cited in this review the neuronal control of arm movement is task-dependent: (a) during skilled hand movements, strong direct cortico-motoneuronal excitation is predominant (dotted lines), while the cervical propriospinal neuronal system is inhibited; (b) during locomotion it is assumed that the brain command is predominantly mediated by interneurons. Cervical and thoraco-lumbar propriospinal systems become coupled and co-ordinate arm and leg movements (dotted lines) [adapted from Dietz 2002 a]. The lower part of the schematic diagram in (b) is adapted from Jankowska and Lundberg (1981).

On the basis of the available evidence, the following hypothesis is put forward (Fig. 4-3). Efficient corticospinal excitation of upper limb motoneurons via propriospinal neurons might be operative during automatically performed movements, such as locomotion. In contrast, during skilled hand movements, strong cortico-motoneuronal (see Box 4-1) input dominates, and transmission through the propriospinal system becomes suppressed (Nicolas et al. 2001). This might explain why stimulating the pyramidal tract or the motor cortex fails to demonstrate the indirect corticospinal projection in the macaque monkey (Maier et al. 1998; Pauvert et al. 1998; Olivier et al. 2001) and in humans (de Noordhout et al. 1999). Such an interaction would be expected if propriospinal neurons in primates are under stronger inhibitory control than in the cat (Alstermark et al. 1999).

6. PRACTICAL CONSEQUENCES

From the observation that basic neuronal control mechanisms underlying locomotor-like movements remain preserved during evolution, the following practical consequences emerge:

1. Knowledge gained about the effects of regenerating pathways on movement performance observed in the rat (Merkler et al. 2001; for review see Raineteau and Schwab 2001) can, at least in part, be transferred to humans. In fact, by comparing the outcome of spinal cord injury in rat and man, using electrophysiological, imaging, and functional measures, a comparable course and extent of recovery was found (Metz et al. 2001).

2. As a consequence of a quadrupedal-like co-ordination of locomotion in humans, it emerges that patients with higher spinal cord lesions would benefit more from neuronal regeneration than those with lower spinal lesions. In high lesions, the integrity of the complex neuronal network underlying locomotion obviously is preserved and "only" a gain regulation is required.

3. The demonstration of an interaction between cervical and thoraco-lumbar neuronal circuits in humans has some relevance in the rehabilitation process of patients with an incomplete spinal cord lesion. Involvement of arm movements in the training, corresponding to spinal cat (De Leon et al. 1999), might lead to a positive effect on the locomotor capacity in these patients.

ACKNOWLEDGEMENT

The Swiss National Science Foundation supported the authors' research.

REFERENCES

Alstermark B, Isa T, Ohki Y, Saito Y (1999) Disynaptic pyramidal excitation in forelimb motoneurones mediated via C(3)-C(4) propriospinal neurones in the Macaca fuscata. J Neurophysiol 82: 3580-3585

Baldissera F, Cavallari P, Leocarni L (1998) Cyclic modulation of the H-reflex in a wrist flexor during rhythmic flexion-extension movements of the ipsilateral foot. Exp Brain Res 118, 427-430

Bartos M, Manor Y, Nadim F, Marder E, Nusbaum MP (1999) Coordination of fast and slow rhythmic neuronal circuits. J Neurosci 19: 6650-6660

Bastiaanse CM, Duysens J, Dietz V (2000) Modulation of cutaneous reflexes by load receptor input during human walking. Exp. Brain Res 135:189-198

Bonnet M, Gurfinkel S, Lipchits MJ, Popov KE (1976) Central programming of lower limb muscular activity in the standing man. Agressologie 17: 35-42

Brooke J D, Cheng J, Collins DF, McIlroy WE, Misiaze KJE, Staines WR (1997) Sensori-sensory afferents conditioning with leg movement: Gain control in spinal reflex and ascending paths. Prog Neurobiol 51: 393-421

Brouwer B, Ashby P (1992) Corticospinal projections to lower limb motoneurones in man. Exp Brain Res 89, 649-654

Brown TG (1914) On the nature of fundamental activity of the nervous centres; together with an analysis of the conditioning of rhythmic activity in progression, and a theory of the evolution of function in the nervous system. J Physiol (Lond) 48: 18-46

Calancie B, Lutton S, Broton JG (1996) Central nervous system plasticity after spinal cord injury in man: Interlimb reflexes and the influence of cutaneous stimulation. Electroenceph Clin Neurophysiol 101: 304-315

Calancie B, Molano, MR, Broton JG (2002) Interlimb reflexes and synaptic plasticity become evident months after human spinal cord injury. Brain 125: 1150-1161

Cazalets JR, Bertrand S (2000) Coupling between lumbar and sacral motor networks in the neonatal rat spinal cord. Eur J Neurosci 12: 2993-3002

Cheng J, Stein RB, Jovanovic K, Yoshida K. Benett DJ, Han Y (1998) Identification, localization, and modulation of neural networks for walking in the mudpuppy (Necturus maculatus) spinal cord. J Neurosci 18: 4295-4304

Debaere P, Swinnen SP, Beatse E, Sunaert S, Van Hecke P, Duysens J (2001) Brain areas involved in interlimb co-ordination: A distributed network. Neuroimage 14: 947-958

De Leon RD, Hodgson JA, Roy RR, Edgerton VR (1999) Retention of hindlimb stepping ability in adult spinal cats after the cessation of step training. J

Delwaide PJ, Crenna P (1984) Cutaneous nerve stimulation and motoneuronal excitability. II: Evidence for non-segmental influences. J Neurol Neurosurg Psychiat 47: 190-196

de Noordhout AM Rapisarda G, Bagacz D, Gerard P, De Pasquo V, Pennisi G, Delwaide PJ (1999) Corticomotoneuronal synaptic connections in normal man: an electrophysiological study. Brain 122: 1327-1340

Dietz V (1992) Human neuronal control of automatic functional movements: interaction between central programs and afferent input. Physiol Rev 72: 33-69

Dietz V (1997) Neurophysiology of gait disorders: present and future applications. (Review) Electroenceph Clin Neurophysiol 103: 333-355

Dietz V (2001) Spinal cord lesion: effects of and perspectives for treatment. (Review) Neural Plasticity 8: 83-90

Dietz V (2002 a) Do human bipeds use quadrupedal coordination? Trends Neurosci 25: 462-467

Dietz V (2002 b) Proprioception and locomotor disorders. Nat Rev Neurosci 3: 781-790

Dietz V, Colombo G, Jensen L, Baumgartner L (1995) Locomotor capacity of spinal cord in paraplegic patients. Ann Neurol 37: 574-582

Dietz V, Fouad K, Bastiaanse C (2001) Neuronal co-ordination of arm and leg movements during locomotion. Eur J Neurosci, 14: 1906-1914

Dietz V, Horstmann GA, Berger W (1989) Interlimb coordination of leg muscle activation during perturbation of stance in humans. J Neurophysiol 62: 680-693

Dietz V, Müller R, Colombo G (2002) Locomotor activity in spinal man: Significance of afferent input from joint and load receptors. Brain 125: 2626-2634

Dietz V, Nakazawa K, Wirz M, Erni T (1999) Level of spinal cord lesion determines locomotor activity in spinal man. Exp Brain Res 128: 405-409

Dietz V, Quintern J, Boos G, Berger W (1986) Obstruction of the swing phase during gait: phase-dependent bilateral leg muscle coordination. Brain Res 384: 166-169

Dietz V, Zijlstra W, Duysens J (1994) Human neuronal interlimb coordination during split-belt locomotion. Exp Brain Res 101: 513-520

Drew T, Rossignol S (1987) A kinematic and electromyographic study of cutaneous reflexes evoked from the forelimb of unrestrained walking cats. J Neurophysiol 57: 1160-1184

Duysens J, Tax AA, Trippel M. Dietz V (1993) Increased amplitude of cutaneous reflexes during human running as compared to standing. Brain Res 613: 230-238

Duysens J, Van de Crommert HWAA (1998) Neural control of locomotion. Part 1: The central pattern generator from cats to humans. (Review) Gait and Posture 7: 131-141

Elftman H (1939) The function of the arms during walking. Hum Biol 11: 529-535

Erni T, Dietz V (2001) Obstacle avoidance during human walking: learning rate and cross-modal transfer. J Physiol (Lond) 534: 303-312

Gans C, Gaunt AS, Webb PW (1997) Vertebrate locomotion. In: W.H. Dantzler (ed.) Handbook of Physiology, Section 13: Comparative Physiology, Oxford Univ Press, New York, pp 55-213

Georgopoulos AP, Grillner S (1989) Visuomotor co-ordination in reaching an locomotion. Science 24: 1209-1210

Gorassini MA, Prochazka A, Hiebert GW, Gauthier MJ (1994) Corrective responses to loss of ground support during walking. I. Intact cats. J. PSG, Stuart DG, Forssberg H, Herman, RM eds) Wenner-Gren International Symposium Series. Vol. 45, Neurobiology of Vertebrate Locomotion, Macmillan, London, pp 505-512

Grillner S (1981) Control of locomotion in bipeds, tetrapods, and fish. In: Brookhart JM, Mountcastle VB. (eds). Handbook of Physiology. The Nervous System. Motor Control. Vol. II. Am Physiol Soc, Washington pp 1179-1236

Grillner S (1986) Interaction between sensory signals and the central networks controlling locomotion in lamprey, dog, fish and cat. In: (Grillner S, Stein)

Grillner S, Deliagina T, Ekeberg O, El Manira A, Hill RH, Lanser A, Orlovsky GN, Wallen P (1995) Neural networks that co-ordinate locomotion and body orientation in lamprey. Trends Neurosci 18: 270-279

Heffner R, Masterton B (1975) Variation in form of the pyramidal tract and its relationship to digital dexterity. Brain Behav Evol 12: 161-200

Herder JG (1785) Ideen zur Philosophie der Geschichte der Menschheit. Bd I,

Hiebert GW, Whelan PJ, Prochazka A, Pearson KG (1996) Contribution of hind limb flexor muscle afferents to the timing of phase transitions in the cat step cycle. J Neurophysiol. 75:1126-1137

Jankowska E, Lundberg A (1981) Interneurones in the spinal cord. Trends Neurosci 4: 230-233

Jones R. (2000) Keeping in step. Nature Rev Neurosci 3: 84-85.

Ito M (1984) The Cerebellum and Neural Control. Raven, New York

Komiyama T, Zehr EP, Stein RB (2000) Absence of nerve specificity in human cutaneous reflexes during standing. Exp Brain Res 133: 267-27238

Lemon RN (1999) Neural control of dexterity: What has been achieved? (Review) Exp Brain Res 128: 6-12

Lundberg A (1980) Half-centres revisited. In: Szentagothai J, Palkovits M, Hamori J (eds) Advanced Physiological Science. Regulatory function of the CNS. Motion and Organization Principles. Vol. 1, Pergamon, New York, pp 155-167

Lundberg A (1999) Descending control of forelimb movements in the cat. Brain Res Bull 50: 323-324

Maier MA, Illert M, Kirkwood PA, Nielsen J, Leon RN (1998) Does a C3-C4 propriospinal system transmit corticospinal excitation in the primate? An investigation in the macaque monkey. J Physiol (Lond) 511: 191-212

Marchand-Pauvert V, Mazevet D, Pierrot-Deseilligny E. Pol S, Prodat-Diehl P (1999) Handedness-related asymmetry in transmission in a system of human cervical premotoneurones. Exp Brain Res 125: 323-334

Marder E (2000) Motor pattern generation. Curr Opin Neurobiol 10, 691-698

Merkler D, Metz GA, Raineteau O, Dietz V, Schwab ME, Fouad K (2001) Locomotor recovery in spinal cord-injured rats treated with an antibody neutralizing the myelin-associated neurite growth inhibitor Nogo-A. J Neurosci 21: 3665-3673

Metz GA, Curt A,van de Meent H; Klusman, Schwab ME, Dietz V (2000) Validation of the weight-drop contusion model in rats: a comparative study of human spinal cord injury. J Neurotrauma 17, 1-17

Nakajima K, Maier MA Kirkwwod PA, Lemon RN (2000) Striking differences in transmission of corticospinal excitation to upper limb motoneurones in two primate species. J Neurophysiol 84: 698-709

Nathan PW, Smith M, Deacon P(1996) Vestibulospinal, reticulospinal and descending propriospinal nerve fibres in man. Brain 119: 1809-1833

Nicolas G, Marchand-Pauvert V,Burke D, Pierrot-Deseilligny E (2001) Corticospinal excitation of presumed cervical propriospinal neurones and its reversal to inhibition in humans. J Physiol (Lond) 533: 903-919

Nielsen J, Petersen, N, Deuschl G, Ballegaard M (1993) Task-related changes in the effect of magnetic brain stimulation of spinal neurones in man. J Physiol (Lond) 471: 223-243

Olivier E, Baker SN, Nakajima K, Brochier T, Lemon RN (2001) Investigation into non-monosynaptic corticospinal excitation of Macaque upper limb single motor units. J Neurophysiol 86:1573-1586

Pang MY, Yang JF (2000) The initiation of the swing phase in human infant stepping: importance of hip position and leg loading. J Physiol (Lond.) 528, 389-404

Pang MY, Yang JF (2001) Interlimb co-ordination in human infant stepping. J Physiol (Lond) 533: 617-625

Pauvert V, Pierrot-Deseilligny E, Rothwell JC (1998) Role of spinal premotoneurones in mediating corticospinal input to forearm motoneurones in man. J Physiol (Lond) 508: 301-312

Pierrot-Deseilligny E (1996) Transmission of the cortical command for human voluntary movement through cervical propriospinal premotoneurones. Prog. Neurobiol. 48: 489-517

Prokop T, Berger W, Zijlstra W, Dietz V (1995) Adaptational and learning processes during human split-belt locomotion: interaction between central mechanisms and afferent input. Exp Brain Res 106: 449

Raineteau O, Schwab ME (2001) Plasticity of motor systems after incomplete spinal cord injury. Nature Rev Neurosci 2: 263-273

Schomburg ED, Petersen N, Barajon I, Hultborn H (1998) Flexor reflex afferents reset the step cycle during fictive locomotion in the cat. Exp Brain Res 122: 339-350

Schubert M, Curt A, Jensen L, Dietz V (1997) Corticospinal input in human gait: modulation of magnetically evoked motor responses. Exp Brain Res 115: 234-246

Swinnen SP, Young DE, Walter CB, Serrien DJ (1991). Control of asymmetric bimanual movements. Exp Brain Res 85: 163-173.

Tax AA, Van Wezel BMH (1995) Bipedal reflex coordination to tactile stimulation of the sural nerve during human running. J Neurophysiol 73: 1947-1964

Thelen E, Ulrich DB, Niles D (1987) Bilateral coordination in human infants: stepping on a split-belt treadmill. J Exp Psychol Hum Percept Perform 13: 405-410

Ting LH, Kautz SA, Brown DA, Zajac FE (2000) Controlateral movement and extensor force generation alter flexion phase muscel co-ordination in pedalling. J Neurophysiol 83: 3351-3365.

Van Wezel BM, Ottenhoff FA, Duysens J (1997) Dynamic control of location-specific information in tactile cutaneous reflexes from the foot during human walking. J Neurosci 17: 3804-3814

Vilensky JA (1987) Locomotor behavior and control in human and non-human primates: comparison with cats and dogs. Neurosci Biobehav Rev 11: 263-274

Wannier T, Bastiaanse C, Colombo G, Dietz V (2001) Arm to leg coordination in humans during walking, creeping and swimming activities. Exp Brain Res 141: 375-379

Yang JF. Stephens MJ, Vishram R (1998) Transient disturbances to one limb produce coordinated, bilateral responses during infant stepping. J Neurophysiol. 79: 2329-2337

Zehr EP, Chua R (2000) Modulation of human cutaneous reflexes during rhythmic cyclical arm movement. Exp Brain Res 135: 241-250

Zehr EP, Collins DF, Chua R (2001 a) Human interlimb reflexes evoked by electrical stimulation of cutaneous nerves innervating the hand and foot. Exp Brain Res 140: 495-504

Zehr EP, Kido A (2001) Neural control of rhythmic, cyclical human arm movement: task dependency, nerve specificity and phase-modulation of cutaneous reflexes. J Physiol (Lond) 537: 1033-1045

Zehr EP, Komiyama T, Stein RB (1997) Cutaneous reflexes during human gait: electromyographic and kinematic responses to electrical stimulation. J. Neurophysiol 77: 3311-3325

Zehr EP. Komiyama T, Stein RB (2001 b) Differential regulation of cutaneous and H-reflexes during leg cycling in humans. J Neurophysiol 85: 1178-1184

Zehr EP, Stein RB (1999) What functions do reflexes serve during human locomotion? Prog Neurobiol 58: 185-205

Zehr EP, Stein RB, Komiyama T (1998) Function of sural nerve reflexes during human walking. J Physiol (Lond) 507: 305-314

Chapter 5

CORTICAL NETWORKS ASSOCIATED WITH LOCOMOTION IN MAN AND PATIENTS WITH HEMIPARETIC STROKE

Ichiro Miyai

Neurorehabilitation Research Institute, Bobath Memorial Hospital, Osaka, 536-0023, Japan

Abstract: A recently developed optical imaging system using near-infrared spectroscopy enabled real-time monitoring of cortical activation during various locomotor tasks. Cortical activation was assessed as increased levels of regional oxygenated hemoglobin. In healthy subjects, walking at 1km/hr was associated by cortical activation that centered in the medial sensorimotor cortices and supplementary motor areas. Walking at higher speed (3 km/hr or 5 km/hr) tended to induce decreased rather than increased activation in the sensorimotor cortices. In patients with hemiparetic stroke, cortical activation patterns during hemiparetic gait were characterized by asymmetrical activation in the sensorimotor cortices and recruitment of other motor-related areas such as the premotor cortices and the prefrontal regions. Importantly activation patterns could be modified by rehabilitative intervention. A facilitation technique, by which therapists assisted patients to walk by pressing the hip forward and backward to ensure the stability of the stance and swing phase of the paretic leg, induced enhanced activation in the motor related areas, particularly that in the premotor area. Partial body weight support during gait training on the treadmill tended to decrease overall activation. It remains undetermined whether these changes in cortical activation patterns are associated with good locomotor recovery in patients with stroke.

Key words: locomotion, near-infrared spectroscopy (NIRS), optical imaging, cerebral cortex, stroke, hemiparesis, gait, functional recovery, neurorehabilitation

1. INTRODUCTION

Experimental studies have indicated that bipedal movements such as walking, are controlled by cerebral cortices including motor neurons in the medial portion of the primary motor cortex (Ferrier, 1876; Penfield, 1950; Leyton and Sherrington, 1917) as well as spinal central pattern generators and multiple motor centers in the brainstem (Armstrong 1988; Drew 1988; Nutt et al., 1993). However there have been few studies concerning the cerebral mechanisms for human gait, mainly due to the limitations of functional neuroimaging techniques for studying dynamic movements. A neuroimaging study using single photon emission computed tomography revealed that the supplementary motor area, medial primary sensorimotor area, the striatum, the cerebellar vermis and the visual cortex were involved in human gait (Fukuyama et al., 1997). The noninvasive and flexible features of the optical imaging technique using near-infrared spectroscopy (NIRS), provide stable real-time monitoring of cortical activation during dynamic as well as static tasks. Although most NIRS studies have replicated previous findings from functional magnetic resonance imaging (fMRI) and positron emission tomography (PET) studies for motor (Hirth et al., 1996, Orbig et al., 1997; Colier et al., 1999), visual (Kato et al., 1993; Meek et al., 1995; Hock et al., 1996; Takahashi et al., 2000) and cognitive tasks (Villringer et al., 1993; Hock et al., 1995; Sakatani et al., 1998; Watanabe et al., 1998; Sato et al., 1999, Hoshi et al., 2000) in humans, NIRS imaging might be most useful in mapping the human brain during locomotor tasks, such as gait (Miyai et al., 2001b) and even running (Suzuki et al., 2002), and in mapping the brain of neonates and children (Bartocci et al., 2000; Taga et al., 200; Isobe, et al., 2001; Hintz, et al., 2001) for which fMRI and PET are ill-suited.

2. THE NIRS IMAGING TECHNIQUE AND ITS APPLICATION TO STUDIES OF HUMAN LOCOMOTION

We are using a 30 to 42-channel NIRS imaging system (OMM-2001, Shimadzu, Kyoto, Japan, Fig. 5-1) depending on the number of detector fibers and light source fibers using continuous wave laser diodes with wavelengths of 780, 805, and 830 nm (Eda et al., 1999). This system detects the absorption spectrum of hemoglobin that depends on its oxygenation state and cortical changes in oxygenated hemoglobin (oxyHb), deoxygenated hemoglobin (deoxyHb) and total hemoglobin (totalHb). Another approach of data acquisition is time-resolved spectroscopy that also assesses the distribution of photon arrival times (Boas et al., 2002; Villringer and Orbig, 2002). In our continuous wave system, each optode with an interoptode distance set to 3.0 cm was placed tightly on the skull using a custom-made holder cap and the optical fibers were suspended from a weight-balancer system to avoid motion-related artifacts during locomotor tasks on the treadmill (Fig. 5-1). An anatomical 3-D MRI scan, performed with marking the optode position on the skull by vitamin D capsules, revealed that the optodes covered an area over fronto-parietal cortices including the primary sensorimotor cortices (SMC), supplementary motor areas (SMA), and premotor cortices (PMC) with 30-channel measurement (Fig. 5-2C), but also including the prefrontal cortices and pre-supplementary motor areas with 36 or 42-channel measurement. Cortical "activation" during locomotor tasks on the treadmill was assessed as increased levels of oxyHb since changes of deoxyHb levels were much less sensitive (Miyai et al., 2001b; Hoshi et al., 2001; Wolf et al., 2002; Strangman et al., 2002). Strangman and colleagues found strong correlations between blood-oxygenation level-dependent (BOLD) fMRI changes and all optical measures including oxyHb, deoxyHb, and totalHb concentrations, with oxyHb providing the strongest correlation, probably due to the superior contrast-to-noise ratio for oxyHb relative to deoxyHb from optical measurements, rather than physiology related to BOLD signal interpretation (Strangman et al., 2002).

3. CORTICAL MAPPING OF HUMAN GAIT

Figure 5.2B shows the time course of changes in hemoglobin oxygenation during walking at 1 km/hr in channel 14 covering the left medial SMC. Levels of oxyHb and totalHb in channels covering the medial SMC started to increase 3 to 5 s after the task onset of walking at 1 km/hr, reached a plateau at 5 to 10 s, and returned to the baseline 3 to 5 s after the end of the task (Fig. 5-2B). Increase of oxyHb and totalHb levels tended to begin earlier as subjects repeated the tasks, probably due to anticipation of starting the tasks. We obtained images depicting increase in cortical oxyHb levels after adapting the linear interpolation to the simultaneously acquired data from neighboring source-detector pairs. Each NIRS map was corrected to match the anatomical location of the source-detector pairs on the brain surface and was overlaid on anatomical MRI surface images (Fig. 5-2A).

Different types of locomotor task activated distinct cortical networks (Fig. 5-2A). Walking on the treadmill at 1 km/hr induced bilateral cortical activation in the medial SMC as well as the SMA. Cortical activation during walking with and without arm swing holding side-rails did not differ although there was apparently less activation in the SMA without arm swing than with arm swing. Simple alternating foot movements were associated with localized activation in the medial SMC. Alternating arm swing without walking activated the lateral SMC. Gait imagery induced more rostral activation centered in the SMA although execution of gait performance and gait imagery appeared to share the similar neuronal networks as supported by PET findings for the changes of regional cerebral blood flow during motor execution and imagery of foot movements in the learning process of the sequence (Lafleur et al., 2002). FMRI findings in the same subjects during foot movements and gait imagery revealed similar activation patterns to those seen in NIRS imaging based on the oxyHb mapping, (Miyai et al., 2001b). Thus it is suggested that coordinated movements such as gait with arm swing and imagery for locomotion might induce enhanced activation in the SMA than less coordinated movements such as gait without arm swing and alternating foot movements do. These findings are in accordance with findings that coordination of wrist and hand movements in different directions was associated with extra activation of the SMA as compared to movements in the same direction (Debaere et al., 2001).

How does locomotor speed affect cortical activation? Unexpectedly, walking at high speed (3 km/hr or 5 km/hr) induced smaller activation in the SMC than walking at low speed (1 km/hr) did (Fig. 5-3) since previous studies showed a positive relationship between force or frequency of finger movements and the amount of SMC activation (Dettmers et al., 1995; Kawashima et al., 1999; Waldvogel et al, 1999). Experimental studies in mammalian quadrupeds have demonstrated that locomotor control is not simply explained by cortical mechanisms and that brain mechanism for gait have a hierarchical structure, including spinal central pattern generators and supraspinal multiple motor centers such as the cerebellum, subthalamic locomotor region, mesencephalic locomotor region, ventral tegmental field, and dorsal tegmental fields (Armstrong 1988; Drew 1988; Nutt et al., 1993). The frontal lobes and basal ganglia loops are involved in higher motor control under complex environmental conditions. Experimental lesions of the motor cortex do not prevent animals from walking on a smooth floor although they impair tasks requiring a high degree of visuomotor coordination such as stepping over an obstacle (Pearson and Gordon, 2000). Our data using an extended 42-channel optical imaging system covering the prefrontal regions, revealed that these areas were more active at 1km/hr than at 3 or 5 km/hr (Fig. 5-3). Furthermore a preliminary result on optical imaging of human running, showed an enhanced prefrontal activation during the acceleration phase of running at 9 km/hr that might require more complex control than walking at the ordinary pace at 3 or 5 km/hr (Suzuki et al., 2002). Each subject was more comfortable and felt making less effort when walking at 3 or 5 km/hr that was the ordinary pace of human walking than at 1 km/hr or running at 9km/hr on the treadmill. Thus these findings might be in accordance with the idea that the main controller of locomotion might shift from the cortical to the subcortical levels (especially to the spinal cord) during automatic locomotion at ordinary pace.

4.   CORTICAL MAPPING OF PATHOLOGICAL GAIT IN HEMIPARETIC STROKE AND ITS CLINICAL RELEVANCE IN NEURO-REHABILITATION

Another way to investigate the neural mechanisms controlling human locomotion is studying patients with gait disorders due to neurological diseases, such as stroke and Parkinson's disease. Hemiparetic gait in patients with stroke is characterized by decreased speed and cadence, asymmetry of swing and stance phase, and abnormal synergic patterns of the affected upper and lower limbs during the whole gait cycles (Roth et al., 1997; Davis, 2000). The latter is primarily due to damage in the primary motor cortex in the unilateral cerebral hemisphere and/or its descending motor pathway, including the corona radiata, internal capsule, cerebral peduncle, and ventral brainstem. We evaluated cortical activation patterns during hemiparetic gait in patients with stroke using the same optical imaging system as described above (Miyai et al., 2001c, 2002b, 2002c). Basic differences of cortical activation patterns between normal gait and hemiparetic gait are 1) asymmetry of activation in the SMC and 2) recruitment of other motor-related areas such as the PMC and the prefrontal regions in patients with stroke (Fig. 5-4). Activation patterns appear to depend on the severity of hemiparesis as well as on the location and size of cerebral lesion. In patients with ambulatory stroke due to capsular lesion, hemiparetic gait was associated with asymmetrical activation in the SMC, with less activation in the affected hemisphere than in the unaffected hemisphere. In patients with severe impairment due to combined damage in the cerebral cortex and subcortical regions including the whole internal capsule and the basal ganglia, recruitment of the PMC - especially in the affected hemisphere - was another characteristic finding during assisted hemiplegic gait (Miyai et al., 2002b).

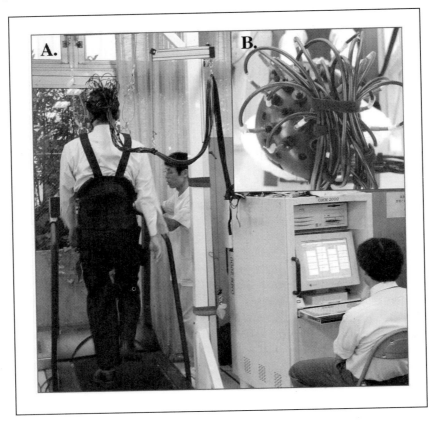

Figure 5-1. Measurement of cortical activation during walking using an optical imaging system.
A. A subject performing a walking task on the treadmill. B. A custom-made head holder cap fixing light source fibers (blue) and detector fibers (red) on the skull. See text for details.
(See color plate section.)

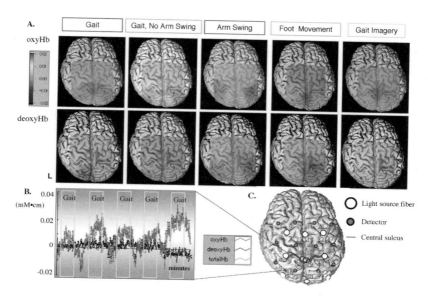

Figure 5-2. Cortical activation during various locomotor tasks (30 channel recording).
A. Different locomotor tasks were associated with distinct cortical activation patterns, as assessed by changes in oxyHb levels in a healthy subject (41 year old, right handed male). There was bilateral activation in the medial SMC and SMA during walking on the treadmill at 1 km/hr, similarly but less in the rostral regions during walking without arm swing, in the lateral SMC during arm swing without walking, in the medial SMC and the SMA during foot movements. Gait imagery activated more rostral regions than foot movements did. L: Left. B. Whole time series of oxyHb (red), deoxyHb (blue), and totalHb (green) levels (mM□cm) during gait at 1 km/hr (n=8) obtained from channel 14 covering the left medial SMC. The time series is composed of 5 task periods (30 s) alternating with 5 rest periods (30 s). This figure shows the average time series for 8 healthy subjects in channel 14, which covers the left medial SMC. There were few changes in deoxyHb levels. C. Location of each optode exposed to the brain surface image and assigned channel numbers.
(See color plate section.)

I. Miyai

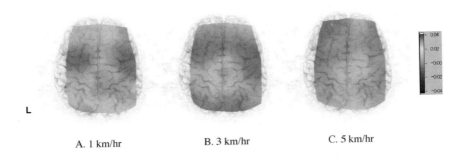

A. 1 km/hr   B. 3 km/hr   C. 5 km/hr

Figure 5-3. Cortical mapping of gait (42 channel recording) at different speeds in a healthy subject (25 years old, right-handed female).
Each mapping is based on changes in oxyHb levels (mM□cm) during walking on the treadmill. Note that activations in the medial SMC and SMA as well as in the prefrontal areas decrease as locomotor speed increases. (*See color plate section.*)

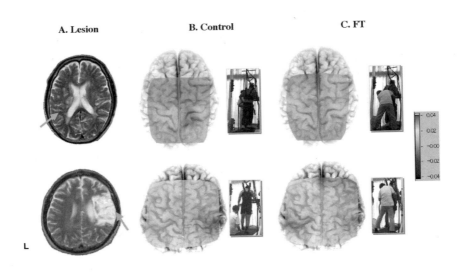

Figure 5-4. Cortical mapping of hemiparetic gait (36 channel recording) in patients with stroke.
Upper row: Cerebral activation during right hemiparetic gait on the treadmill in a patient with subcortical stroke (53 year old right handed male, 53 days poststroke) in the left corona radiata (A: arrow). During unassisted gait (B. 0.2 km/hr), SMC activation was asymmetrical with less activation in the affected hemisphere than in the unaffected hemisphere. During assisted gait by a therapist who facilitated swing and stance of the paretic leg by providing sensorimotor stimulation to the pelvis (C. FT: facilitation technique), SMC activation appears to be symmetrical with enhanced activation in the premotor and prefrontal areas. Lower row:

Cerebral activation during left hemiplegic gait in a patient (58 year old right handed male, 102 days poststroke) with diffuse infarction in the territory of the right middle cerebral artery (A: arrow). The patient needed partial body weight support (20% of body weight) to perform the locomotor tasks (0.2 km/hr). During assisted gait by a therapist that mechanically supported the swing of the paretic leg (B), mild activations are observed in the premotor and prefrontal regions of the affected hemisphere and in the parietal regions of the unaffected hemisphere. Enhanced activation in these regions and the unaffected SMC are observed during FT (C). Note that there is no activation in the damaged cortical regions. (Miyai et al., 2001b, Copyright© 2001, Elsevier Science, Reprinted by Permission Elsevier Science). (*See color plate section.*)

Importantly, activation patterns could be modified by rehabilitative interventions including therapeutic handling, such as a facilitation technique and partial body weight support (BWS) during locomotor training. During the facilitation technique, therapists (a) pressed the hip forward and backward to prevent hyperextension of the knee in the stance phase, (b) assisted flexion of the knee for the initiation of the swing phase with the weight well over the sound leg, and (c) prevented the pelvis from being hitched up as the hemiparetic leg moved forward (Bobath, 1978; Davis, 2000). A neurophysiological study demonstrated that this technique induced a more balanced walking pattern with a prolonged single stance period of the affected leg, an unobstructed hip movement, enhanced weight acceptance and a faster gait as immediate effects (Hesse et al., 1998). A crucial role of proprioceptive afferent inputs from the hip joints in the generation of locomotor activity, has been also shown in patients with complete spinal cord injury, trained with partial BWS on the treadmill (Dietz et al., 2002).

Real-time NIRS measurements during treadmill training in patients with hemiparetic stroke revealed that the facilitation technique (a) induced enhanced overall activation in motor-related cortical areas, particularly in the PMC of the affected hemisphere (Fig. 5-4) and (b) that it improved asymmetry of cortical activation (Fig. 5.5) along with improvement of gait asymmetry and increased cadence (Miyai et al., 2002b). These observations suggest that this technique immediately modifies cerebral activation patterns during pathological walking. Dietz et al. also demonstrated that interlimb coordination during walking depended on a supraspinal input since unilateral locomotion on the treadmill with partial BWS (70 %) in patients with spinal cord injury was associated with a normal pattern of leg muscle EMG activity only restricted to the moving leg, while in the healthy subjects a bilateral EMG activation occurred during unilateral locomotion. Thus improved gait performance might be associated with sensory inputs from the hip joints during assisted walking using a facilitation technique that modifies motor execution at the cortical levels. It is likely that some of these activation patterns in multiple motor related areas might lead to restoration of hemiparetic gait in patients with stroke.

There is growing evidence that the PMC and SMA play crucial roles in controlling human locomotion. In healthy subjects, right parieto-premotor activation has been shown to be involved in execution of limb-independent antiphase movements of hands or feet (de Jong et al., 2002). Moreover, the early phase of learning of an explicitly known sequence of foot movements, was associated with increases in regional cerebral blood flow in the bilateral dorsal premotor cortex and cerebellum as well as in the left inferior parietal lobule (Lafleur et al., 2002). The PMC and SMA are involved in the purposeful modification and initiation of locomotion through connections with the brainstem, basal ganglia, cerebellum, and spinal cord (Armstrong et al., 1988; Drew et al., 1988; Nutt et al., 1993). These areas participate in control of the contralateral proximal musculature and bilateral axial musculature (Freund et al., 1985; Halsband et al., 1993). Parallel motor pathways among the primary motor cortex, PMC, SMA and the final effectors in the spinal cord are known to exist (Fries et al., 1993; Fink et al., 1997). Neuroimaging studies, using single photon emission computed tomography in patients with Parkinson's disease, showed reduced activity in SMA during hypokinetic gait (Hanakawa et al., 1999b) as compared with healthy subjects, suggesting that these areas are associated with generating steady gait cycles in human. They also observed an enhanced activity of the PMC when hypokinetic gait was improved by visual cues (Hanakawa et al., 1999a).

An important role of the PMC in locomotor recovery is supported by a clinical observation that damage in the PMC might predict poor mobility outcome in patients with diffuse stroke in the territory of the middle cerebral artery (Seitz et al, 1998; Miyai et al., 1999). Patients with capsular stroke with MRI evidence for Wallerian degeneration of the pyramidal tract below the level of lesion, exhibited comparable final locomotor outcome to those without evidence for Wallerian degeneration although the former showed delayed improvement (Miyai et al., 1998b). Furthermore, the location of lesion rather than the volume of lesion could be related to mobility outcome after stroke. Multiple damage in the basal ganglia-thalomocortical motor circuits including the SMC, PMC, and SMA (Alexander and Crutcher, 1990), might result in better outcome than single-locus damage in these structures after stroke (Miyai et al., 1997, 2000b). More specifically, the greater volume of damage in the networks might possibly act to stimulate more effective reorganization of the intact structures. Thus, both functional neuroimaging studies and lesion studies indicate that the SMC and PMC as well as the primary motor cortex and its descending pathway might control, either in a parallel or hierarchical manner, human gait over subcortical and spinal centers.

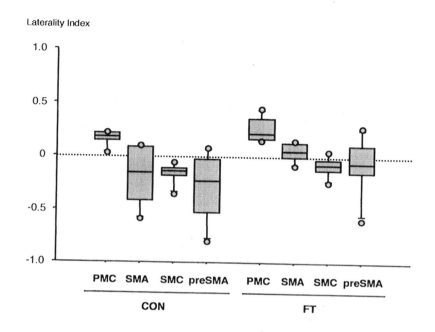

Figure 5-5. Changes of Laterality index (LI) in the motor-related areas during hemiplegic gait with different types of rehabilitative intervention (n=6).

Asymmetry in cortical activation during hemiplegic gait was assessed in six patients with severe stroke (4 males, 2 females; 4 with right and 2 with left hemiplegia; 57 years old and 3 months poststroke on average). LI was defined as (ΔoxyHb in the affected hemisphere - ΔoxyHb in the unaffected hemisphere) / (ΔoxyHb in the affected hemisphere + ΔoxyHb in the unaffected hemisphere). Positive LI indicates more activation in the affected hemisphere than in the unaffected hemisphere and negative LI indicates the reverse. An ANOVA revealed that there were significant main effects, both for type of rehabilitation ($F[1, 20] = 12.810$, $p = 0.0019$) and site of region ($F[3, 20] = 6.102$, $p = 0.0040$). There was no significant interaction between these factors. Namely, the facilitation technique (FT) improved asymmetry of cortical activation during hemiparetic gait. LI was significantly greater in the PMC than in the SMA ($p < 0.05$), SMC ($p < 0.005$), and preSMA ($p < 0.001$). The lower, middle, and upper horizontal lines of the boxes represent 25th, 50th, and 75th percentiles, respectively. The vertical lines extend from the 10th to the 90th percentiles (Miyai et al., 2002b, Copyright© 2002, Wiley-Liss, Inc., Reprinted by Permission John Wiley & Sons, Inc.).

Locomotor training on a treadmill with partial BWS by means of an overhead harness with a pelvic belt and thigh has the advantage that it is easy to describe and therefore replicable in any rehabilitation facility. Recent studies have shown that treadmill training with partial BWS is effective in improving mobility outcome in patients with spinal cord injury (Visintin et al., 1989; Wernig et al., 1992; Dobkin 1999), stroke (Hesse et al., 1994, 1995; Visintin et al., 1998), cerebral palsy (Schindl et al., 2000), and Parkinson's disease (Miyai, et al., 2000a, 2002a). The favorable outcome in treadmill training with partial BWS might be attributed to BWS rather than to treadmill walking since all patients initially tolerated a higher treadmill speed when walking on the treadmill with BWS than without BWS. A controlled trial, comparing the effect of treadmill training with BWS and without BWS in stroke, also showed greater efficacy of the former therapy than the latter (Visintin et al., 1989). However the mechanisms for the functional improvement remain unclear. Enhancement of the central pattern generator has been postulated as a mechanism for the efficacy of treadmill training with partial BWS. Our preliminary observation revealed that BWS tended to decrease overall activation (Fig. 5-6). Thus it is also possible that improvement of hemiparetic gait during treadmill training with partial BWS might be associated with improved efficacy of SMC activation.

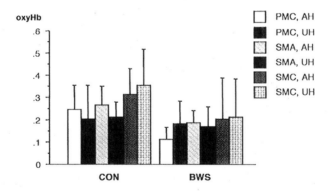

Figure 5-6. Effect of body weight support (BWS) on cortical activation during hemiparetic gait.
Cortical activation was measured during treadmill training with and without BWS (10 %) in 5 hemiparetic patients due to initial subcortical stroke (4 males, 1 female; 2 with right and 4 with left hemiparesis; 57 years old on average, and 2 months poststroke). For purposes of quantification of the activation, changes of oxyHb during the task period were integrated and then subtracted from integrated values of the changes during the rest period (n=5). Two-factorial repeated measures ANOVA demonstrated a significant main effect for BWS [$F(1,24)$ = 32.739, $p < 0.0001$]. Post-hoc test revealed that BWS induced significantly smaller

activation (p<0.0001) than control (CON). There was no main effect for site of region or no significant interaction between BWS and site of region, suggesting that BWS generally reduced cortical activation (Miyai et al., 2001b). Error bars are mean ± SD. AH: affected hemisphere, UH: unaffected hemisphere.

5. CEREBRAL MECHANISMS FOR LOCOMOTOR RECOVERY VERSUS HAND RECOVERY FOLLOWING STROKE

In stroke, restoration of gait is the major goal of neurorehabilitation. More than 70 % of patients with stroke are expected to be able to walk after inpatient rehabilitation (Miyai et al., 1998a; Yagura et al., 2003) whereas functional prognosis of the paretic hand is mostly determined by the initial impairment (Nakayama et al., 1994; Duncan et al., 2002; Yagura et al., 2003). Thus locomotor function appears to be less vulnerable to the brain damage than hand function. This might be partly due to the hieratical control of human locomotion as described above and the properties of the functional anatomy, controlling the distal and proximal muscles of the extremities. More specifically, axial and proximal muscles are controlled by bilateral corticospinal pathways whereas distal muscles are principally controlled by contralateral corticospinal pathways (Davidoff, 1990). This implies that the limbs might receive discordant efferent commands from both hemispheres and neural-crosstalk (Swinnen, 2002) might play an important role in locomotor recovery.

Changes of cortical activation involving multiple motor-related areas during hemiparetic gait as discussed above, are similar to those reported in functional neuroimaging studies regarding cerebral mechanisms for recovery of hand function in patients with stroke. Although patients generally made remarkable recoveries at the time of the studies in most neuroimaging studies using finger movement tasks, reorganization after stroke involved several candidate regions, including the peri-infarct area in the primary motor cortex of the injured hemisphere, the PMC and SMA in the damaged hemisphere, and some combination of these areas in the undamaged hemisphere (Chollet et al., 1991; Weiller et al., 1992, 1993; Cramer et al., 1997; Cao et al., 1998)

Several functional neuroimaging studies have shown that the ratio of contralateral to ipsilateral SMC and/or PMC activity during paretic hand movement increased as the paretic hand regained function (Marshall et al., 2000; Calautti et al., 2001; Miyai et al., 2001a, Feydy et al., 2002). This finding was confirmed by observation from Carey et al. (2002) that was designed to clarify the specific effect of rehabilitative training on brain activation. They showed that improved motor performance after intensive finger tracking training was accompanied by brain reorganization in patients with chronic stroke. The reorganization was characterized by a shift of cortical activation from predominantly ipsilateral to the affected hand to contralateral in SMC, and PMC. An increase in the hand area of the primary motor cortex in the damaged hemisphere after rehabilitative training was also shown in neurophysiological studies of primates with experimental cortical lesions (Nudo et al., 1996) and patients with chronic stroke (Liepert et al., 2000). Furthermore, a preliminary attempt to expand the hand area by combining rehabilitative training and deafferentation of proximal muscle by regional anesthesia in the upper brachial plexus, revealed some improvement of hand function in patients with chronic stroke (Muellbacher et al., 2002). Thus the capacity for reorganization in the primary motor cortex might be essential for good recovery of hand function after stroke.

Involvement of the PMC in the cerebral mechanisms of motor recovery of the affected hand is supported by the observation that improved finger movements in patients with middle cerebral artery infarction was associated with bilateral activation of the PMC (Seitz et al., 1988). Patients with capsular infarct and Wallerian degeneration on MRI appeared to require greater cortical reorganization than those without Wallerian degeneration to reach the similar functional level, since the former group showed persistent ipsi-lesional fMRI activation during grasping tasks of the paretic hand (Miyai, et al., 2001a). The emerging importance of the ipsi-lesional PMC in motor recovery finds additional support in the in vivo experimental literature (Liu et al., 1999).

In contrast, the ipsilateral corticospinal tract appears to play an important role in functional recovery of the paretic hand in cases who suffered from brain damage during the perinatal period and in some exceptional cases with adult stroke. Eyre and colleagues showed that ipsilateral responses to transcranial magnetic stimulation showed significantly shorter onsets than contralateral responses but similar thresholds and amplitudes in neonates. From 3 to 18 months, ipsilateral responses were significantly smaller and had significantly higher thresholds and longer onset latencies than contralateral responses. Accordingly, subjects with hemiplegic cerebral palsy exhibited ipsilateral responses with onsets, thresholds and amplitudes similar to those of contralateral responses, whereas small and late ipsilateral responses were observed after transcranial magnetic stimulation of the intact cortex in subjects with adult stroke (Eyre et al., 2001). Only few adult cases with stroke represent ipsi-lesional hemiparesis because of the predominant uncrossed corticospinal pathways (Hosokawa et al., 1996, Terakawa et al., 2000). However, the clinical relevance of the ipsilateral pathways in functional recovery in adult stroke is still suggested by the observation of two patients with stroke whose recovered limbs became re-paralyzed when they developed a new stroke in the other hemisphere (Fisher, 1992). Furthermore, bilateral hand movements have shown to enhance activation of the primary motor cortex in the affected hemisphere, as compared with unilateral movements of the affected hand after stroke (Staines, et al., 2001). Since gait performance is bipedal by nature, multiple structures including the PMC and SMA as well as bilateral SMC appear to be involved in locomotor recovery as discussed above. Thus impairment of locomotion due to damage in some of these structures, is more likely compensated for than impairment of hand function for a longer period after the insult.

6. SUMMARY

A comparison of the findings from neuroimaging studies regarding hand recovery after stroke and the current optical imaging studies, indicates that improved asymmetry in the SMC activation and recruitment of other motor related areas, especially the ipsi-lesional PMC, might be closely related to restoration of gait in hemiparetic stroke. However it remains to be determined what type of activation is associated with favorable locomotor recovery unless further longitudinal studies correlate real-world functional outcome with changes in cortical activation patterns. A preliminary serial optical imaging study showed that improvement of the asymmetry in the sensorimotor activation significantly correlated with improvement of asymmetry in gait parameters (Miyai et al., 2002c). These approaches might contribute to the establishment of brain-based as well as evidence-based strategies for neurorehabilitation.

REFERENCES

Alexander RL, Crutcher MD. Functional architecture of basal ganglia circuits: neural substrates of parallel processing. Trends Neurosci 1990;13:266-271.

Armstrong DM. The supraspinal control of mammalian locomotion. J Physiol 1988;405:1-37.

Bartocci M, Winberg J, Ruggiero C, Bergqvist LL, Serra G, Lagercrantz H. Activation of olfactory cortex in newborn infants after odor stimulation: a functional near-infrared spectroscopy study. Pediatr Res 2000;48:18-23.

Boas DA, Franceschini MA, Dunn AK, Strangman G: Noninvasive imaging of cerebral activation with diffuse optical tomography, In Frostig ed, In vivo optical imaging of brain function, CRC press, Boca Raton, p193-221, 2002.

Bobath, B. 1978. Adult hemiplegia. Evaluation and treatment, 2nd ed. William Heinemann Medical Books Ltd, London, England.

Calautti C, Leroy F, Guincestre JY, Marie RM, Baron JC. Sequential activation brain mapping after subcortical stroke: changes in hemispheric balance and recovery. Neuroreport 2001;12:3883-3886.

Cao Y, D'Olhaberriague L, Vikingstad EM, Levine SR, Welch KM. Pilot study of functional MRI to assess cerebral activation of motor function after poststroke hemiparesis. Stroke 1998;29:112-122.

Carey JR, Kimberley TJ, Lewis SM, Auerbach EJ, Dorsey L, Rundquist P, et al. Analysis of fMRI and finger tracking training in subjects with chronic stroke. Brain 2002;125:773-788.

Chollet F, DiPiero V, Wise RJ, Brooks DJ, Dolan RJ, Frackowiak RS. The functional anatomy of motor recovery after stroke in humans: a study with positron emission tomography. Ann Neurol 1991;29:63-71.

Colier WN, Quaresima V, Oeseburg B, Ferrari M. Human motor-cortex oxygenation changes induced by cyclic coupled movements of hand and foot. Exp Brain Res 1999;129:457-461.

Cramer SC, Nelles G, Benson RR, Kaplan JD, Parker RA, Kwong KK, et al. A functional MRI study of subjects recovered from hemiparetic stroke. Stroke 1997;28:2518-2527.

Davidoff RA. The pyramidal tract. Neurology 1990;40:332-339.

Davis PM. Steps to follow, 2nd ed. New York: Springer, 2000.

Debaere F, Swinnen SP, Beatse E, Sunaert S, Van Hecke P, Duysens J. Brain areas involved in interlimb coordination: a distributed network. Neuroimage 2001;14:947-958.

de Jong BM, Leenders KL, Paans AM. Right parieto-premotor activation related to limb-independent antiphase movement. Cereb Cortex 2002;12:1213-1217.

Dettmers C, Fink GR, Lemon RN, Stephan KM, Passingham RE, Silbersweig D, et al. Relation between cerebral activity and force in the motor areas of the human brain. J Neurophysiol 1995;74:802-815.

Dietz V, Muller R, Colombo G. Locomotor activity in spinal man: significance of afferent input from joint and load receptors. Brain 2002;125:2626-2634.

Dobkin BH. An overview of treadmill locomotor training with partial body weight support: a neurophysiologically sound approach whose time has come for randomized clinical trials. Neurorehabil Neural Repair 1999;13:157-165.

Drew T. Motor cortical cell discharge during voluntary gait modification. Brain Res 1988;457:181-187.

Duncan PW, Lai SM, Keighley J. Defining post-stroke recovery: implications for design and interpretation of drug trials. Neuropharmacology 2000;39:835-841.

Eda H, Sase I, Seiyama A, Tanabe HC, Imaruoka T, Tsunazawa Y, Yanagida T. Optical Topography System for Functional Brain Imaging: Mapping Human Occipital Cortex During Visual Stimulation. Proc of Inter-Institute Workshop on In Vivo Optical Imaging at the NIH. pp. 93-99, 2000. Optical Society of America.

Eyre JA, Taylor JP, Villagra F, Smith M, Miller S. Evidence of activity-dependent withdrawal of corticospinal projections during human development. Neurology 2001;57:1543-1554.

Ferrier, D. 1876. The Functions of the Brain. Smith, Elder, London.

Feydy A, Carlier R, Roby-Brami A, Bussel B, Cazalis F, Pierot L, et al. Longitudinal study of motor recovery after stroke: recruitment and focusing of brain activation. Stroke 2002;33:1610-1617.

Fink GR, Frackowiak RS, Pietrzyk U, Passingham RE. Multiple nonprimary motor areas in the human cortex. J Neurophysiol 1997;77:2164-2174.

Fisher CM. Concerning the mechanism of recovery in stroke hemiplegia. Can J Neurol Sci 1992;19:57-63.

Frackowiak RSJ. The cerebral basis of functional recovery. In Frackowiak RSJ, Friston KJ, Frith CD, Dolan RJ, Mazziotta JC, eds, Human brain function, Academic press, San Diego, p275-299, 1997.

Freund HJ, Hummelsheim H. Lesions of premotor cortex in man. Brain 1985;108:697-733.

Fries W, Danek A, Scheidtmann K, Hamburger C. Motor recovery following capsular stroke. Role of descending pathways from multiple motor areas. Brain 1993;116:369-382.

Fukuyama H, Ouchi Y, Matsuzaki S, Nagahama Y, Yamauchi H, Ogawa M, et al. Brain functional activity during gait in normal subjects: a SPECT study. Neurosci Lett 1997;228:183-186.

Halsband U, Ito N, Tanji J, Freund HJ. The role of premotor cortex and the supplementary motor area in the temporal control of movement in man. Brain 1993;116:243-266.

Hanakawa T, Fukuyama H, Katsumi Y, Honda M, Shibasaki H. Enhanced lateral premotor activity during paradoxical gait in Parkinson's disease. Ann Neurol 1999;45:329-336.

Hanakawa T, Katsumi Y, Fukuyama H, Honda M, Hayashi T, Kimura J, et al. Mechanisms underlying gait disturbance in Parkinson's disease: a single photon emission computed tomography study. Brain 1999;122:1271-1282.

Hesse S, Bertelt C, Schaffrin A, Malezic M, Mauritz KH. Restoration of gait in nonambulatory hemiparetic patients by treadmill training with partial body-weight support. Arch Phys Med Rehabil 1994;75:1087-1093.

Hesse S, Bertelt C, Jahnke MT, Schaffrin A, Baake P, Malezic M, et al. Treadmill training with partial body weight support compared with physiotherapy in nonambulatory hemiparetic patients. Stroke 1995;26:976-981.

Hesse S, Jahnke MT, Schaffrin A, Lucke D, Reiter F, Konrad M. Immediate effects of therapeutic facilitation on the gait of hemiparetic patients as compared with walking with and without a cane. Electroencephalogr Clin Neurophysiol 1998;109:515-522.

Hintz SR, Benaron DA, Siegel AM, Zourabian A, Stevenson DK, Boas DA. Bedside functional imaging of the premature infant brain during passive motor activation. J Perinat Med 2001;29:335-343.

Hirth C, Obrig H, Villringer K, Thiel A, Bernarding J, Muhlnickel W, et al. Non-invasive functional mapping of the human motor cortex using near-infrared spectroscopy. Neuroreport 1996;7:1977-1981.

Hock C, Muller-Spahn F, Schuh-Hofer S, Hofmann M, Dirnagl U, Villringer A. Age dependency of changes in cerebral hemoglobin oxygenation during brain activation: a near-infrared spectroscopy study. J Cereb Blood Flow Metab 1995;15:1103-1108.

Hock C, Villringer K, Muller-Spahn F, Wenzel R, Heekeren H, Schuh-Hofer S, et al. Decrease in parietal cerebral hemoglobin oxygenation during performance of a verbal fluency task in patients with Alzheimer's disease monitored by means of near-infrared spectroscopy (NIRS)--correlation with simultaneous rCBF-PET measurements. Brain Res 1997;755:293-303.

Hoshi Y, Oda I, Wada Y, Ito Y, Yutaka Y, Oda M, et al. Visuospatial imagery is a fruitful strategy for the digit span backward task: a study with near-infrared optical tomography. Brain Res Cogn Brain Res 2000;9:339-342.

Hoshi Y, Kobayashi N, Tamura M. Interpretation of near-infrared spectroscopy signals: a study with a newly developed perfused rat brain model. J Appl Physiol 2001;90:1657-1662.

Hosokawa S, Tsuji S, Uozumi T, Matsunaga K, Toda K, Ota S. Ipsilateral hemiplegia caused by right internal capsule and thalamic hemorrhage: demonstration of predominant ipsilateral innervation of motor and sensory systems by MRI, MEP, and SEP. Neurology 1996;46:1146-1149.

Isobe K, Kusaka T, Nagano K, Okubo K, Yasuda S, Kondo M, et al. Functional imaging of the brain in sedated newborn infants using near infrared topography during passive knee movement. Neurosci Lett 2001;299:221-224.

Kato T, Kamei A, Takashima S, Ozaki T. Human visual cortical function during photic stimulation monitoring by means of near-infrared spectroscopy. J Cereb Blood Flow Metab 1993;13:516-520.

Kawashima R, Inoue K, Sugiura M, Okada K, Ogawa A, Fukuda H. A positron emission tomography study of self-paced finger movements at different frequencies. Neuroscience 1999;92:107-112.

Lafleur MF, Jackson PL, Malouin F, Richards CL, Evans AC, Doyon J. Motor Learning Produces Parallel Dynamic Functional Changes during the Execution and Imagination of Sequential Foot Movements. Neuroimage 2002;16:142-157.

Liepert J, Bauder H, Wolfgang HR, Miltner WH, Taub E, Weiller C. Treatment-induced cortical reorganization after stroke in humans. Stroke 2000;31:1210-1216.

Liu Y, Rouiller EM. Mechanisms of recovery of dexterity following unilateral lesion of the sensorimotor cortex in adult monkeys. Exp Brain Res 1999;128:149-159.

Marshall RS, Perera GM, Lazar RM, Krakauer JW, Constantine RC, DeLaPaz RL. Evolution of cortical activation during recovery from corticospinal tract infarction. Stroke 2000;31:656-661.

Meek JH, Elwell CE, Khan MJ, Romaya J, Wyatt JS, Delpy DT, et al. Regional changes in cerebral haemodynamics as a result of a visual stimulus measured by near infrared spectroscopy. Proc R Soc Lond B Biol Sci 1995;261:351-356.

Miyai I, Blau AD, Reding MJ, Volpe BT. Patients with stroke confined to basal ganglia have diminished response to rehabilitation efforts. Neurology 1997;48:95-101.

Miyai I, Reding M. Stroke Recovery and Rehabilitation. In Cerebrovascular Disease: Pathology, Diagnosis, and Management. Ginsberg MD, Bogousslavsky J, Eds. Blackwell Scientific Publications, Malden, MA, 1998, 2043-2056.

Miyai I, Suzuki T, Kii K, Kang J, Kubota K. Wallerian degeneration of the pyramidal tract does not affect stroke rehabilitation outcome. Neurology 1998;51:1613-1616.

Miyai I, Suzuki T, Kang J, Kubota K, Volpe BT. Middle cerebral artery stroke that includes the premotor cortex reduces mobility outcome. Stroke 1999;30:1380-1383.

Miyai I, Fujimoto Y, Ueda Y, Yamamoto H, Nozaki S, Saito T, et al. Treadmill training with body weight support: its effect on Parkinson's disease. Arch Phys Med Rehabil 2000;81:849-852.

Miyai I, Suzuki T, Kang J, Volpe BT. Improved functional outcome in patients with hemorrhagic stroke in putamen and thalamus compared with those with stroke restricted to the putamen or thalamus. Stroke 2000;31:1365-1369.

Miyai I, Suzuki T, Mikami A, Kubota K, Volpe BT. Patients with capsular infarct and Wallerian degeneration demonstrate persistent regional premotor cortex activation on functional MRI. J Stroke Cereb Dis 2001;10:210-216.

Miyai I, Tanabe HC, Sase I, Eda H, Oda I, Konishi I, et al. Cortical mapping of gait in humans: a near-infrared spectroscopic topography study. Neuroimage 2001;14:1186-1192.

Miyai I, Yagura H, Kubota K, Suzuki T, Oda I, Konishi I, et al. Cortical activation patterns during hemiparetic gait are altered by rehabilitative intervention. A near-infrared spectoroscopic topography study. Soc Neurosci Abstr. 2001;27: 831.2.

Miyai I, Fujimoto Y, Yamamoto H, Ueda Y, Saito T, Nozaki S, et al. Long-term effect of body weight-supported treadmill training in Parkinson's disease: a randomized controlled trial. Arch Phys Med Rehabil 2002;83:1370-1373.

Miyai I, Yagura H, Oda I, Konishi I, Eda H, Suzuki T, et al. Premotor cortex is involved in restoration of gait in stroke. Ann Neurol 2002;52:188-194.

Miyai I, Yagura H, Oda I, Konishi I, Suzuki T, Kubota K. Cortical reorganization associated with locomotor recovery in stroke. An optical imaging study. Soc Neurosci Abst 2002;28:664.12.

Muellbacher W, Richards C, Ziemann U, Wittenberg G, Weltz D, Boroojerdi B, et al. Improving hand function in chronic stroke. Arch Neurol 2002;59:1278-1282.

Nakayama H, Jorgensen HS, Raaschou HO, Olsen TS. Recovery of upper extremity function in stroke patients: the Copenhagen Stroke Study. Arch Phys Med Rehabil 1994;75:394-398.

Nudo RJ, Wise BM, SiFuentes F, Milliken GW. Neural substrates for the effects of rehabilitative training on motor recovery after ischemic infarct. Science 1996;272:1791-1794.

Nutt JG, Marsden CD, Thompson PD. Human walking and higher-level gait disorders, particularly in the elderly. Neurology 1993;43(2):268-279.

Obrig H, Hirth C, Junge-Hulsing JG, Doge C, Wenzel R, Wolf T, et al. Length of resting period between stimulation cycles modulates hemodynamic response to a motor stimulus. Adv Exp Med Biol 1997;411:471-480.

Roth EJ, Merbitz C, Mroczek K, Dugan SA, Suh WW. Hemiplegic gait. Relationships between walking speed and other temporal parameters. Am J Phys Med Rehabil 1997;76:128-133.

Pearson K, Gordon J. Locomotion. In: Kandel ER, Schwartz JH, Jessell TM (eds) Principles of neural science, 737-755, McGraw-Hill, New York, 2000.

Sakatani K, Xie Y, Lichty W, Li S, Zuo H. Language-activated cerebral blood oxygenation and hemodynamic changes of the left prefrontal cortex in poststroke aphasic patients: a near-infrared spectroscopy study. Stroke 1998;29:1299-1304.

Sato H, Takeuchi T, Sakai KL. Temporal cortex activation during speech recognition: an optical topography study. Cognition 1999;73:B55-66.

Schindl MR, Forstner C, Kern H, Hesse S. Treadmill training with partial body weight support in nonambulatory patients with cerebral palsy. Arch Phys Med Rehabil 2000;81:301-306.

Seitz RJ, Hoflich P, Binkofski F, Tellmann L, Herzog H, Freund HJ. Role of the premotor cortex in recovery from middle cerebral artery infarction. Arch Neurol 1998;55:1081-1088.

Staines WR, McIlroy WE, Graham SJ, Black SE. Bilateral movement enhances ipsilesional cortical activity in acute stroke: a pilot functional MRI study. Neurology 2001;56:401-404.

Strangman G, Culver JP, Thompson JH, Boas DA. A quantitative comparison of simultaneous BOLD fMRI and NIRS recordings during functional brain activation. Neuroimage 2002;17:719-731.

Swinnen SP. Intermanual coordination: from behavioural principles to neural-network interactions. Nat Rev Neurosci 2002;3:348-359.

Suzuki M, Miyai I, Ono T, Yagura H, Oda I, Konishi I, Eda H, Tanabe HC, Kochiyama T, Kubota K. Running induces prefrontal activation. An optical imaging study. Soc Neurosci Abst 2002;28:854.10.

Taga G, Konishi Y, Maki A, Tachibana T, Fujiwara M, Koizumi H. Spontaneous oscillation of oxy- and deoxy- hemoglobin changes with a phase difference throughout the occipital cortex of newborn infants observed using non-invasive optical topography. Neurosci Lett 2000;282:101-104.

Takahashi K, Ogata S, Atsumi Y, Yamamoto R, Shiotsuka S, Maki A, et al. Activation of the visual cortex imaged by 24-channel near-infrared spectroscopy. J Biomed Opt 2000;5:93-96.

Terakawa H, Abe K, Nakamura M, Okazaki T, Obashi J, Yanagihara T. Ipsilateral hemiparesis after putaminal hemorrhage due to uncrossed pyramidal tract. Neurology 2000;54:1801-1805.

Villringer A, Planck J, Hock C, Schleinkofer L, Dirnagl U. Near infrared spectroscopy (NIRS): a new tool to study hemodynamic changes during activation of brain function in human adults. Neurosci Lett 1993;154:101-104.

Villringer A, Obrig H: Near-infrared spectroscopy and imaging, In Toga, Mazziotta eds, Brain mapping. The methods, 2nd ed, Academic Press, San Diego, p141-158, 2002.

Visintin M, Barbeau H. The effects of body weight support on the locomotor pattern of spastic paretic patients. Can J Neuro Sci 1989;16:315-325.

Visintin M, Barbeau H, Korner-Bitensky N, Mayo NE. A new approach to retrain gait in stroke patients through body weight support and treadmill stimulation. Stroke 1998;29:1122-1128.

Waldvogel D, van Gelderen P, Ishii K, Hallett M. The effect of movement amplitude on activation in functional magnetic resonance imaging studies. J Cereb Blood Flow Metab 1999;19:1209-1212.

Watanabe E, Maki A, Kawaguchi F, Takashiro K, Yamashita Y, Koizumi H, et al. Non-invasive assessment of language dominance with near-infrared spectroscopic mapping. Neurosci Lett 1998;256:49-52.

Weiller C, Chollet F, Friston KJ, Wise RJ, Frackowiak RS. Functional reorganization of the brain in recovery from striatocapsular infarction in man. Ann Neurol 1992;31:463-472.

Weiller C, Ramsay SC, Wise RJ, Friston KJ, Frackowiak RS. Individual patterns of functional reorganization in the human cerebral cortex after capsular infarction. Ann Neurol 1993;33:181-189.

Wernig A, Müller S. Laufband locomotion with body weight support improved walking in persons with severe spinal cord injuries. Paraplegia 1992;30:229-238.

Wolf M, Wolf U, Toronov V, Michalos A, Paunescu LA, Choi JH, et al. Different time evolution of oxyhemoglobin and deoxyhemoglobin concentration changes in the visual and motor cortices during functional stimulation: a near-infrared spectroscopy study. Neuroimage 2002;16:704-712.

Yagura H, Miyai I, Seike Y, Suzuki T, Yanagihara T. Benefit of In-patient Multidisciplinary Rehabilitation up to 1 Year after Stroke. Arch Phys Med Rehab 2003, in press.

PART II

BIMANUAL COORDINATION

NEURAL MECHANISMS AND BEHAVIORAL PRINCIPLES

Chapter 6

ELECTROPHYSIOLOGICAL APPROACHES TO BIMANUAL COORDINATION IN PRIMATES

Opher Donchin[1], Simone Cardoso de Oliveira[2]

[1] *Dept. Biomedical Engineering, Johns Hopkins University, Baltimore, USA,* [2]*Institut für Arbeitsphysiologie an der Universität Dortmund, Dortmund, Germany*

Abstract: Neuronal activity in primates has been used to address two different questions. The first is what part, or parts of the brain mediate bimanual coordination. The second is how the activity of different parts of the brain is coordinated to achieve bimanual coordination. The answers to both questions are preliminary, but we outline the evidence for a widespread network using both rate coding and temporal coding to mediate bimanual coordination. We also address the relevance of monkey studies to human bimanual coordination.

Key words: Electrophysiology, Local field potential (LFP), single unit, primary motor cortex (MI), Supplementary motor area (SMA), premotor cortex, parietal cortex, rate coding, temporal coding, population vector, monkey

1. INTRODUCTION

The neurophysiological approach, using extracellular intracortical microelectrodes to record neuronal activity within the brain of behaving primates (see Box 6-1), has been used to ask two major questions about bimanual coordination. The first is "what part of the brain is responsible for bimanual coordination?" To answer this question, researchers have sought modulation of neural activity that is specific to bimanual movements, and considered this to be evidence that the neurons generating the modulated activity are involved in bimanual coordination.

Box 6-1: Intracortical electrode recording techniques

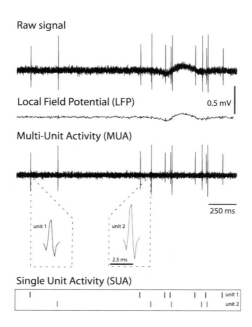

Raw signal

Local Field Potential (LFP)   0.5 mV

Multi-Unit Activity (MUA)

250 ms

unit 1   unit 2

2.5 ms

Single Unit Activity (SUA)

unit 1
unit 2

Thin, sharp metal electrodes are introduced through a small hole made in the skull of the animal. These electrodes are insulated along their shafts, usually with either glass or enamel, leaving only the tip exposed. The electrodes are not used to penetrate neurons. Instead, they record extracellular voltage changes, referenced to a ground electrode usually attached to the animals skull. These voltage changes are caused by currents into or out of neurons that arise during both synaptic activation and action potentials. Synaptic activation produces a much slower signal than action potentials, so the two components of the signal can be separated by appropriate filtering. The local field potential (LFP) is acquired by filtering the signal from 1-140 Hz, and is thought to reflect primarily synaptic activity. The spiking activity is usually filtered from 300-10,000 Hz. Typically, an electrode is only sensitive to neurons in a volume of about $0.5 - 1$ mm^3 near its tip. This may include many neurons, and so the spiking activity recorded from a single electrode is usually called multi unit activity (MUA). However, the shape of extracellularly recorded action potentials depends on the size and distance of the recorded cell, so spike shapes and amplitudes from different cells can be isolated (see upper and lower spike shapes) and the activity of individual neurons recorded, resulting in so-called single unit activity (SUA). All data generated by simulation.

The stronger the modulation, the stronger the evidence is considered to be. If one region of the brain shows modulation specific to bimanual movements and another region does not show such modulation, this is considered evidence that the first area is involved in bimanual coordination and the second area is not. Of course, the logic of this approach is relatively weak, and so such evidence will not be considered ultimately conclusive without converging evidence from lesion studies, stimulation studies, or anatomical studies.

The second question is "how is bimanual coordination achieved?" To answer this question, it is not enough to find neurons whose activity is modulated during bimanual movements. One must discover which parameters of neuronal activity are modulated, and then relate this modulation to parameters of the movement. For instance, it has been suggested that temporal correlation between the activity of different neurons may serve to bind the activity of different neurons together. One may similarly hypothesize that correlations between neurons in different cortical hemispheres bind the actions of the limbs to produce bimanual coordination.

While current research has not provided full answers to either of these questions, this chapter will summarize the current partial answers to each of the two questions in turn. Our focus is on behaving primate neurophysiology, but we will briefly review other relevant literature. For a more extensive discussion of findings from other techniques, the readers are referred to the other chapters of this book, especially the chapter on imaging by Wenderoth et al. (Chapter 8).

2. WHICH BRAIN AREAS ARE INVOLVED IN BIMANUAL COORDINATION?

Early electrophysiological research focused on the supplementary motor area (SMA, see Box 6-2) – located in the medial aspect of the frontal cortex – as a strong candidate for mediating bimanual coordination. This focus grew out of early clinical reports of bimanual deficits following lesions of medial frontal cortex (for review see Brust, 1996) and an influential study of bimanual deficits following similar, controlled, lesions in primates (Brinkman, 1984). Furthermore, the SMA possesses dense interhemispheric connections that could facilitate interactions between the activities of neurons controlling both arms (Rouiller et al., 1994, Wiesendanger et al., 1996). EEG (Lang et al., 1990) and imaging studies have shown that the SMA of humans is activated during interlimb coordination. Since bimanual coordination is a complex task especially refined in primates and higher motor cortical areas are particularly evolved in primates, it seems natural to

---

Box 6-2: Glossary of brain areas possibly involved in bimanual coordination

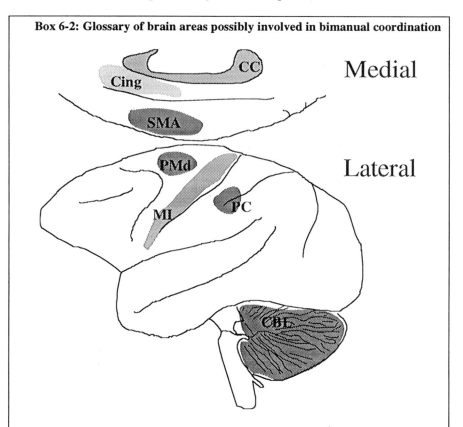

Motor Cortex (MI): Also called primary motor cortex because it is the main cortical area directly controlling movements. Damage to this area induces paralysis or paresis of the affected body part. MI, like other motor cortical areas, has a topographic representation of the body. In MI, electrical stimulation at very low thresholds (5-10 μA) elicits muscle twitches or small movements in corresponding body parts. Neuronal activity in MI is closely related to the characteristics of upcoming movements.

Supplementary Motor Area (SMA): Supplementary motor area. Stimulation of this area also evokes movements, but higher thresholds are necessary (40-80 μA) and evoked movements are less consistent, and can also be multi-jointed or bilateral. Lesions of this area do not cause paresis, but frequently specific deficits in bimanual coordination have been described. The hypothesis that the SMA is involved in bimanual coordination is supported by dense interhemispheric connectivity. It has also been suggested that SMA is involved in self-generated movements and postural control (Weisendanger et al., 1996).

Cingulate Cortex (Cing): Cortical area that classically was considered to be part of the limbic system. Recent studies showed that neurons in this area also show clear movement-related activity (Dum & Strick, 2002).

Dorsal Premotor Area (PMd: Secondary motor area, with less dense direct corticospinal projections, and from which movements can be elicited only at high current intensities.

Parietal Cortex (PC): Cortical area considered to belong to the associative visual cortex. Neurons in this area respond to stimuli that elicit movements. Lesions in this area produce deficits in controlling goal-directed movements under visual control.

Basal Ganglia (BG, not shown): Basal part of the forebrain subserving a complex set of limbic, cognitive and motor functions. Diseases affecting the basal ganglia cause a number of major motor disorders including Tourette syndrome, Parkinson's and Huntington's disease. The basal ganglia receive their input from the cortex and are connected back to cortical areas through projections to the thalamus.

Corpus Callosum (CC): Thick band of fibers connecting the two cortical hemispheres. Because each hemisphere reacts to and controls mainly the contralateral half of the body, severing these connections causes typical 'split-brain' syndromes, e.g., having trouble reaching with the left hand to an object presented in the right visual hemifield. Also, coordination between the two arms is altered.

Spinal Cord (not shown): The nervous tissue within the spine contains all the neurons that have direct connections to muscles. No voluntary or reflexive movement is possible without the spinal cord or when the connection between the spinal cord and the muscles is severed (as, e.g, in the case of hemi- or tetraplegia). Spinal mechanisms are strongly implicated in interlimb coordination of gait for all species studied so far.

Cerebellum (CBL): Large brain area overlying the brainstem. The cerebellum is known to mediate involuntary motor functions as well as motor coordination and motor learning.

assume that higher motor cortical areas are involved in this task in bimanual coordination.

In the first effort to test this hypothesis electrophysiologically, a monkey was trained to press buttons with the fingers of either hand separately or both hands together (Tanji et al., 1988). A large fraction of neurons in SMA were active during the bimanual movements and not during movements of either hand separately, suggesting that there are some neurons that seem to be 'specific' for bimanual movements. Neuronal activity in SMA specific to bimanual movements has now been described by a number of groups using different tasks, although this specificity has been defined slightly differently by different groups. Donchin et al. (1998) used a task in which the monkey performed unimanual and bimanual movements involving the whole arm moving in different directions (Fig. 6-1). These authors were not only interested in units exclusively activated during bimanual movements. Instead, they defined a 'bimanual specificity' index characterizing the extent to which bimanual activity could be explained by the activity of the neuron during unimanual movements. Figure 6-2 demonstrates this notion. The figure shows two examples of 'bimanual specific' units where the specificity arises either from increased activity or decreased activity during the bimanual movements. Donchin et al. (1998) reported that more than 50% of neurons in SMA showed bimanual specificity. In another task involving whole-arm movements, where the monkey was required to open a drawer with one hand and retrieve a raisin from it with the other hand, such bimanual specific SMA activity has also been described (Kermadi et al., 1998), although a different study on the same task reported that only a small percentage of neurons was exclusively activated during bimanual movements (Kazennikov et al., 1999). The drawer pulling task is particularly interesting because the monkey performs it naturally and interlimb synchronization is achieved without specific training (Wiesendanger et al., 1994). In the tasks used by Donchin (1998) and Tanji (1988), extensive training was required before the monkeys could perform the required movements simultaneously. It is possible, even likely, that this training had an effect on the neural activity.

Figure 6-1. A bimanual reaching task for primates. The monkey controls two cursors (see crosses) on a computer monitor using two manipulanda. The task is to move the two cursors simultaneously from central origins (centers of the circle) to peripheral targets (lighter circles,

left targets for the left hand, right targets for the right hand). The choice of targets allowed for movements of the two hands in the same direction or in different directions.

Despite these encouraging single-unit studies, a number of attempts to replicate the human bimanual deficits after SMA lesion in monkeys have failed (Kazennikov et al.,1998, Gribova et al.,1998). One possibility is that studies of human lesions involved damage to areas beyond the SMA. Indeed, even in the experimental lesions in monkeys it is possible that damage extended beyond the SMA and spread into additional areas. It may be possible, therefore, that SMA is not the only area involved in bimanual coordination, and a more widely distributed network, involving multiple cortical and subcortical areas, is involved in this task. This picture is also compatible with recent imaging work on human subjects (see chapter of Wenderoth et al., Chapter 8). In the following section, we will review the electrophysiological evidence supporting this hypothesis.

In contrast to the secondary motor cortical areas, the **primary motor cortex** (MI, see Box 6-2) was long seen as a relatively simple structure containing essentially 'upper motor neurons' that directly control the muscles. In this view, it was assumed that MI in each hemisphere was responsible for controlling the movements of the contralateral limb. Despite this long-standing understanding, every single unit study of MI since 1966 (Evarts, 1966) has reported that a significant proportion of the neurons in each hemisphere are activated during movements of the ipsilateral arm (e.g.: Tanji et al., 1988, Donchin et al., 1998, 2002, Kermadi et al., 1998, Cisek et al., 2003). One study has even suggested that there is a specific region of MI which has an especially dense representation of the ipsilateral hand (Aizawa, 1990). However, this finding has yet to be replicated and most researchers report a mixture of ipsilaterally and contralaterally activated neurons all over MI. It is worth noting, however, that many studies used whole arm movements, while the Aizawa (1990) study used movements of the hand and fingers. Perhaps distal and proximal representations of the ipsilateral limb are organized differently. Indeed, whole arm studies finding ipsilateral activation in MI must contend with the possibility that these findings are tainted by postural adjustments or spurious contralateral activation (Cisek et al., 2003) and the ipsilateral activation during distal movements in the Aizawa study could be considered more compelling evidence of an ipsilateral representation in MI.

A few studies compared the extent of bilateral activation in MI and SMA, but the results of these comparisons have been inconclusive. Some studies have reported that SMA has more neurons related to ipsilateral movements (Tanji, 1988; Wiesendanger, 1996) while others have reported a similar percentage in the two areas (Donchin et al., 1998, Kermadi et al., 1998).

Thus, while it is not clear whether the SMA is more ipsilaterally activated than MI, the single unit studies do seem to indicate that MI plays some role in the generation of ipsilateral movements. This idea has also found support in human subjects, as reviewed by Chen et al. (1997). To summarize this evidence briefly, fMRI confirms ipsilateral activation of MI in humans for movements of the elbow, wrist and fingers (Alkadhi et al., 2002), strokes or transcranial-magnetic stimulation (TMS) induced inactivation of MI cause deficits in interlimb coordination on the ipsilateral side of the lesion (Debaere et al., 2001b, Chen et al., 1997), and TMS-induced activation of MI causes ipsilateral muscle activation (Wassermann et al., 1994).

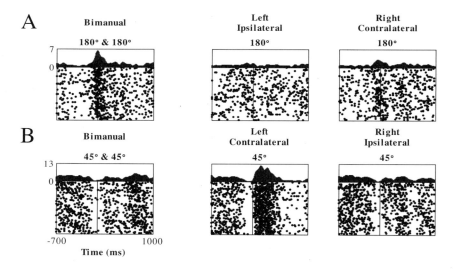

Figure 6-2. Bimanual specific activity of two neurons in SMA. A) A neuron that is more active during bimanual movements; B) A neuron that is less active during bimanual movements. Directions on top of each plot indicate movement directions of one or both arms. Dots indicate action potentials, where each horizontal lines represents the activity recorded in a single movement. The histogram on top of each display shows activity averaged over all repetitions. The time axis in each repetition was aligned to movement onset (at 0 ms).

If each MI can be activated during movements of both arms, it seems possible that this area may also be involved in bimanual coordination. While an early study reported that bimanual specific units were restricted to the SMA (Tanji, 1988), more recent studies found bimanual specific units in similar quantities in both MI and SMA (Donchin et al., 1998; Donchin et al., 2002; Kermadi et al., 1998). Also, local field potential amplitude in bimanual movements was increased compared to unimanual movements in both SMA and MI (Donchin et al., 2001). These more recent findings indicate that MI may well be part of a network participating in bimanual control. The differences between these latter results and those of Tanji (1988) could arise for a number of reasons. The most probable, it seems, is that MI is involved in bimanual coordination in certain tasks but not in others. The task used by Tanji (1988) required only small movements of the fingers, while the tasks used by Donchin (1998) and Kermadi (1998) involved larger, whole limb movements. Maybe MI plays a larger role in ipsilateral movements of the whole limb than in ipsilateral movements of the fingers (Brinkman and Kuypers, 1973), and perhaps this distinction extends to bimanual movements as well. It is also possible that postural adjustments or uncontrolled changes in the way the component unimanual movements are performed occur more frequently with proximal movements. These may lead to an appearance of bimanual specific activity (Kazenikov et al., 1999). Donchin et al. (2002) addressed this possibility directly, performing a number of statistical analyses to test whether differences in movement details might explain the difference between unimanual and bimanual neural activation. Their conclusion was that differences in the movements could not explain the bimanual activations. On the basis of these results, it seems safe to assume that bimanual specific activity in MI reflects neuronal processes that are specifically related to bimanual coordination. A replication of these results by other groups and an adaptation of the methods to different bimanual tasks would help to further prove this point.

However, other cortical areas may also play a role in bimanual coordination. For instance, we suggested that early lesion results implicating SMA in bimanual coordination may have arisen because of additional damage to neighboring cortical areas. These neighboring areas in medial and frontal cortex may thus be implicated in bimanual control. Indeed, lesions involving the **cingulate cortex** do result in bimanual deficits (Stephan et al., 1999). Further, the single-unit study of Kermadi et al. (2000) found substantial bimanual-specific activity also in the cingulate cortex (see Box 6-2). fMRI has also quite recently shown bimanual activation of cingulate cortex in humans (Immisch et al., 2001).

In addition, the study of Kermadi et al. (2000) also reported bimanual specific activity in the **dorsal premotor cortex** (see Box 6-2), an area where Tanji (1988) had previously shown that cells specifically responded to bilateral finger movements. This area has also been implicated in human studies. For example, a PET study (Sadato et al., 1997) showed activity in this area that seemed related to the complexity of the bimanual synergy. Similarly, lesions involving the dorsal premotor cortex lead to deficits in bimanual rhythm production (Halsband et al., 1993).

Kermadi et al. (2000) also showed bimanual specific activity is the **posterior parietal cortex** (see Box 6-2), an area that had been previously unexplored using electrophysiology and bimanual tasks. Again, this is consistent with data in humans where posterior parietal cortex of humans is activated during bimanual movements (Nair et al., 2003), and damage to the parietal cortex of humans impairs bimanual coordination (Serrien et al., 2001). While posterior parietal is classically considered the home of the visual associative areas, its relation to eye-hand coordination and reaching movements (e.g., Batista et al., 1999) seem to indicate that it plays a role in motor function as well.

There is also the possibility that subcortical structures may be involved in bimanual coordination. For instance, patients suffering from diseases originating from the **basal ganglia** (see Box 6-2), such as those with Parkinson's disease (Johnson et al., 1998, Van den Berg et al., 2000, Serrien et al., 2000) and Huntington's disease (eg., Johnson et al., 2000) show deficits in bimanual coordination. This has lead to the suggestion that the basal ganglia are involved in bimanual coordination. However, the output of the basal ganglia indirectly also affects, through thalamocortical loops, a number of cortical areas, including M1 and premotor cortex, and has especially heavy projections to the SMA. Therefore, deficits observed may be an indirect effect caused by disruption of the cortical network. Recent evidence of single unit activity specific to bimanual movements (Orlov et al., 1999; Wannier et al., 2002) might be considered a more direct indication that the basal ganglia play a role in the coordination of bimanual movements. Additionally, because neurosurgical techniques are often used to treat Parkinson's, electrophysiological evidence from human beings is also available. In humans, the amplitude of oscillatory activity in the globus pallidum (the output nucleus of the basal ganglia) is strongly correlated with bimanual task duration (Brown et al., 2002). This convergent evidence from monkey and human studies strongly suggests that the basal ganglia play a role in bimanual coordination.

The **cerebellum** (see Box 6-2) is another motor control area that should be considered a potential candidate for involvement in bimanual control. Cerebellar injury in humans has been shown to result in deficits in bimanual coordination (Brown et al., 1993; Serrien & Wiesendanger, 2000), and recent imaging studies revealed an activation of the cerebellum during bimanual movements (eg., Tracy et al., 2001, for more references, see the chapter of Wenderoth et al., Chapter 8). However, we are not aware of any single unit experiments that test for a cerebellar role in bimanual coordination.

Finally, its clear involvement in the bilateral control of gait (see the chapters of Dietz, Chapter 4) makes the **spinal cord** (see Box 6-2) an area that might well have a role during voluntary bimanual movements. However, techniques for spinal cord single unit recording in awake behaving primates have only recently been developed (Prut and Fetz, 1999), and data about non-locomotor bimanual activations in primates are still missing.

Taken together, the above mentioned evidence seems to indicate that coding of bimanual movements is most probably not accomplished by a single structure, but by a distributed network of cortical and subcortical regions (Swinnen, 2002). This assumption is also supported by imaging studies in human subjects, showing widespread bimanual-related activation. Such a multi-level organization also offers the possibility of some functional specialization, i.e. that certain aspects of bimanual interactions and bimanual control are dealt with in certain areas. Researchers have yet to compare different bimanual tasks in single unit studies, but this may well be the key to sorting out the functional roles of the different areas implicated in bimanual control.

3.   POSSIBLE NEURONAL MECHANISMS OF BIMANUAL COORDINATION

The movements of the two hands couple spontaneously. This constrains bimanual behavior by making some forms of bimanual coordination easier than others (see Chapters 9 and 10 of Spijkers & Heuer, and Ivry et al., respectively). With training, spontaneous coupling can be partially overcome, and a wider bimanual behavioral repertoire can develop. What are the neurophysiological mechanisms underlying the spontaneous coupling, and how is it regulated in a task-dependent way? While there are a number of possibilities, we explore the hypothesis that interhemispheric interactions through the corpus callosum (see Box 6-2) are at the root of at least some aspects of bimanual coupling, and, further, that modulating these interactions enables different modes of bimanual coordination.

This hypothesis is supported by reported abnormalities of bimanual coordination in patients with impaired callosal function. Abnormalities caused by callosal section have been clearest in spatial coordination (Preilowski 1972, Tuller & Kelso 1989, Franz et al., 1996, Eliassen et al., 1999, Stephan et al., 1999), but there have also been reports of deficits in temporal coupling, as well (Eliassen et al., 2000, Kennerley et al., 2002). Also, an intact corpus callosum seems to be necessary to acquire new bimanual skills (Franz et al., 2000). A recent EEG study on human subjects has provided a functional explanation for this finding. Andres et al. (1999) have suggested that the interhemispheric communication through the corpus callosum may be reflected in correlations between the activities of the two hemispheres. They investigated learning in a task that required fusion of two unimanually learned sequences into a bimanual one. They found that during the initial learning process, interhemispheric task-related coherence of sensorimotor cortices transiently increased and decreased again when learning was complete. Thus, the corpus callosum may mediate interhemispheric communication necessary for bimanual learning.

The hypothesis is also supported by recent electrophysiological evidence. Cardoso de Oliveira et al (2001) recorded simultaneously from multiple electrodes in the primary motor cortex of both hemispheres of monkeys performing bimanual and unimanual whole-arm reaching movements (Fig. 6-3). The bimanual reaching movements were either symmetric – both hands moving the same distance in the same direction – or unsymmetric – different distances or different directions of movement. The study analyzed synchronization of the local field potential (LFP, Box 6-1) on electrodes within the same hemisphere and in both hemispheres. Frequently, interhemispheric correlation increased for LFPs when the monkey was beginning a movement. Most interestingly, these increases were stronger for symmetric movements than for unsymmetric or unimanual movements. These stronger increases in interhemispheric interactions during symmetric bimanual movements coincided with higher correlations of movement velocities in the symmetric movements. Thus, interactions between the hemispheres may directly contribute to the coupling between the arms. EEG studies in humans support the notion that interhemispheric synchronization may have some direct connection to bimanual performance. Serrien & Brown (2002) showed for cyclical movements that increasing cycling frequency induces a progressive decrease in bimanual coupling. They found that this decrease in bimanual coupling was accompanied by decreased task-related interhemispheric coherence between the sensorimotor cortices. Furthermore, when subjects learned to move the arms at different frequencies, the ability to decouple the arms correlated with decreased task-related interhemispheric coherence in sensorimotor and midline cortical areas (Serrien & Brown, 2003).

So, physiological evidence suggests that temporal synchronization of activity in the cortical hemispheres functions to bind the movements of the two arms. This idea is similar to the binding hypothesis proposed for resolving ambiguities in visual scenes (see Box 6-3). The hypothesis suggests that neuronal assemblies exploit the temporal relations between neuronal activities as a coding dimension that can be used in addition to and in parallel with the firing rates of individual neurons. Here, however, it is the different components of the movement that are being bound together rather than the objects in the visual field. Indeed, one of the compelling aspects of the recent findings is that they suggest that sensory and motor areas of cortex rely on similar mechanisms to solve different problems. This is intuitively appealing because researchers have long believed that the similarity of circuitry across all cortical areas must be related to functional similarities between different regions that are related to very different parts of the system.

Box 6-3: Temporal coding and the binding hypothesis

It is now well accepted that firing rate – the number of action potentials generated by neurons per unit time – is a basic component of neural coding. However, the temporal relation between the activity of different neurons could also be an important information carrier. This idea has received growing interest over the last two decades. One suggestion, first made by a group recording in visual cortex, is that while firing rates record the details of the visual scene, the temporal relations among the neurons provide information about how the different parts of this visual scene are connected to each other. This group showed that, when two bars are perceived as parts of a single bar (because they move in parallel and in the same directions and speed), the neuronal activity in visual cortex was more tightly temporally coupled than when the they were perceived as separate objects (Kreiter & Singer, 1992). This observation led to the hypothesis that when a number of visual features (e.g., see the outer marks on the above sketch) are perceived as being part of the same object (1 vase), the activity of the neurons responding to them should be synchronized, while if they are seen as parts of different objects (2 faces), their activities should be temporally independent of each other. In this way, synchronous neuronal activity could solve the so called 'binding problem', by binding together activities of neurons that code for stimulus features that form a perceptual unity.

In the motor system, an analogous 'binding' problem occurs: How are individual movement components that are simultaneously executed bound into a single, coordinated movement pattern? A mechanism that generates precise temporal relations between neuronal activities seems especially attractive to account for the generation of precise temporal relations between movement components in a coordinated movement.

Is it possible that bimanual coordination is reflected both in the bimanual specific changes in single unit firing rate and in temporal coupling? Since a temporal code and a firing rate code rely on independent dimensions of neuronal activity, it seems reasonable that the two could coexist (Riehle et al., 1997). It remains to be seen whether the synchronization and the bimanual-specific activity code different aspects of bimanual coordination, or whether they are redundant.

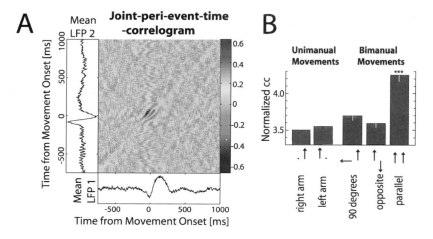

Figure 6-3. Dynamic interactions detected in the LFPs from different hemispheres during a bimanual movement. A. In the joint-peri-event-time-correlogram, the correlation coefficients between two LFPs are shown for all possible combinations of time bins in the two signals. The diagonal of the rectangular plot indicates the synchronous correlation over time. Around movement onset, a strong, positive correlation between the two LFPs arises. B. Average normalized correlation coefficients for time bins with such movement-related increases of correlation. Arrows show examples of the different movement types for which the data were evaluated separately (for details, see Cardoso de Oliveira et al., 2001). Three asterisks indicate that bimanual parallel movements show significantly stronger movement-related correlation increases than unimanual or bimanual opposite or 90 degrees movements. [Figure adapted from Cardoso de Oliveira et al. , 2001].

Both the bimanual specific activity and neural representations of the movements of each arm seem to rely on firing rate. Does this lead to a problem in coding the individual movements of each arm? Steinberg et al (2002) found that the bimanual specific changes in firing rate do not prevent the two hemispheres from successfully using a rate coding strategy to

Box 6–4: Population vector: How discharge rates of neuronal populations encode movement parameters

Many studies have shown that the rate of action potentials per time unit (discharge rate) of single neurons in motor cortex is related to parameters of an upcoming movements, such as force, joint torque, direction, velocity and acceleration (e.g., Kakei et al., 1999). The most extensively studied movement parameter, however, is movement direction. Neurons in motor cortex display typically a cosine shaped tuning to movement directions, with highest discharge rates at the preferred direction (PD) and lowest rates at the opposite direction. The movement direction of an upcoming movement can be inferred by the 'population vector' method (Georgopoulos et al., 1982). To this end, the activities of single neurons are interpreted as vectors pointing into the preferred direction of each cell, with the length of each vector corresponding to the momentary discharge rate. The sum of all these vectors results in the population vector (PV) that points in the direction of the upcoming movement.

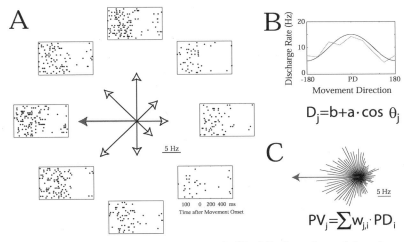

$$D_j = b + a \cdot \cos \theta_j$$

$$PV_j = \sum w_{j,i} \cdot PD_i$$

A. Dot diplays (plotted in the same way as in Fig.6.2) show the activity of a model neuron during movements in 8 different directions, arrow magnitude shows mean activities for each direction, and the arrow pointing towards the left indicates the preferred direction. B. Directional tuning curve for the neuron in A. Disrupted line: data from A, smoothe line: fitted cosine function. C. The activity of many neurons is displayed as vectors pointing into the PD, with a length of the vector indicating the discharge rate (w). In order to create a PV for a given movement direction, the vector sum of all these vectors is calculated. Data generated by simulation.

represent the directions of movement for both arms simultaneously. That is, it is possible for population vectors computed from the two hemispheres to produce good predictions of symmetric and unsymmetric bimanual movements as well as unimanual movements all using a single code (Steinberg et al., 2002, Box 6-4, Fig. 6-4). Thus, it is possible for the rate changes to exist, without compromising a rate coding strategy for direction of movement. An alternative is that the bimanual specific changes in rate are a reflection, on the single unit level, of the changes necessary to produce interhemispheric synchronization at the network level. This possibility remains to be explored as multiple electrode recording techniques make it possible to include increasingly large numbers of neurons and, thus, characterize the single unit behavior and the network behavior simultaneously.

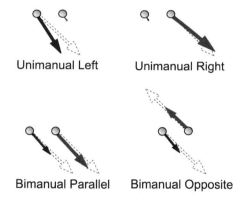

Figure 6-4. Population vectors (thick, solid arrows) for different unimanual and bimanual movement types (indicated by black dotted arrows). Two different populations of neurons (one for each arm) were formed by assigning each cell to the arm to which it responded most strongly.

4.   CAN STUDIES ON NON-HUMAN PRIMATES
      TEACH US ABOUT THE MECHANISMS OF
      BIMANUAL COORDINATION IN HUMANS?

   Electrophysiological techniques in monkeys are a sort of 'gold standard' in understanding the functional role of neural activity. However, it is important to ask whether the behavior in question is the same in monkeys and humans. Can we draw inferences from the studies reviewed here about bimanual coordination in humans? It is certainly possible that the mechanisms through which monkeys coordinate their limbs are different from ours. There are a number of reasons to think this might be the case.

   Perhaps the most striking of these reasons is that motor control in humans seems to be significantly more lateralized than it is in monkeys. While individual monkeys may have hand preferences for specific tasks, these preferences do not seem to be as strong as they are in humans (Ettlinger, 1988, MacNeilage et al., 1987, Westergaard et al., 2000), and species-wide lateralization in monkeys still has not been conclusively demonstrated despite numerous efforts to do so (Hopkins et al., 1992, Rigamonti et al. 1998, Westergaard et al., 2001). This difference in laterality is not the only reason to assume that bimanual coordination may be different in monkeys and humans. It also seems that biomechanical differences in the upper limbs (Christel & Billard, 2002) and also differences in typical body posture (Westergaard et al., 1998) can result in differences in motor coordination in general. Furthermore, particularly with control of the distal musculature, humans are capable of more precise voluntary control over individual muscles than monkeys. Again, this could play a critical role in the mechanisms of coordination.

   There is another difficulty in comparing electrophysiological experiments in monkeys to data collected from human subjects. Electrophysiology is almost exclusively done on over-trained monkeys while human subjects are usually only tested once. Over-training is used in monkey experiments in order to facilitate stable behavioral conditions for the days and months of data collection. Humans are only tested once for almost exactly the same reason: to prevent learning on one day from influencing behavior on the next. Gribova et al. (2002) have, in fact, demonstrated subtle differences in motor performance in overtrained monkeys as compared to naïve human subjects: while in humans bimanual movements took more time than unimanual ones, movement times of bimanual movements in monkeys were even shorter than unimanual ones. There is mounting evidence that training has profound effects on the motor system (Karni, 1998, Rioult-Pedotti, 2000). Thus, one might suspect that over-training could influence the mechanisms of bimanual coupling and coordination, leading to difficulties interpreting the physiological findings. This possibility has to be taken into account while trying to extrapolate from electrophysiological recordings in monkeys during a bimanual activity to the neuronal basis of the same behavior in man.

On the other hand, recent studies that compared the performances of humans and monkeys in the same task have revealed considerable similarities. Basic psychophysics of bimanual coordination can be replicated in monkeys, at least for arm movements. Thus, monkeys show the same temporal and spatial bimanual binding that people do. For instance, natural movements like opening a drawer and reaching into to it in order to retrieve a reward have some markedly similar properties in monkeys and humans (Kazennikov et al., 2002, Perrig et al., 1999, Kazennikov et al., 1994). In both monkeys and humans, there is significant trial-to-trial variability in the timing of the movements, but specific points of the movement (such as the end of the drawer pull and the end of the reach towards the drawer) are always highly synchronized. These points of natural synchronization are preserved across species.

In a more structured reaching task, another study compared temporal aspects of bimanual coupling in monkeys and humans and also found marked similarities (Gribova et al., 2002). It was found that, in both monkeys and humans, reaction times in bimanual movements were not longer than the reaction times for the slower arm in unimanual movements. In both species, movement initiation was highly correlated between the arms, but temporal correlation progressively declined after movement initiation. These behavioral findings seemed to be a manifestation of observable neural processes in the monkey. For instance, decorrelation of movements over time was reflected in a net decorrelation of the neuronal population activity in MI and SMA. Such a similarity between inter-manual coupling and activity coupling in the two hemispheres was also observed in humans. Serrien & Brown (2002) found that, in a cyclic task, progressive deterioration of coupling between the arms with increasing cycling frequency was related to decreasing task-related coherence between the primary sensorimotor cortices. Furthermore, during learning of a cyclical task in which the hands had to be moved at different frequencies, task-related coherence between the primary sensorimotor and midline areas decreased, while coupling between the arms also continuously decreased. Thus, it seems that, while monkeys' bimanual coordination is probably not identical to human bimanual coordination, it is also not completely different. The use of converging evidence is probably the best way to determine which findings are relevant to both species and which are not.

5. SUMMARY

To summarize, the last years have revealed many details about the electrophysiological basis of bimanual coordination. Bimanual specific activity has been found in a number of cortical and also subcortical areas, suggesting that a widespread network of multiple brain areas is involved in this task. Future research will have to clarify whether different areas control different aspects of bimanual coordination. Convergent evidence from monkey and man supports the idea that the interactions between the movements of the two arms may employ temporal interactions between the hemispheres of motor cortex. Although species-specific differences may exist between human and non-human primates, electrophysiological studies in monkeys may offer insight into the neuronal basis of bimanual coordination in humans.

REFERENCES

Aizawa H, Mushiake M, Inase M, Tanji J (1990): An Output Zone of the Monkey Primary Motor Cortex Specialized for Bilateral Hand Movement. Exp.Brain Res 82:219-221

Alkadhi H, Crelier GR, Boendermaker SH, Hepp-Reymond MC, Kollias SS. (2002): Somatotopy in the ipsilateral primary motor cortex. Neuroreport 13:2065-2070.

Andres FG, Mima T, Schulman AE, Dichgans J, Hallett M, Gerloff C (1999): Functional coupling of human cortical sensorimotor areas during bimanual skill acquisition. Brain 122: 855-870

Batista AP, Buneo CA, Snyder LH, Andersen RA (1999): Reach plans in eye-centered coordinates. Science 285:257-260

Brinkman C (1984): Supplementary motor area of the monkey's cerebral cortex: short- and long-term deficits after unilateral ablation and the effects of subsequent callosal section. J Neurosci 4: 918-929

Brinkman J, Kuypers HG. (1973) Cerebral Control of Contralateral and Ipsilateral Arm, Hand and Finger Movements in the Split-Brain Rhesus Monkey. Brain 96:653-674.

Brown RG, Jahanshahi M, Marsden CD. (1993): The execution of bimanual movements in patients with Parkinson's, Huntington's and cerebellar disease. J Neurol Neurosurg Psychiatry 56: 295-297

Brown P, Williams D, Aziz T, Mazzone P, Oliviero A, Insola A, Tonali P, Lazzaro Di V (2002): Pallidal activity recorded in patients with implanted electrodes predictively correlates with eventual performance in a timing task. Neurosci Lett 330:188-192

Brust JCM (1996): Lesions of the supplementary motor area. In H.O. Lüders, Advances in Neurology, Vol. 70: Supplementary Sensorimotor area (pp. 237-247). Philadelphia: Lippincott-Raven Publishers

Cardoso de Oliveira S, Donchin O, Gribova A, Bergman H, Vaadia E (2001): Dynamic interactions between and within cortical hemispheres during bilateral and unilateral arm movements. Eur J Neurosci 14:1881-1896

Chen R, Cohen L, Hallett M (1997): Role of the ipsilateral motor cortex in voluntary movement. Can J Neurol Sci 24:284-291

Christel MI, Billard A (2002) Comparison between macaques' and humans' kinematics of prehension: the role of morphological differences and control mechanisms. Behav Brain Res 131:169-184

Cisek P, Crammond DJ, Kalaska JF (2003): Neural activity in primary motor and dorsal premotor cortex in reaching tasks with the contralateral versus ipsilateral arm. J Neurophysiol 89:922-942

Debaere F, Swinnen SP, Beatse E, Sunaert S, Hecke PV, Duysens J (2001a): Brain areas involved in interlimb coordination: a distributed network. NeuroImage 14:947-958

Debaere F, Van Assche D, Kiekens C, Verschueren SM, Swinnen SP (2001b): Coordination of upper and lower limb segments: deficits on the ipsilesional side after unilateral stroke. Exp .Brain Res 141:519-529

Donchin O, Gribova A, Steinberg O, Bergman H, Vaadia E (1998): Primary motor cortex is involved in bimanual coordination. Nature 395:274-278.

Donchin O, Gribova A, Steinberg O, Bergman H, Cardoso de Oliveira S, Vaadia E (2001): Local field potentials related to bimanual movements in the primary and supplementary motor cortices. Exp Brain Res 140:46-55

Donchin O, Gribova A, Steinberg O, Mitz AR, Bergman H, Vaadia E (2002): Single-unit activity related to bimanual movements in the primary and supplementary motor cortices. J Neurophysiol 88:3498-3517

Dum RP, Strick PL (2002): Motor areas in the frontal lobe of the primate. Physiol Behav 77:677-682.

Eliassen JC, Baynes K, Gazzaniga MS (1999): Direction information coordinated via the posterior third of the corpus callosum during bimanual movements. Exp Brain Res 128:573-577

Eliassen JC, Baynes K, Gazzaniga MS (2000): Anterior and posterior callosal contributions to simultaneous bimanual movements of the hands and fingers. Brain 123: 2501-2511

Ettlinger G (1998): Hand Preference, Ability, and Hemispheric Specialization: In How Far Are These Factors Related in the Monkey? Cortex 24:389-398

Evarts EV (1966): Pyramidal tract activity associated with a conditioned hand movement in the monkey. J Neurophysiol 29:1011-1027

Franz EA, Eliassen JC, Ivry RB, Gazzaniga MS (1996): Dissociation of spatial and temporal coupling in the bimanual movements of callosotomy patients. Psychol Sci 7:306-310

Franz EA, Waldie KE, Smith MJ (2000): The effect of callosotomy on novel versus familiar bimanual actions: a neural dissociation between controlled and automatic processes? Psychol Sci 11: 82-85

Georgopoulos AP, Kalaska JF, Caminiti R, and Massey JT (1982): On the relations between the direction of two-dimensional arm movements and cell discharge in primate motor cortex. J Neurosci 2:1527-1537

Goerres GW, Samuel M, Jenkins H, Brooks DJ (1998): Cerebral control of unimanual and bimanual movements: an H_2 ^{15}O PET study. NeuroReport 9: 3631-3638

Gribova A, Cardoso de Oliveira S, Donchin O, Steinberg O, Bergman H, Vaadia E (1998): Bilateral lesions of the supplementary motor area (SMA): movement deficits and compensatory changes in other motor areas. Neurosci Lett Suppl 51: S17

Gribova A, Donchin O, Bergman H , Vaadia E, Cardoso de Oliveira S (2002): "Timing of bimanual movements in human and non-human primates in relation to neuronal activity in primary motor cortex and supplementary motor area". Exp Brain Res 146:322-335

Halsband U, Ito N, Tanji J, Freund HJ (1993): The role of premotor cortex and the supplementary motor area in the temporal control of movement in man. Brain 116:243-266.

Hopkins WD, Washburn DA, Berke L, Williams M (1992): Behavioral asymmetries of psychomotor performance in rhesus monkeys (Macaca mulatta): a dissociation between hand preference and skill. J Comp Psychol 106:392-397

Immisch I, Waldvogel D, van Gelderen P, Hallett M (2001): The role of the medial wall and its anatomical variations for bimanual antiphase and in-phase movements. Neuroimage 14:674-684

Jaencke L, Peters M, Himmelbach M, Nösselt T, Shah J, Steinmetz H (2000): fMRI study of bimanual coordination. Neuropsychologia 38:164-174

Johnson KA, Cunnington R, Bradshaw JL, Phillips JG, Iansek R, Rogers MA (1998): Bimanual coordination in Parkinson's disease. Brain 121:743-753

Johnson KA, Bennett JE, Georgiou N, Bradshaw JL, Chiu E, Cunnington R, Iansek R (2000): Bimanual co-ordination in Huntington's disease. Exp Brain Res 134:483-489

Kakei S, Hoffman DS, Strick PL (1999): Muscle and movement representations in the primary motor cortex. Science 285:2136-2139

Karni A, Meyer G, Rey-Hipolito C, Jezzard P, Adams MM, Turner R, Ungerleider LG (1998): The acquisition of skilled motor performance: fast and slow experience-driven changes in primary motor cortex. PNAS 95:861-868

Kazennikov O, Wicki U, Corboz M, Hyland B, Palmeri A, Rouiller EM, Wiesendanger M (1994): Temporal structure of a bimanual goal-directed movement sequence in monkeys. Eur J Neurosci 6:203-210

Kazennikov O, Hyland B, Wicki U, Perrig S, Rouiller EM, Wiesendanger M (1998): Effects of lesions in the mesial frontal cortex on bimanual co-ordination in monkeys. Neurosci 85:703-716

Kazennikov O, Hyland B, Corboz M, Babalian A, Rouiller EM, Wiesendanger M (1999): Neural activity of supplementary and primary motor areas in monkeys and its relation to bimanual and unimanual movement sequences. Neurosci 89:661-667

Kazennikov O, Perrig S, Wiesendanger M (2002): Kinematics of a coordinated goal-directed bimanual task. Behav Brain Res 134:83-91

Kennerley SW, Diedrichsen J, Hazeltine E, Semjen A, Ivry RB (2002) Callosotomy patients exhibit temporal uncoupling during continuous bimanual movements. Nat Neurosci 5:376-81.

Kermadi I, Liu Y, Tempini A, Calciati E, Rouiller EM (1998): Neuronal activity in the primate supplementary motor area and the primary motor cortex in relation to spatio-temporal bimanual coordination. Somatosens Mot Res 15:287-308

Kermadi I, Liu Y, Rouiller EM (2000): Do bimanual motor actions involve the dorsal premotor (PMd), cingulate (CMA) and posterior parietal (PPC) cortices? Comparison with primary and supplementary motor cortical areas. Somatosens Mot Resarch 17:255-271

Kreiter AK, Singer W. (1992): Oscillatory Neuronal Responses in the Visual Cortex of the Awake Macaque Monkey. Eur J Neurosci 4:369-375

Lang W, Obrig H, Lindinger G, Cheyne D, Deecke L (1990): Supplementary motor area activation while tapping bimanually different rhythms in musicians. Exp Brain Res 79:504-514

MacNeilage PF, Studdert-Kennedy MG, Lindblom B (1987): Primate Handedness Reconsidered. Behav Brain Sci 10:247-303

Nair DG, Purcott KL, Fuchs A, Steinberg F, Kelso JA. (2003): Cortical and cerebellar activity of the human brain during imagined and executed unimanual and bimanual action sequences: a functional MRI study. Brain Res Cogn Brain Res 15:250-260

Orlov AA, Selezneva EV, Afanas'ev SV, Tolkunov BF (1999): Correlates of sequential elements of bimanual behavior in the neuronal activity of the neostriatum in monkeys. Neurosci Behav Physiol 29:53-59

Perrig S, Kazennikov O, Wiesendanger M. (1999): Time structure of a goal-directed bimanual skill and its dependence on task constraints. Behav Brain Res 103:95-104

Preilowski BFB (1972): Possible contribution of the anterior forebrain commissures to bimanual motor coordination. Neuropsychologia 10:267-277

Prut Y, Fetz EE (1999): Primate spinal interneurons show pre-movement instructed delay activity. Nature 401:590-594

Riehle A, Gruen S, Diesmann M, Aertsen A (1997): Spike synchronization and rate modulation differentially involved in motor cortical function. Science 278:1950-1953

Rigamonti MM, Previde EP, Poli MD, Marchant LF, McGrew WC (1998): Methodology of motor skill and laterality: new test of hand preference in Macaca nemestrina. Cortex 34:693-705

Rioult-Pedotti MS, Friedman D, Donoghue JP (2000): Learning-induced LTP in neocortex. Science 290:533-536

Rouiller EM, Babalian A, Kazennikov O, Moret V, Yu XH, Wiesendanger M. (1994): Transcallosal connections of the distal forelimb representations of the primary and supplementary motor cortical areas in macaque monkeys. Exp Brain Res 102:227-243

Sadato N, Yonekura Y, Waki A, Yamada H, Ishii Y (1997): Role of the supplementary motor area and the right premotor cortex in the coordination of bimanual finger movements. J Neurosci 17:9667-9674

Serrien DJ, Bogaerts H, Suy E, Swinnen S (1999): The identification of coordination constraints across planes of motion. Exp Brain Res 128:250-255

Serrien DJ, Wiesendanger M (2000): Temporal control of a bimanual task in patients with cerebellar dysfunction. Neuropsychologia 38:558-565

Serrien DJ, Steyvers M, Debaere F, Stelmach GE, Swinnen SP (2000): Bimanual coordination and limb-specific parameterization in patients with Parkinson's disease. Neuropsychologia 38: 1714-1722

Serrien DJ, Brown P (2002): The functional role of interhemispheric synchronization in the control of bimanual timing tasks. Exp Brain Res 147:268-272

Serrien DJ, Brown P (2003): The integration of cortical and behavioural dynamics during initial learning of a motor task. Eur J Neurosci 17:1098-1104

Serrien DJ, Cassidy MJ, Brown P (2003): The importance of the dominant hemisphere in the organization of bimanual movements. Hum Brain Mapp 18:296-305.

Steinberg O, Donchin O, Gribova A, Cardoso de Oliveira S, Bergman H, Vaadia E(2002): Neuronal Populations in Primary Motor Cortex Encode Bimanual Arm Movements. Eur J Neurosci 15:1371-1380

Stephan KM, Binkofski F, Halsband U, Dohle C, Wunderlich G, Schnitzler A, Tass P, Posse S, Herzog H, Sturm V, Zilles K, Seitz RJ, Freund HJ (1999): The role of ventral medial wall motor areas in bimanual co-ordination. A combined lesion and activation study. Brain 122:351-36

Swinnen S. (2002): Intermanual coordination: From behavioral principals to neural-network interactions. Nature Rev. Neurosci. 3: 351

Tanji J, Okano K, Sato KC (1988): Neuronal activity in cortical motor areas related to ipsilateral, contralateral, and bimanual digit movements of the monkey. J Neurophysiol 60: 325-343

Tracy JI, Faro SS, Mohammed FB, Pinus AB, Madi SM, Laskas JW. (2001): Cerebellar mediation of the complexity of bimanual compared to unimanual movements. Neurology 57:1862-1869

Tuller B & Kelso JAS (1989): Environmentally-specified patterns of movement coordination in normal and split-brain subjects. Exp Brain Res75:306-316

Ullen F, Forssberg H, Ehrsson HH (2003): Neural networks for the coordination of the hands in time. J Neurophysiol 89:1126-1135

Van den Berg C, Beek PJ, Wagenaar RC, Wieringen PCW (2000): Coordination disorder in patients with Parkinson's disease: a study of paced rhythmic forearm movements. Exp Brain Res 134:174-186

Wannier T, Liu J, Morel A, Jouffrais C, Rouiller EM (2002) : Neuronal activity in primate striatum and pallidum related to bimanual motor actions. Neuroreport 13:143-147

Wasserman EM, Pascual-Leone A, Hallet M (1994): Cortical motor representation of the ipsilateral hand and arm. Exp Brain Res 100:121-132

Westergaard GC, Kuhn HE, Suomi SJ (1998): Bipedal posture and hand preference in humans and other primates. J Comp Psychol 112:55-64

Westergaard GC, Liv C, Haynie MK, Suomi SJ (2000): A comparative study of aimed throwing by monkeys and humans. Neuropsychologia, 38:1511-1517

Westergaard GC, Lussier ID, Higley JD (2001): Between-species variation in the development of hand preference among macaques. Neuropsychologia, 39:1373-8

Wiesendanger M, Wicki U, Rouiller E (1994): Are there unifying structures in the brain responsible for interlimb coordination. In: Swinnen et al., (Eds.), Interlimb Coordination: Neural, Dynamical and cognitive constraints (pp. 179-207). New York: Academic

Wiesendanger M, Rouiller EM, Kazennikov O, Perrig S (1996): Is the supplementary motor area a bilaterally organized system? Adv Neurol 70:85-93

Figure 5-1 (see p. 114)

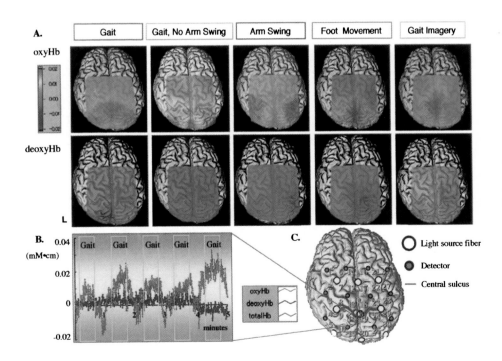

Figure 5-2 (see p. 115)

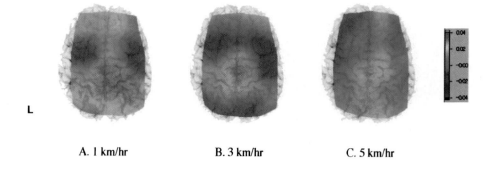

A. 1 km/hr   B. 3 km/hr   C. 5 km/hr

Figure 5-3 (see p. 116)

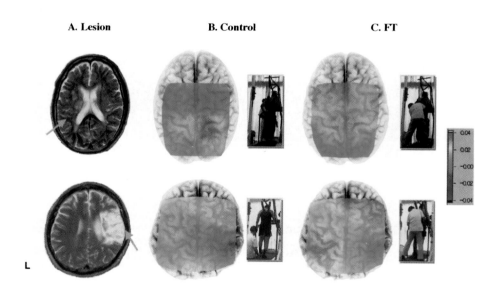

Figure 5-4 (see p. 116)

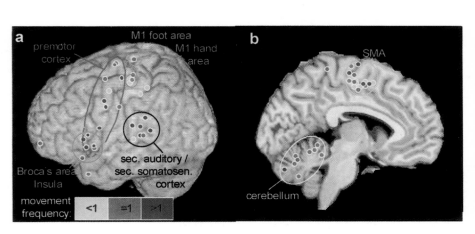

Figure 8-2 (see p. 197)

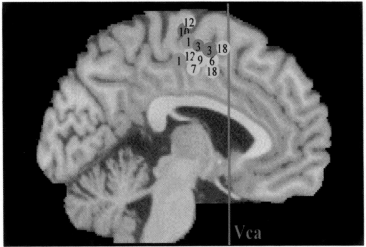

	Limbs	Comparison	MNI-coordinates		
12	bimanual	antiphase vs inphase finger-thumb opposition	-12	-13	68
12		parallel vs mirror finger abduction-adduction	14	-9	52
7	bimanual	assymetric vs symmetric finger tapping	4	-9	51
9	bimanual	assymetric vs unimanual finger tapping	-4	-7	52
18	bimanual	antiphase vs inphase finger tapping	-16	0	60
18		2:3 vs inphase finger tapping	-8	-4	52
3	bimanual	param. response to inc. incompatibility	4	-4	54
3			-4	-10	56
10	bimanual	param. response to inc. frequency in anti-phase	2	-13	66
1	hand-foot	non-isodirectional vs isodirectional coordination	6	-12	63
1			9	-15	54
6	unimanual	non-synergistic vs synergystic finger coordination	-4	-4	52

Figure 8-3 (see p. 199)

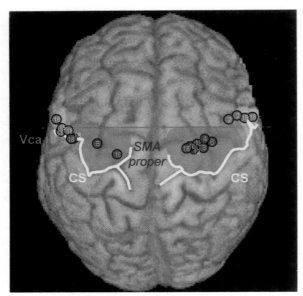

Figure 8-4 (see p. 203)

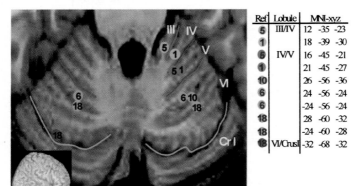

Ref	Lobule	MNI-xyz
5	III/IV	12 -35 -23
1		18 -39 -30
6	IV/V	16 -45 -21
1		21 -45 -27
10		26 -56 -36
6		24 -56 -24
6		-24 -56 -24
18		28 -60 -32
18		-24 -60 -28
18	VI/CrusI	-32 -68 -32

Figure 8-5 (see p. 205)

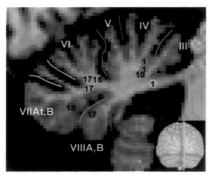

Ref	Lobule	MNI-xyz
1	II-IV	9 -42 -27
1		6 -51 -18
1		6 -51 -15
10		0 -52 -20
18	VI	8 -64 -24
17		10 -68 -25
17		-2 -69 -24
17	VII/VIII	0 -68 -34
18		0 -76 -32

Figure 8-6 (see p. 207)

Chapter 7

THE MODULATION OF EXCITABILITY IN CORTICOSPINAL PATHWAYS DURING RHYTHMIC MOVEMENT

Winston D. Byblow[1], Gwyn N. Lewis[1,2], James W. Stinear[1], Richard G. Carson[3]

[1] *Human Motor Control Laboratory, Department of Sport and Exercise Science, University of Auckland, Auckland, New Zealand,* [2] *Sensory Motor Performance Program, Rehabilitation Institute of Chicago,* [3] *Perception and Motor Systems Laboratory, School of Human Movement Studies, The University of Queensland, Brisbane, Australia*

Abstract: We report on several studies that have examined changes in corticospinal excitability during rhythmic movement of the upper limbs. During passive wrist flexion-extension there is a cyclic potentiation and inhibition of corticospinal excitability. The potentiation appears to be associated with a release of intracortical inhibition directed toward the same muscle, and is also associated with an overall depression of segmental excitability, as determined by H-reflex. There appears to be suppression of intracortical inhibition when homologous forearm muscles are shortened and lengthened synchronously, whereas less suppression of intracortical inhibition is evident when the movement pattern is asynchronous. H-reflex modulation does not distinguish between the two patterns and hence, differences likely reflect a cortical phenomenon. The composition of muscle synergies is shown to impact directly upon between-limb neural coupling and cortical input to spinal motoneurons appears to be strongly modulated by changes in the functional context of opposite limb muscles. The clinical implications of this research are described in a pilot study of patients recovering from monohemispheric stroke, and results show promise for interventions that exploit changes along the neuroaxis associated with interlimb movement.

Key words: human; movement upper limb; coordination; corticospinal excitability; transcranial magnetic stimulation; H-reflex; motor cortex; intracortical inhibition; bimanual; stroke

1.   INTRODUCTION

In this chapter, we seek to describe the nature of the changes in corticospinal excitability that occur during rhythmic movement of the upper limbs. Our specific aim is to demonstrate the means by which such changes mediate interactions between the hands during bimanual coordination. We describe recent findings on neurologically intact populations and on patients with brain injury (following stroke) investigated under active, passive or bilateral movement protocols. Beyond furthering our basic knowledge of the neural processes that mediate interlimb coordination, we also seek to illustrate how the understanding of the mechanisms through which movement-mediated modulation occurs within the CNS has led us toward new insights into possible rehabilitation strategies for neurologically impaired populations.

The chapter consists of two complementary themes. In the first theme, we will focus upon the contribution of peripheral afferent feedback to the crossed facilitation/inhibition of motor pathways that arises during active and passive movements. In the second theme, we will turn our attention to factors that influence the degree of coupling between the limbs by examining the changes in corticomotor excitability that result from (voluntary) movements on the opposite side of the body. We will consider not only the extent to which bilateral interactions are contingent upon the level of activation of the muscles of the opposite limb, but also the impact of the biomechanical context in which the movements occur. Prior to the discussions presented in these two themes, we will first provide an introduction to the fundamental investigative techniques employed in our laboratories, and an outline of the general experimental methodology and equipment.

2.   EXPERIMENTAL METHODOLOGIES

Most studies investigating motor control mechanisms associated with rhythmic movements have focused on examining motor pathway and reflex excitability during various phases of the movement cycle, identifying the levels of the motor system at which changes in neuronal state arise, and establishing the influence of peripheral sensory information on the ongoing motor output. Although many of the techniques for neuronal investigation have been utilized for a number of years, advances in technology and expertise in recent times have resulted in the development of new methods and procedures with which to probe the motor and sensory systems in greater detail, allowing a more complete analysis of the neuronal circuitry to be obtained.

Most early investigations involving cyclic movement examined neuronal excitability at the spinal level using the H-reflex technique, first described by Hoffman (1922). The H-reflex is the increase in electromyographic (EMG) activity evoked by electrical stimulation of autogenic fibres of a mixed nerve. At certain stimulus intensities two waves of EMG activity are detectable post-stimulation: an M-wave, representing direct stimulation of the motoneurons; and the longer latency H-wave, representing stimulation of Ia afferent fibres and the consequent largely monosynaptic (Burke et al. 1984) transmission of the signal to homonymous motoneurons via the spinal cord. The amplitude of the H-reflex provides a general indication of changes in motoneuronal excitability. It is most easily obtained from the soleus and quadriceps muscle in the lower limb, and in the flexor carpi radialis (FCR) muscle in the upper limb (Pierrot-Deseilligny, 1997). There are, however, a number of caveats associated with the use of this technique, which bear upon the interpretation of experimental data. One important consideration is that the response amplitude is not a simple correlate of the state of the motoneuron pool; it also reflects the excitability of the synapse between Ia afferent and the α-motoneuron pool, which is subject to presynaptic inhibition (PSI) of the sensory fibre terminals. Even when the post-synaptic state of the motoneurons remains stable, any descending inhibitory or excitatory input which alters PSI will consequently alter the size of the test reflex. A further presynaptic influence on the Ia terminals arises from post-activation depression, the reduction in transmitter release that occurs in fibres that have been previously activated (Capaday and Stein, 1987). Although these factors complicate the interpretation of results obtained using the H-reflex, a variety of methodological manipulations have been developed in conjunction with this technique that have enabled researchers to estimate the relative contribution of these mechanisms to alterations in H-reflex size. When used judiciously therefore, the H-reflex may help to provide a clearer understanding of the role of the spinal cord circuitry in the production of movement.

In recent years there has been a steadily increasing use of the technique of transcranial magnetic stimulation (TMS) to examine the role of supraspinal pathways during the production of movement. Pioneered in the 1980's by Barker and colleagues (Barker et al. 1985), TMS applied over the motor cortex preferentially stimulates cortico-cortico fibres and the deep tangential fibres of the grey matter (Day et al. 1989), leading to indirect stimulation of the corticospinal tract (see Box 7-1). The benefits of the indirect nature of the activation invoked by TMS in a research context become clear

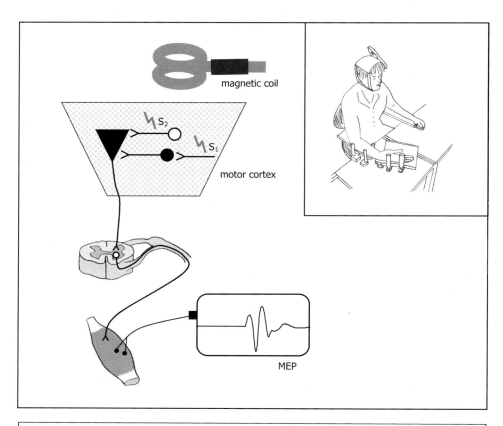

Box 7-1. Principles of TMS.

When the coil is discharged over the scalp, a rapidly changing magnetic field induces a flow of current that excites the neural tissue in the motor cortex situated below. The impulse is transmitted mainly via the corticospinal pathway to the motoneuron pool of the muscle of interest (e.g., FCR) , where it is recorded as a motor evoked potential (MEP) in the electromyography trace. A variety of evidence indicates that the descending corticospinal tract is activated transynaptically via intracortical interneurons. TMS is therefore able to probe changes in excitability of the descending pathway at both cortical and motoneuronal levels. With paired-pulse TMS, two stimuli are delivered through the same coil at short interstimulus intervals. The first stimulus (S1) is subthreshold for activation of the corticospinal tract and activates low threshold inhibitory interneurons (●) and these are expressed mostly at interstimulus intervals of 2-5 ms where the test stimulus (S2), is reduced in size in comparison to a non-conditioned stimulus. ○ denotes an excitatory interneuron, ▼ denotes a corticospinal tract fibre. Inset. TMS applied during passive movement. A subject with hand in the motor-driven apparatus such that the wrist is aligned with the axis of rotation of the apparatus. TMS is applied via coil situated over the motor regions of the scalp on the contralateral side. The subject wears a cap over the head which marks a grid pattern for marking precisely the location the optimal position for stimulating the muscle of interest. Surface EMG is collected from muscles of interest in forearm.

when comparisons are made with transcranial electric stimulation (TES). At normal stimulus intensities, a greater proportion of the corticospinal neurons activated by TES are depolarized directly by the electric stimulus, resulting in a so-called direct or D-wave (e.g. Di Lazzaro et al. 1998a; Di Lazzaro et al. 1999). In contrast, a large proportion of the muscular response to TMS is brought about by trans-synaptic excitation of corticospinal cells, which results in one or more "indirect" descending volleys or I-waves (Di Lazzaro et al. 1998a; Di Lazzaro et al. 1998c; Edgley et al. 1991; Mazzocchio et al. 1994). Magnetic stimulation, as well as being painless, therefore generates responses that are more strongly influenced by the excitability of the motor cortex than those arising from TES.

Although responses evoked by cortical stimulation and those obtained via the H-reflex technique share the final common pathway of the spinal motoneuron, the two techniques are distinguished with respect to the factors that can mediate variations in the size of the test response. Responses to TMS reflect the excitability of the corticomotor pathway as a whole. The evidence that has been obtained to date indicates that monosynaptic corticospinal projections to α-motoneurons are not subject to PSI. Corticospinal transmission to proximal upper limb motoneurons is however also mediated by di- and oligo-synaptic pathways that receive excitatory and inhibitory input from peripheral sensory fibres (see Pierrot-Deseilligny 1996 for a review). While the functional significance of these pathways remains the subject of considerable debate, it appears presently that their impact upon the excitability of α-motoneurons is diminished in the absence of concurrent descending motor activity (Nicolas et al., 2001). In contrast to the spinal H-reflex therefore, responses to TMS are influenced to a greater degree by postsynaptic inputs to spinal motoneurons, and by the state of higher level (i.e. cortical) elements of the corticospinal tract.

Prior to our discussion of the alterations in corticomotor excitability that occur in response to active and passive movements of the upper limb, we will outline the basic procedures that remain consistent across a range of our experiments. Passive movement of the wrist joint is induced using a custom-built manipulandum consisting of a hand-piece mounted on a steel framed table (Box 7-1). The proximal end of the hand-piece is mounted on a rotating shaft located coaxially with the wrist joint, enabling free rotation of the hand in the horizontal plane. An AC Servo motor is mounted underneath the unit and is coupled to the shaft of the manipulandum, such that movement amplitude and frequency of the hand-piece can be specified. For protocols involving active movements or bimanual tasks, a second manipulandum is utilised. In most respects this unit is a replication of the first, the exception being that there is no motor, *i.e.* motion of the hand-piece can only be generated actively by the subject.

The subject is seated comfortably in front of the manipulandum and the target hand is inserted and secured into the hand-piece so that the wrist joint is aligned coaxially with the shaft of the manipulandum. EMG electrodes are placed 2 cm apart on the belly of the target muscle (typically flexor carpi radialis; FCR). The optimal site for cortical stimulation is determined by applying suprathreshold magnetic stimuli through a figure-of-eight coil over the cortex contralateral to the driven hand (Box 7-1). Test stimulus intensity for TMS is normally set to a common percentage of each individual's threshold for eliciting a motor evoked potential (MEP) in the target muscle. For paired TMS, the conditioning stimulus intensity is set at or just below 90% of active threshold, i.e. an intensity at which stimulation of the corticomotor tract should not occur (Abbruzzese et al. 1999; Di Lazzaro et al. 1998c; Ziemann et al. 1996).

In experiments involving the collection of H-reflex responses, stimulation is applied to the median nerve at the cubital fossa. Stimuli (1 ms duration) are typically delivered at an intensity to elicit an H-reflex of about 10% M_{max} with the wrist joint in a neutral position and the forearm muscles at rest. The amplitude of the M-wave is repeatedly checked to ensure a constant intensity of stimulation.

During active and passive movement of the hand-piece, the timing of cortical/peripheral stimuli is pre-determined relative to hand-piece angle ($\pm$ less than 1°) and/or time ($\pm$ less than 1 ms). The system is programmed to deliver stimuli in eight distinct phases of the movement cycle (peak flexion to peak flexion). The precise time of stimulus delivery is determined by dividing the movement cycle into eight even intervals of time and establishing the centre of each time period (Fig. 7-1). Several stimuli are delivered per trial, with one or more occurring in each phase of the movement cycle. For each condition, numerous movement trials are completed in order to collect a minimum of eight responses per cycle phase.

3.   THEME I: MODULATION OF CORTICOSPINAL EXCITABILITY IN RESPONSE TO MOVEMENT-ELICITED AFFERENCE

In this section we will discuss the results of studies in which we employed the techniques outlined above to examine the state of the motor circuitry during rhythmic movement of the ipsilateral upper limb. It is necessary to be familiarised with the basic findings from these unimanual investigations so that the effects associated with inter-limb studies, described in Theme II, can be more fully understood and interpreted.

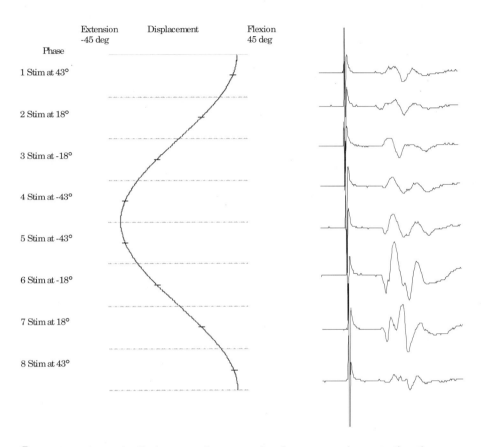

Figure 7-1. Left. Wrist angle displacement for one cycle of movement demonstrating the eight cycle phases for applying stimulation. Stimuli were delivered in the centre of each time period at the displacement indicated. Right. Motor evoked potentials from FCR of one subject at each phase during 0.6 Hz movement.

There have been many previous studies which have demonstrated a modulatory pattern of spinal- and cortical excitability during rhythmic lower limb movements, such as walking or cycling (e.g. Brooke et al. 1992; Capaday and Stein 1986; Schubert et al. 1997). In one of the earliest investigations to focus on changes in spinal reflex excitability during gait, Capaday and Stein (1986) reported that the magnitude of the soleus H-reflex increased during the stance phase in a ramp-like manner, approaching or exceeding standing values, and was then markedly inhibited during the swing phase. We have observed a comparable modulatory effect in the upper limb during rhythmic wrist flexion/extension movements (Carson et al., 1999). The results of this initial study demonstrated that during the movement cycle, there is a parallel facilitation of FCR H-reflex amplitude, and of potentials evoked in FCR in response to TMS, during the final stages of wrist extension and in the initial period of wrist flexion.

In both the upper- and the lower-limbs, phasic modulation of H-reflex and TMS evoked response amplitude tracks the pattern of muscle activation that is present in the target muscle during active voluntary movements. Thus, an interesting question that arises is whether the observed modulation of spinal reflex pathways is of central or peripheral origin. Indeed, a number of previous authors have suggested that the alterations in spinal excitability evident in active rhythmic movements of the lower limb, may be attributable to a centrally generated pattern of modulation, rather than to movement related afference (Lavoie et al. 1997; Schneider et al. 2000). On the basis of studies of the soleus spinal reflex pathway during walking, Schneider and colleagues (2000) proposed that a task-dependent and centrally specified pattern of pre-synaptic input serves to enhance the gain of the spinal (stretch) reflex during the stance phase of gait, during which there occurs activation of the plantar-flexors (soleus and gastrocnemius).

An alternative, although not mutually exclusive, possibility is that presynaptic inhibitory influences arising from peripheral input act to modulate the gain of the reflex pathway (Brooke et al., 1995). In an extensive series of papers, Brooke and colleagues provided a thorough and comprehensive analysis of the influence of movement elicited afference on the excitability of various motor pathways, by investigating the modulation of responses that occurs during passive movements of the lower limb (see Brooke et al. 1997 for review). The use of a passive movement paradigm diminishes the efferent contribution to changes in the excitability of the motor pathways. Necessarily therefore, any residual modulations are more likely to be of peripheral origin. In order to enhance our understanding of the changes in excitability that occur during rhythmic movements of the upper limb, we recently implemented a similar protocol, in which joint motion was imposed upon quiescent musculature.

3.1   Modulations in Corticomotor Excitability during Passive Movement

In an initial experiment, we examined responses to TMS in the FCR muscle, when the wrist was moved passively through an amplitude of either

50° or 100° at a frequency of 1.0 Hz (Carson et al., 2000b). The responses were evoked in the eight phases of the movement cycle, in the fashion described previously. The imposed displacement of the wrist joint led to profound alterations in the size of the evoked potential. Specifically, MEP amplitudes were potentiated when the target muscle was shortening, and inhibited somewhat during the phases of muscle lengthening. This modulation was quite robust, and evident to some degree in all of the subjects tested. As a means of illustrating the degree of MEP modulation that was observed, we have expressed the response amplitudes recorded during the passive movement relative to those obtained when the wrist joint was statically constrained in a neutral position. It is apparent that during wrist extension, the MEPs are equivalent to, or slightly smaller than those obtained at rest. During wrist flexion, when there is the greatest degree of facilitation, the responses are on average almost three times the size of those elicited in the static condition.

In the same experiment (Carson et al. 2000b), we also elicited spinal H-reflexes. During the passive movement cycle, there was a large overall depression of the amplitude of the H-reflex elicited in FCR, relative to resting control conditions. It was also apparent that the extent of this depression was not constant throughout the movement cycle. Specifically, there was less depression of the H-reflex as the wrist moved from extension into flexion. The findings are thus similar to those reported previously for the lower limb. In studies involving passive unimanual pedalling, McIlroy and colleagues (1992) and Cheng and colleagues (1995) also reported similar modulations of soleus H-reflex amplitude at corresponding phases of the pedalling cycle.

3.2   Ia Afferents Likely Mediate Response Amplitude During Movement

It is therefore clear that movement elicited afference, in the absence of descending motor activity, modulates the excitability of motor pathways in a predictable and reproducible manner, and that this modulation occurs in both the upper and lower limbs. There are several lines of evidence that implicate muscle stretch receptors in this regard. The finding that the modulations in excitability that occur during passive wrist flexion-extension are closely linked to changes in wrist displacement, rather than the muscle activation profiles seen during the equivalent active movement, provides an indication that the sensory receptors involved are likely to be those which are sensitive to alterations in muscle length. To further investigate this possibility, an additional study was conducted (Lewis et al. 2001) in which we examined responses to TMS when three distinct velocities of passive movement were imposed using 90° amplitude movements at 0.2, 0.6, 1.0 Hz. It was hypothesised that if alterations in corticomotor excitability arose through Ia input, modulations in response amplitude would be greater at the higher movement frequency. The results of the study largely supported this hypothesis in that, at the higher movement velocity, response amplitude was greater at the facilitated cycle phases and smaller at the inhibited cycle phases compared to the slow movement speed (Lewis et al. 2001) (Fig. 7-2). This enhanced differentiation of response amplitude appears consistent with the greater modulation of Ia output that is associated with cyclic movements at high velocity, although the relationship between joint angular velocity and MEP amplitude observed in the experimental data was not strictly linear. A later study conducted at an even slower movement frequency of 0.05 Hz revealed further reductions in the modulation of MEP amplitude compared to movement driven passively at 0.2 Hz (Lewis and Byblow, 2002).

Similar modulations in evoked potentials have been demonstrated in the soleus muscle during passive cycling motions (Cheng et al. 1995; McIlroy et al. 1992). As either the frequency of pedaling or the range of stretch of the knee extensors (obtained by lengthening the crank) was increased, the size of the H-reflex response was reduced. In fact, Cheng and colleagues (1995) found that for a given rate of stretch in the knee extensors a similar magnitude of response was elicited, even when obtained through differing frequency and crank length combinations. This provides strong evidence that the gain of the spinal reflex pathway at least is mediated by input from muscle receptors that appear to be most sensitive to the rate of change of muscle length.

Contributions from joint and cutaneous receptors to alterations in pathway excitability cannot be discounted. During periods of muscle shortening the unloading of tension in the muscle may serve to reduce Ib inhibition from Golgi tendon organs, increasing the resting excitability of the motoneuron. As Ib afferents are less influential during passively induced movement compared to active movements, and make a minimal contribution to the resting potential in relaxed conditions (Burg et al. 1973), the influence of these joint receptors is likely to be relatively small. Similarly, although cutaneous and joint receptors are sensitive to alterations in joint angle, it has been demonstrated that the majority respond at the limits of joint motion (Burke et al. 1988) unless an exaggerated stretch of skin is imposed (Collins et al., 2000). This makes it unlikely that these receptors account for the marked modulation of MEP and H-reflex amplitudes that occurs during the cycling and wrist movement tasks described above.

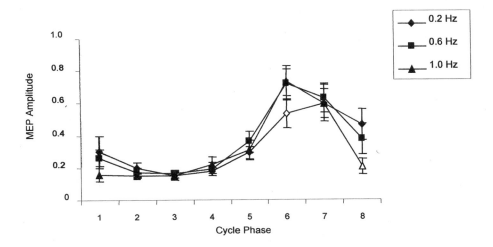

Figure 7-2. Group averages of FCR MEP amplitude obtained during passive movement at three different frequencies of movement (from Lewis et al., 2001). Response amplitudes from each subject were normalised to the peak amplitude obtained from that subject in passive movement trials. Bars represent one standard error. Hollow symbols indicate a significant difference in amplitude from the remaining two frequencies. Reproduced with permission from Elsevier Science.

3.3   Motor pathway excitability is also modulated at different static positions

An interesting feature apparent when examining forearm flexor responses elicited at different frequencies of wrist joint movement is a slight phase shift in peak potentiation as movement frequency is reduced. Whereas peak facilitation generally occurs at the initial phases of wrist flexion at the higher movement frequencies, this gradually shifts to the later periods of flexion and early extension at very slow movement rates (Lewis and Byblow, 2002). One potential explanation for this finding is that, at the lower movement speeds, type IIa receptor output is more influential, reflecting a greater influence of static muscle position rather than a dynamic effect of shortening *per se*. Is it possible that most, or even all, of the modulation in pathway excitability during movement could be accounted for by changes in static position of the wrist joint? We investigated this prospect by eliciting responses in static conditions with the wrist constrained at different angles (Lewis et al., 2001). Four wrist joint angles were chosen to correspond to the position of the wrist joint during stimulation under passive movement conditions. A marked facilitation of response amplitude in the FCR muscle was noted when the wrist was placed in a flexed posture, *i.e.* when the target muscle was in a shortened position (Fig. 7-3). Again, this feature was prominent in all subjects and has been confirmed by subsequent studies using TMS (Lewis and Byblow, 2002). It is apparent, therefore, that motor pathway excitability is influenced by static changes in target muscle length.

To take this feature into account in the passive movement results, MEP amplitudes obtained during movement were normalised to the size of the evoked response obtained when the wrist joint was in the equivalent static position (Lewis et al., 2001). When this was completed, clear modulations in response amplitude remained during the passive movement trials. The pattern of modulation was similar to that displayed in the non-normalised data, with responses strongly potentiated during the phases of muscle shortening. This confirms that at least some of the effects on the motor pathway arise as a consequence of the dynamic effect of movement.

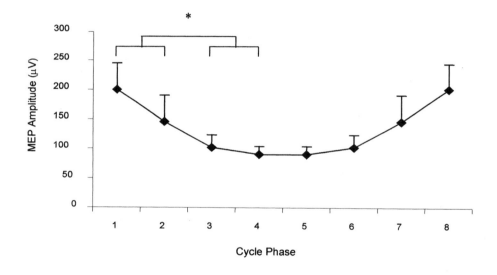

Figure 7-3. Group averages of MEP amplitude in static conditions from Lewis et al. (2001). Results of the four static positions (see also Figure 7-1) have been mirrored to enable easier comparisons to passive movement data.

3.4 Neural mechanisms of motor pathway modulation during passive movement

One caveat of passive movement paradigms and the consequent quiescent nature of the musculature is the inability to determine the state of the neural inputs influencing the motor pathways. Indeed, there are a number of means by which peripheral afferents can influence the excitability of corticospinal pathways during rhythmic movement. Polysynaptic corticospinal pathways, mediated by spinal interneurons and propriospinal neurons, operate in parallel with the direct monosynaptic cortico-motoneuronal projections (e.g. Burke et al., 1994). Alterations in the excitability of these interneuronal circuits can arise from a variety of sources, including peripheral input from the moving limbs (Pierrot-Deseilligny, 1996). Corticospinal input to motoneurons is also relayed via a population of segmental premotoneurons intercalated in disynaptic pathways that receive excitation and inhibition from peripheral afferents (Pauvert et al. 1998; see also Burke et al. 1994). In so much as these networks appear to have widely distributed projections, they suggest several means by which changes in peripheral afferent input may modulate transmission of the descending voluntary command.

One of the more intriguing questions in relation to the neural mechanisms of response modulation is the level/s of the neuroaxis at which these influences arise. There are numerous techniques available based upon TMS that enable the (indirect) examination of intracortical circuits, as opposed to probing the descending corticomotor pathway as a whole. Paired-pulse protocols comprising a subthreshold conditioning stimulus delivered prior to a suprathreshold test stimulus result in inhibition or facilitation of the test MEP amplitude when short (2-6 ms) or long (10-15 ms) interstimulus intervals are applied, respectively (Kujirai et al., 1993). The inhibitory and facilitatory effects are thought to reflect the activation of low-threshold intracortical circuits with the conditioning stimulus, which then impact upon neurons of the corticospinal tract and influence response amplitude accordingly (See Box 7-1). As the subthreshold conditioning stimulus does not penetrate below the level of the cortex, the extent of inhibition or facilitation of the test response provides an indication of intracortical excitability. If cortical circuits are influenced by the ascending afferent information elicited by passive movement, it would be expected that the level of facilitation or inhibition would be modulated during the movement cycle.

We have examined both intracortical inhibition (ICI) and facilitation (ICF) of the corticospinal pathway to the FCR muscle during passive wrist movement using interstimulus intervals optimal for each. Interestingly, we found that the extent of ICF is not altered during the passive movement cycle, whereas significant shifts in the level of ICI have been detected (Carson et al., 2000b). Specifically, less ICI was evident at cycle phases 5 and 6, corresponding to the initial period of wrist flexion, compared to that obtained in static conditions. This effect on ICI has been found to be restricted to passive movement imposed at high movement frequencies (1.0

Hz) (Lewis et al. 2001). Previous research with paired-pulse protocols has revealed that deafferentation, (either functional ischaemic blockage [Ziemann et al., 1998], or permanent amputation [Chen et al., 1998]), reduces the extent of ICI in musculature proximal to the blockage. We therefore propose that the reduction in spindle output during the wrist flexion phase of passive movement, as the induced movement in the FCR muscle alters from lengthening to shortening, may lead to a similar suppression of intracortical inhibitory circuits. The periods of disinhibition corresponded approximately to the phases of peak MEP facilitation, thus implying that both cortical and spinal mechanisms may contribute to the enhancement in pathway excitability at this time. These findings should be interpreted with caution, however, as the extent of short-interval ICI is known to be influenced by the amplitude of the non-conditioned response (Abbruzzese et al. 1999; Ridding et al. 1995; Ziemann et al. 1996). The modulatory nature of the test response itself therefore complicates the formation of more definitive conclusions using this procedure. Further investigations of this issue are ongoing.

There is further supporting evidence for alterations in cortical excitability during passive movement that has been provided by additional techniques. Similar to experiments conducted in the lower limb (see Brooke et al. 1995), we have imposed passive joint movement upon musculature activated at low levels of contraction, thus maintaining the state of the motoneurons at a fixed level. If the modulations in excitability of the corticomotor pathway arise solely at the spinal level through direct connections to the motoneurons, it would be expected that modifications of response amplitude during movement the pattern of response modulation would be altered while the target muscle is activated. In the experimental paradigm adopted subjects were required to maintain a $6 \pm 2\%$ maximum voluntary contraction in the FCR or ECR muscle during imposed wrist joint movement. To assist the subjects in consistently maintaining this low level of activation, a frequency of movement was employed that was much slower than previous investigations (0.05 Hz) and high gain visual feedback of EMG was provided. Responses were also analyzed during movement in conditions of muscle quiescence. The results indicated a comparable *pattern* of response amplitude modulation during the movement cycle in relaxed and activated conditions. This finding was evident in both the FCR and ECR muscles, and

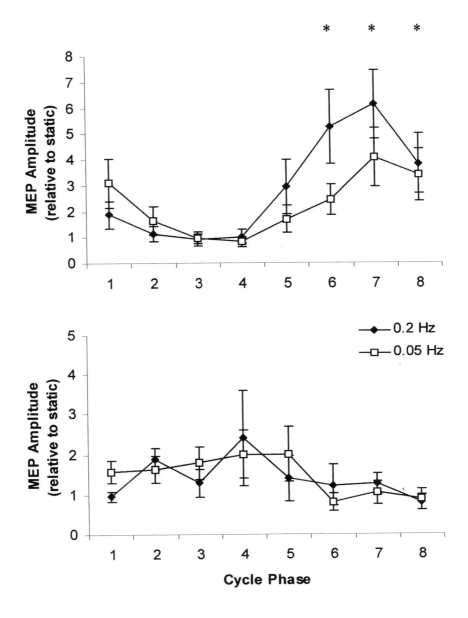

Figure 7-4. FCR (top) and ECR (bottom) MEP amplitudes during passive movement at 0.05 and 0.2 Hz from Lewis & Byblow (Experiment 1, 2002). Values are expressed relative to the MEP amplitude obtained in static conditions (1.0). * indicates phases significantly different from unity (static MEP amplitude). Error bars represent 1 SEM. Reproduced with permission from Elsevier Science.

provides some evidence that afference arising from the induced movement influences the corticomotor pathway at a pre-motoneuronal level. A marked reduction in the extent of MEP modulation was apparent during trials with muscle activation, suggesting that the supraspinal component to response modulation may be small (Fig. 7-4).

To summarize, the evidence derived from a number of complementary research protocols suggests that the marked modulation in the excitability of the corticomotor pathway that occurs during passive limb movement is mediated at both spinal and supraspinal levels. Although direct evidence of a cortical element is lacking, there is a convergence of findings from studies involving indirect methodologies that point to a demonstrable cortical component. This would indicate that afference arising from movement has the potential to influence motor pathways at a number of levels of the neuroaxis.

The likely role for modulation in intracortical excitability in particular has important implications for rehabilitation protocols directed toward injured limb/hemisphere systems due to the powerful influence of intracortical inhibitory networks within the primary motor cortex that may mediate functional reorganisation following injury (Jacobs and Donoghue 1991; Liepert et al. 2000). This intriguing possibility is discussed further in Theme II.

4.   THEME II: MODULATION OF CORTICOSPINAL EXCITABILITY IN RESPONSE TO VOLUNTARY MOVEMENT OF THE OPPOSITE LIMB

What are the mechanisms that mediate interactions between the limbs during rhythmic voluntary movement? It is now well established that the contraction of muscles on one side of the body leads to an increase in the excitability of the (contralateral) homologous motor pathways (e.g. Carson et al. 1999; Muellbacher et al. 2000; Stedman et al. 1998; Stinear et al. 2001; Ziemann and Hallett 2001). The precise mechanisms of this crossed-facilitation remain a matter of debate. The finding that the potentiation of cortically elicited motor responses during forceful contractions of the opposite hand is preserved in patients with agenesis of the corpus callosum suggests that at least some of the spread of excitation takes place at a subcortical level (Gerloff et al. 1998). Interestingly, the elevation of motor evoked potentials of biceps brachii observed in an above elbow amputee when the muscles of the opposite hand were contracted indicates that afferent input from the muscle is not necessarily required for crossed-facilitation to occur (Hess et al. 1986). In this theme we will consider the role of pathways defined at a number of levels of the neuroaxis that may mediate interactions between the limbs during rhythmic voluntary movement. The bimanual interactions are built upon the context of passive movement as described in Theme 1. In the next section, modulation of corticomotor excitability in a passive limb is examined while the contralateral limb is voluntarily activated.

4.1 Homologous wrist flexor shortening suppresses intracortical inhibition

We have recently extended the investigations of corticomotor excitability during passive wrist flexion-extension (Lewis et al. 2001) to include the assessment of the modulation of ICI during synchronous and asynchronous patterns of bimanual coordinated movement. Our interest in the modulation of ICI arose from the interactions of crossed facilitation in corticomotor excitability (cited above), and from evidence of an association between the down-regulation of ICI and cortical reorganisation. The important implication of this relationship is that release of ICI may be an important precursor to cortical plasticity following injury such as stroke. For example, in motor cortex (M1) of rats, Jacobs and colleagues induced disinhibition by administering a gamma-aminobutyric acid (GABA) receptor antagonist (bicuculline methobromide) directly into the forelimb representation in M1 (Jacobs and Donoghue, 1991). Administering the GABA antagonist resulted in forelimb movements not only being elicited when that area was electrically stimulated, but also when neighbouring vibrissae representations were stimulated. The expansion of the cortical area from which movements could be elicited suggests that latent intracortical connections were unmasked with this procedure. Furthermore, in human subjects, Ziemann and colleagues (2001) examined practice-dependant cortical plasticity by inducing GABA-related cortical disinhibition using an ischemic nerve block applied to the forearm. The results of their experiment suggested that the down-regulation of cortical inhibition facilitated practice-dependent plasticity during extended performance of upper limb ballistic movements.

This evidence led us to examine the regulation of ICI in human M1 during different patterns of rhythmical bimanual movements performed in active and passive contexts (Stinear and Byblow, 2002). It is well known that mirror symmetric kinematic patterns of upper limb movement are performed more reliably than all other patterns (Cohen, 1970; Kelso 1984). In our study, the corticomotor excitability of FCR and ECR was examined during synchronous (mirror symmetric) and asynchronous bimanual wrist flexion-extension movements. A tracking paradigm was employed, whereby one wrist was passively and rhythmically flexed and extended by a computer-controlled servomotor at 1 Hz, while the participants produced active voluntary flexion and extension movements of the opposite wrist at the same frequency. We chose an active-passive paradigm for three reasons. First, we have previously demonstrated that the dynamics of this tracking task share many similarities with those in which both limbs are moved voluntarily (Stinear and Byblow, 2001). Second, it was expected that maintaining the test muscles in a passive state would reveal task-related modulation of corticomotor excitability that would otherwise be masked by voluntary activation (Ridding et al., 1995). Third, during flexion and extension of the wrist, the muscles primarily engaged are agonist-antagonist pairs subject to reciprocal inhibition. As such, the excitability of the pathway to the agonist is expected to be high at the same time as the excitability of the pathway to the antagonist is expected to be low. In our task, the synchronous pattern of wrist flexion-extension required that voluntary contraction of a given muscle occurred simultaneously with the passive shortening of the homologous effector. That is, both limbs reached peak flexion or extension at the same time. The asynchronous pattern required timing of the voluntary contractions such that the limb moved actively lagged the motion of the passively driven limb. Specifically, the limb moved voluntarily reached peak flexion 1/6th of a movement cycle *after* the passively driven test limb. The extent of ICI was assessed using two transcranial magnetic stimuli, a subthreshold conditioning stimulus followed 2 ms later by a suprathreshold stimulus (Kujirai et al. 1993), as the driven test limb passed through a wrist-forearm angle of 0°. TMS was delivered at this joint angle to evoke responses in FCR during flexion, and to evoke responses in ECR during extension. The main finding from the experiment was that synchronous bimanual movement down-regulated ICI in M1 contralateral to the passively driven limb to a greater extent than asynchronous movement (Fig. 7-5).

Down-regulation of ICI, or disinhibition, was also demonstrated clearly when the target limb alone was being driven, consistent with our earlier findings (Lewis et al., 2001). These results suggest that cortical inhibition assists in the maintenance of asynchronous patterns of homologous FCR muscle activation. A similar effect of movement condition was not revealed for ECR. This latter finding may be an indication that the cortex maintains greater individual control of wrist flexors than wrist extensors. Such an interpretation is consistent with previously reported anatomical and physiological differences between wrist flexors and extensors (Cheney et al.,

1991). In order to conclude effects were mediated cortically, segmental-level excitability was examined using H-reflex techniques.

4.2   Homologous wrist flexor shortening modulates spinal level excitability

In this study, H-reflexes were potentiated just after peak extension and when the muscle was beginning to shorten, and inhibited at these same wrist angles during the extension phase during which the FCR is being stretched. The largest responses obtained during the passive movement cycle were

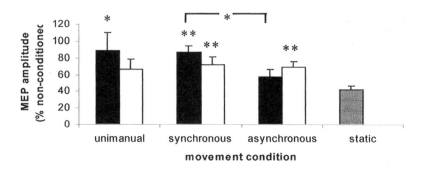

Figure 7-5. Cortical disinhibition in the representation of FCR from Experiment 1 of Stinear & Byblow (2002). Mean MEP amplitudes of conditioned responses expressed as a percent of non-conditioned responses. Filled bars represent mid-flexion responses, open bars represent mid-extension responses, the hatched bar represents static responses. Error bars represent 1 standard error of the mean. *, $p = < 0.05$; **, $p = < 0.01$, from one-tailed Student's t-tests, df 1,7. The bracketed star indicates there is a difference between synchronous and asynchronous averages during flexion. The remaining levels of significance are from movement condition means compared with static mean. Reproduced with permission from Cambridge University Press.

similar in size to those obtained when the limb was at rest with the wrist at 0°. During movements of the opposite limb, H-reflex amplitudes were also depressed relative to control values, although to a lesser extent than that observed during movements of the target limb. The extent of the depression was also influenced by the amplitude of the movement. In conditions in which the opposite limb was moved through 100 degrees, the H-reflexes were smaller than those elicited when the limb was driven passively through 50 degrees. To assess whether the marked suppression of ICI evident in cortical representations of FCR during synchronous mirror symmetric patterns of wrist flexion-extension described in the previous section was due to a spinal effect, we again assessed motoneuronal excitability using H-reflexes recorded from FCR (Stinear and Byblow, 2002). The H-reflexes were evoked when the wrist angle was passing through 0° during either unimanual flexion, or during flexion of the test limb when the same two synchronous and asynchronous patterns of bimanual movement described previously were being performed. The findings supported those of the Carson et al. (1999) study with respect to phasic modulation of response amplitude that occurred during the movement cycle, and the overall suppression of response amplitude relative to when the limb was at rest. The responses were larger during flexion than extension, and similar in size to the those reported in the Carson et al. (1999) study in the unimanual condition when the wrist angle was passing 0° during flexion (i.e. ~50% of control response size). Importantly in the more recent study, there was no effect of pattern. That is, the responses obtained in the synchronous pattern were not distinguished from those recorded during asynchronous movements. This result supports our conclusion that the effect of *pattern* observed in the ICI study was unlikely to have been mediated spinally.

4.3   Contraction of wrist effectors modulates ipsilateral corticomotor excitability

Another question we asked following the above two studies was to what extent the active movement of the contralateral limb in isolation may have contributed to the pattern of corticomotor excitability (and disinhibition) observed in the passively driven test limb during the two patterns of synchronous and asynchronous bimanual movement. To explore this issue we examined corticomotor excitability of a static limb at rest during active flexion and extension of the contralateral limb. Responses to TMS using the short-interval paired-pulse paradigm were recorded from FCR and ECR of the limb held static, as the active contralateral limb cycled through flexion-extension. It was evident that rhythmical flexion-extension of one wrist phasically modulated the excitability of corticomotor pathways to FCR and ECR of the contralateral limb maintained at rest (Carson et al., 1999). This phasic modulation was evident for both non-conditioned and conditioned responses. The pattern of modulation was consistent with the pattern we have previously observed in FCR pathways when the test limb was driven passively (e.g. Carson et al. 2000b; Lewis et al. 2001). What makes this finding interesting is that the excitability of corticomotor pathways to muscles *at rest* was up-regulated by voluntary contraction of contralateral homologous muscles. This effect is thought to occur as a result of transcallosal inhibitory interneurons suppressing the action of intracortical inhibitory interneurons (Schnitzler et al., 1996). Our study confirmed these findings in wrist effector representations.

To what extent did the active movement of the contralateral limb in isolation contribute to the pattern of corticomotor excitability in the static limb at rest? We found that at cycle phases analogous to the synchronous and asynchronous patterns of movement, the marked disinhibition associated with the synchronous pattern was not demonstrated for the contralateral-active hand moving alone. This finding supports the interpretation that the marked disinhibition for the synchronous pattern of movement was the result of the *simultaneous shortening* of homologous FCR muscles, and was unlikely to be the sole result of voluntary contraction of the contralateral FCR.

4.4 Levels of interlimb neural coupling are influenced by task context

It has recently become apparent that the functional role of specific muscles in relation to an ongoing movement pattern determines the nature of the interaction between homologous motor pathways. That is, the modulation of corticospinal input that occurs is not simply contingent upon the level of recruitment of the homologous muscle of the opposite limb. It also depends upon the manner in which that muscle is engaged.

In seeking to examine this issue, we were prompted initially by an interesting pattern of results that emerged during our previous investigations of sensorimotor coordination. If during pronation and supination of the forearm, the axis of rotation is adjacent to the radius, synchronisation of supination with the beat of a metronome is more stable than a pattern in which pronation is synchronised with the metronome. In contrast, if the rotations are about an axis that is to the ulnar side of the forearm, patterns of movement requiring pronation on-the-beat are more stable than patterns requiring supination on-the-beat (Carson et al., 2000a). In the context of bimanual movements, these constraints exert a powerful influence upon the dynamics of interlimb coordination (Byblow et al., 1994). When the position of the axis of rotation is equivalent for the left and the right limbs, transitions from antiphase (left limb pronating and right limb supinating simultaneously, and vice-versa) to inphase (both limbs pronating simultaneously, and supinating simultaneously) patterns of coordination are observed. In marked contrast, when the position of the axis of rotation for the left and right limb is contradistinct, transitions from inphase (symmetric) to antiphase (parallel) patterns of coordination predominate (Carson et al., 2000a).

What are the mechanisms that affect such profound changes in the coupling between the limbs? Mechanical context determines the changes in muscle length and moment arm that occur during movement, and thus regulates the capacity of individual muscles to contribute to net joint torque. The central nervous system (CNS) displays sensitivity to these factors in the composition of muscle synergies. In circumstances in which the radius can rotate freely about the ulna, the FCR muscle exhibits a moment arm for pronation that is of a magnitude similar to those observed for pronator teres and pronator quadratus (Ettema et al., 1998). During movements executed in this mechanical context, FCR is activated strongly as the forearm moves from supination into pronation. In contrast, when the axis of rotation is displaced towards the radius the muscle is disengaged (Carson et al., 2000a). The question that thus arises is whether alterations in the composition of muscle synergies impact directly upon the neural coupling between the limbs.

In a study designed to examine this issue (Carson and Riek 2000), we sought to determine whether the degree of crossed facilitation of motor pathways, resulting from the phasic recruitment of muscles in the arm, is simply contingent upon the level of activation of the muscles, or whether it is also sensitive to the mechanical context in which the movements are performed. We required that volunteers coordinate either maximum pronation or maximum supination of the left arm with the beats of a metronome. We also manipulated the mechanical context in which the movements were performed. In separate blocks of trials, the external axis of rotation was located either adjacent to the radius, or adjacent to the ulna. To a first approximation, the EMG activity in active muscles reflects the level of excitation of the spinal motoneurons. EMG recordings were therefore obtained from the FCR, ECR, pronator teres (PT), and biceps brachii (BB) muscles of both arms. In order to obtain a measure of the excitability of the homologous motor pathways, we employed TMS of the motor cortex to elicit MEPs in the muscles of the limb contralateral to the stimulation. As we have noted previously, the magnitude of the MEP reflects both the level of cortical input, and the excitability of elements post-synaptic to the motor cortex, such as spinal premotoneurons and interneurons, that may be regulated by peripheral afferent input (Di Lazzaro et al., 1998b). Motor potentials were evoked in the quiescent muscles of the right limb by TMS of the left motor cortex during the rhythmic movements of the left arm.

The amplitude of the EMG activity recorded from the muscles of the left limb was strongly contingent upon the mechanical context in which the movements were performed (see also Carson et al., 2000a). In addition, for three of the four muscles examined (FCR, PT & BB), the degree of association between the amplitude of the motor potential evoked in the right limb and the level of EMG recorded contiguously from the homologous muscle of the left limb, was influenced profoundly by the location of the external axis of rotation. When the axis of rotation was adjacent to the ulna, there was a positive linear relationship between the magnitude of the potentials evoked in FCR and PT, and the amplitude of the EMG activity recorded from the homologous muscles of the opposite limb. In contrast, when the axis of rotation was adjacent to the radius, the sign of these relationships was reversed (Fig. 7-6). The degree of linear association noted for BB was reliably higher when the axis of rotation was adjacent to the radius, than when the axis was adjacent to the ulna. These data suggest that the transmission of cortical input to the spinal motoneurons can be modulated strongly by changes in the functional context in which the skeletal musculature of the opposite limb is engaged (Carson & Riek 2000).

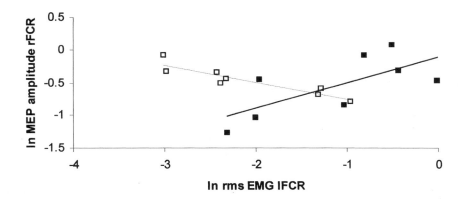

Figure 7-6. Results of a single participant performing the "supination-on-the-beat" task (Carson & Riek, 2000). Participants grasped and moved a manipulandum with the left hand while the right hand remained stationary. The relationship between the amplitude of the motor potential evoked in the right flexor carpi radialis (rFCR) and the r.m.s. amplitude of EMG activity recorded contiguously from the left FCR (lFCR) is shown. Each point represents the mean of 12 responses. The open symbols and dashed line of best fit represent mean values from each movement phase when axis of the manipulandum was aligned with the radius. The filled squares and solid line-of-best-fit depict trials conducted in the same mode of coordination, when the axis of rotation of the manipulandum was adjacent to the left ulna.

In task contexts in which a particular muscle contributes significantly to the focal action, there appears to be a positive relationship between the amplitude of the EMG activity recorded from the muscle and the level of the corticospinal input to the homologous muscle of the opposite limb. In other postures, in which the mechanical contribution of the muscle is diminished, the relationship between the level of EMG and the amplitude of the evoked potential recorded in the opposite limb is diminished and may even be reversed. In a recent study that also bears directly upon this issue, Sohn et al. (2003) obtained MEPs in several muscles of the left arm during discrete self-initiated movements of individual fingers of the right hand. It was noted that during the period immediately following EMG onset, there was a diffuse inhibition of responses evoked by TMS in the muscles of the opposite arm. Significantly however, the extent of this inhibition was lower in the homologous muscle than in other adjacent muscles. There was also a marked increase in inter-hemispheric facilitation in the motor cortex ipsilateral to the voluntary contractions, the expression of which was confined to the homologous muscle of the left hand. No changes in inter-hemispheric inhibition were observed. It thus appears that when a muscle acts as a prime mover in a specific task context, there exist excitatory relations between the cortical representation of the muscle and that of its homologue. In circumstances in which the muscle is not thus engaged, the relationship is largely inhibitory.

5.   CLINICAL IMPLICATIONS

A question of practical interest is whether the knowledge gained from the several experiments described thus far has any application in a clinical setting. To explore this question, in the next section we briefly describe a recent study designed to test the hypothesis that bimanual wrist flexion-extension applied as a long-term therapy to patients recovering from stroke can induce changes in corticomotor excitability that are associated with functional recovery of the affected upper limb.

5.1   Simultaneous bilateral movement disinhibits motor cortex post-stroke

The efficacy of bilateral movement therapy applied to functional restoration of upperlimb following stroke has been examined in only a limited number of studies (Mudie et al., 1996; Whittal et al., 2000). There may be an association between the down-regulation of ICI (as described in section 4.1) and cortical reorganisation which is potentially of importance to functional recovery post-stroke. To examine this issue we applied long-term bimanual flexion-extension as a therapy and used TMS to examine the excitability of affected and unaffected pathways to FCR and ECR in a group of post-stroke hemiparetic patients with a range of residual motricity and chronicity from one-month post-stroke through several years post-stroke, and a mix of cortical and subcortical lesions. A purpose-built manipulandum was located in each patient's home. The manipulandum was used to induce passive flexion-extension of the affected (paretic) wrist by active flexion-extension of the unaffected wrist. Five patients practised synchronous movement and four practised asynchronous movement. Six ten-minute sessions of rhythmical flexion-extension were practised each day for a period of four weeks. Independent clinical assessments of the patients were made using the upper limb section of the Fugl-Meyer motricity scale (Fugl-Meyer et al. 1975) before (PRE) and after (POST) the four-week period of intervention. At these times TMS-derived maps of FCR and ECR cortical representations were also constructed to provide measures of corticomotor excitability (map volume). Five of nine subjects in this pilot study improved their upper limb motricity in response to this novel active-passive bimanual movement therapy. Unaffected cortical map volume decreased, indicating a reduction in corticomotor excitability, especially for a sub-group of five patients who had a post-intervention increase in motricity. No change in unaffected map volume was revealed for the four patients who did not improve their post-intervention motricity (Fig. 7-7).

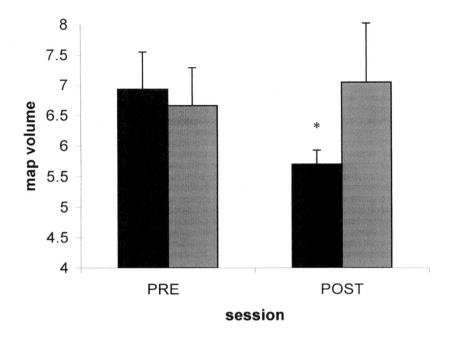

Figure 7-7. "Balancing" between hemispheres following bimanual intervention during stroke recovery. **Mean muscle unaffected (n = 10)** map volumes at pre-intervention (PRE) and post-**intervention (POST), for the five of nine** patients with motricity (Fugl-Meyer) scores that increased by > 10% (filled bars); and mean muscle unaffected (n = 8) map volumes, for the four of nine subjects with FM scores that increased by < 10% (hatched bars). Error bars represent one standard error of the mean. The asterisk indicates significant difference between T2 and POST. *, p = < 0.05. From Stinear (2003) and Stinear & Byblow (submitted).

No consistent shifts in cortical map centre of gravity were revealed. These findings suggest that high-dose passive bimanual movement therapy may have potential to improve motricity by promoting "balancing" of between-hemisphere corticomotor excitability. A direct causal link between the therapy, changes in corticomotor excitability and functional recovery, could not be made from this pilot study. However, there appears to be enough evidence to justify the assessment of the rehabilitative effects of this novel intervention, designed from advances in understanding interlimb coordination, in future studies where the issues of sample homogeneity (lesion site, level of chronicity), and causal links can be addressed.

6. SUMMARY

In this chapter we have presented a brief account of the changes in corticospinal excitability that occur during rhythmic movement of the upper limbs. It is clear that movement elicited afference modulates corticomotor excitability. During passive wrist flexion-extension we have described a cyclic potentiation and inhibition of MEP amplitudes during FCR shortening and lengthening respectively. This modulation was associated with an overall depression of the FCR H-reflex, with less depression being evident during the transition from extension to flexion. Flexor MEP amplitudes were also facilitated when the wrist was statically constrained in flexion compared with extension. The potentiation of corticomotor excitability during muscle shortening was associated with a release of intracortical inhibition, not intracortical facilitation. We therefore propose that a reduction in spindle output during flexor shortening leads to a suppression of intracortical inhibitory circuitry.

We also described the suppression of intracortical inhibition when homologous wrist flexors were shortening synchronously during bilateral wrist flexion-extension. When the muscles were shortening asynchronously less suppression of intracortical inhibition was evident. However, the amplitudes of H-reflexes did not distinguish the two patterns of movement. Therefore, the modulation of afference-driven corticomotor excitability is likely a cortical phenomenon.

We asked whether alterations in the composition of muscle synergies impact directly upon between-limb neural coupling and found that cortical input to spinal motoneurons was strongly modulated by changes in the functional context of opposite limb muscles. It appears that when a prime mover is engaged in a task an excitatory relationship exists between homologous cortical representations. This relationship is largely inhibitory if the prime mover is not engaged.

Some clinical implications of our understanding of the modulation of corticomotor excitability during bilateral upper limb movements were examined in patients recovering from monohemispheric stroke. We conducted a pilot study designed to provide evidence in support of the idea that there may be an association between the down-regulation of intracortical inhibition, cortical reorganisation and functional recovery after stroke. Patients' voluntary unaffected wrist flexion-extension was used to drive the affected wrist through passive flexion-extension. This novel intervention enhanced motricity and decreased the hyper-excitability of the unaffected hemisphere in a sub-group of patients. The application of bilateral rhythmical movement in a rehabilitation setting shows promise and is worthy of further investigation.

ACKNOWLEDGMENTS

The following funding organizations are gratefully acknowledged for the support provided for the authors and the research: The New Zealand Foundation for Research Science and Technology; The Neurological Foundation of New Zealand; The National Health and Medical Research Council of Australia; The Australian Research Council, and Staff Research Grants from The University of Auckland and The University of Queensland.

REFERENCES

Abbruzzese G, Assini A, Buccolieri A, Marchese R, Trompetto C (1999) Changes of intracortical inhibition during motor imagery in human subjects. Neurosci Lett 263:113-116

Barker AT, Jalinous R, Freeston IL (1985) Non-invasive stimulation of human motor cortex. Lancet ii:1106-1107

Brooke JD, McIlroy WE, Collins DF (1992) Movement features and H-reflex modulation I. pedalling versus matched controls. Brain Res 582:78-84

Brooke JD, Cheng J, Misiaszek JE, Lafferty K (1995) Amplitude modulation of the soleus H reflex in the human during active and passive stepping movements. J Neurophysiol 73:102-111

Brooke JD, Cheng J, Collins DF, McIlroy WE, Misiaszek JE, Staines WR (1997) Sensori-sensory afferent conditioning with leg movement: gain control in spinal reflex and ascending paths. Prog Neurobiol 51:393-421

Burg D, Szumski AJ, Struppler A, Velho F (1973) Afferent and efferent activation of human muscle receptors involved in reflex and voluntary contraction. Exp Neurol 41:754-768

Burke D, Gandevia SC, McKoen B (1984) Monosynaptic and oligosynaptic contribution to human ankle jerk and H-reflex. J Neurophysiol 52:438-448

Burke D, Gandevia SC, Macefield G (1988) Reponses to passive movement of receptors on joint, skin and muscle of the human hand. J Physiol 402:347-361

Burke D, Gracies JM, Mazevet D, Meunier S, Pierrot-Deseilligny E (1994) Non-monosynaptic transmission of the cortical command for voluntary movement in man. J Physiol 480:191-202

Byblow WD, Carson RG, Goodman D (1994) Expressions of asymmetries and anchoring in bimanual coordination. Hum Move Sci 13:3-28

Capaday C, Stein RB (1986) Amplitude modulation of the soleus H-reflex in the human during walking and standing. J Neurosci 6:1308-1313

Capaday C, Stein RB (1987) Difference in the magnitude of the human soleus H reflex during walking and running. J Physiol 392:513-522

Carson RG, Riek S (2000) Musculo-skeletal constraints on corticospinal input to upper limb motoneurones during coordinated movements. Hum Move Sci 19:451-474

Carson RG, Riek S, Bawa P (1999) Electromyographic activity, H-reflex modulation and corticospinal input to forearm during active and passive rhythmic movements. Hum Move Sci 18:307-343

Carson R, Riek S, Smethurst C, Parraga J, Byblow W (2000a) Neuromuscular-skeletal constraints upon the dynamics of unimanual and bimanual coordination. Exp Brain Res 131:196-214

Carson RG, Byblow WD, Riek S, Lewis GN, Stinear JW (2000b) Passive movement alters the transmission of corticospinal input to upper limb motoneurons. Abstracts for 30th Annual Meeting of Society for Neuroscience, New Orleans, USA

Chen R, Corwell B, Yaseen Z, Hallett M, Cohen LG (1998) Mechanisms of cortical reorganization in lower-limb amputees. J Neurosci 18:3443-3450

Cheney PD, Fetz EE, Mewes K (1991) Neural mechanisms underlying corticospinal and rubrospinal control of limb movements. Prog Brain Res 87:213-252

Cheng J, Brooke JD, Misiaszek JE, Staines WR (1995) The relationship between the kinematics of passive movement, the stretch of extensor muscles of the leg and the change induced in the gain of the soleus H reflex in humans. Brain Res 672:89-96

Cohen L (1970) Interaction between limbs during voluntary activity. Brain 93:259-272

Collins DF, Refshauge KM, Gandevia SC (2000) Sensory integration in the perception of movements at the human metacarpophalangeal joint. J Physiol 529:505-515

Day BL, Dressler D, Maertens de Noordhout A, Marsden CD, Nakashima K, Rothwell JC, Thompson PD (1989) Electric and magnetic stimulation of human motor cortex: surface EMG and and single motor unit responses. J Physiol 412:449-473

Di Lazzaro V, Oliviero A, Profice P, Saturno E, Pilato F, Insola A, Mazzone P, Tonali P, Rothwell J (1998a) Comparison of descending volleys evoked by transcranial magnetic and electric stimulation in conscious humans. Electroencephalogr clin Neurophysiol 109:397-401

Di Lazzaro V, Restuccia D, Oliviera A, Profice P, Ferrara L, Insola A, Mazzone P, Tonali P, Rothwell JC (1998b) Effects of voluntary contraction on descending volleys evoked by transcranial stimulation in conscious humans. J Physiol 508:625-633

Di Lazzaro V, Restuccia D, Oliviera A, Profice P, Ferrara L, Insola A, Mazzone P, Tonali P, Rothwell JC (1998c) Magnetic transcranial stimulation at intensities below active motor threshold activates inhibitory circuits. Exp Brain Res 119:265-268

Di Lazzaro V, Rothwell J, Oliviera A, Profice P, Insola A, Mazzone P, Tonali P (1999) Intracortical origin of the short latency facilitation produced by pairs of threshold magnetic stimuli applied to human motor cortex. Exp Brain Res 129:494-499

Ettema GJC, Styles G, Kippers V (1998) The moment arms of 23 muscle segments of upper limb with varying elbow and forearm positions: implications for motor control. Hum Move Sci 17:201-220

Fugl-Meyer AR, Jaasko L, Leyman I, Olsson S, Steglind S (1975) The post-stroke hemiplegic patient 1. a method for evaluation of physical performance. Scand J Rehabil Med 7:13-31

Gerloff C, Cohen LG, Floeter MK, Chen R, Corwell B, Hallet M (1998) Inhibitory influence of the ipsilateral cortex on responses to stimulation of the human cortex and pyramidal tract. J Physiol 510:249-259

Hess C, Mills K, Murray N (1986) Magnetic stimulation of the human brain: facilitation of motor responses by voluntary contraction of ipsilateral and contralateral muscles with additional observations on an amputee. Neurosci Lett 71:235-240

Hoffmann P (1922) Die eigenreflexe (sehnenreflexe) menschlicher muskeln. Springer, Berlin

Jacobs KM, Donoghue JP (1991) Reshaping the cortical motor map by unmasking latent intracortical connections. Science 251:944-947

Kelso JAS (1984) Phase transitions and critical behaviour in human bimanual coordination. Am J Physiol Regul Integr Comp Physiol 246:R1000-1004

Kujirai T, Caramia MD, Rothwell JC, Day BL, Thompson PD, Ferbert A, Wroe S, Asselman P, Marsden CD (1993) Corticocortical inhibition in human motor cortex. J Physiol 471:501-519

Lavoie BA, Devanne H, Capaday C (1997) Differential control of reciprocal inhibition during walking versus postural and voluntary motor tasks in humans. J Neurophysiol 78:429-438

Lewis GN, Byblow WD (2002) Modulations in corticomotor excitability during passive upper-limb movement: Is there a cortical influence? Brain Res 943:263-275

Lewis GN, Byblow WD, Carson RG (2001) Phasic modulation of corticomotor excitability during passive movement of the upper limb: effects of movement frequency and muscle specificity. Brain Res 900:282-294

Liepert J, Bauder H, Miltner WHR, Taub E, Weiller C (2000) Treatment-induced cortical reorganization after stroke in humans. Stroke 31:1210-1216

Mazzocchio R, Rothwell JC, Day BL, Thompson PD (1994) Effect of tonic voluntary activity on the excitability of human motor cortex. J Physiol 474:261-267

McIlroy WE, Collins DF, Brooke JD (1992) Movement features and H-reflex modulation. II. Passive rotation, movement velocity and single leg movement. Brain Res 582:85-93

Mudie H, Matyas T (1996) Upper extremity retraining following stroke: effects of bilateral practice. J Neurol Rehabil 10:167-184

Muellbacher W, Facchini S, Boroojerdi B, Hallet M (2000) Changes in motor cortex excitability during ipsilateral hand muscle activation in humans. Clin Neurophysiol 111:344-349

Nicolas G, Marchand-Pauvert V, Burke D, Pierrot-Deseilligny E (2001) Corticospinal excitation of presumed cervical propriospinal neurones and its reversal to inhibition in humans. J Physiol 533:903-919

Pauvert V, Pierrot-Deseilligny E, Rothwell JC (1998) Role of spinal premotoneurones in mediating corticospinal input to forearm motonerones in man. J Physiol 508:301-312

Pierrot-Deseilligny E (1996) Transmission of the cortical command for human voluntary movement through cervical propriospinal premotoneurons. Prog Neurobiol 48:489-517

Pierrot-Deseilligny E (1997) Assessing changes in presynaptic inhibition of Ia afferents during movement in humans. J Neurosci Methods 74:189-199

Ridding MC, Taylor J, Rothwell JC (1995) The effect of voluntary contraction on cortico-cortical inhibition in human motor cortex. J Physiol 487:541-548

Schneider C, Lavoie B, Capaday C (2000) On the origin of the soleus H-reflex modulation pattern during human walking and its task-dependent differences. J Neurophysiol 83:2881-2890

Schnitzler A, Kessler KR, Beneke R (1996) Transcallosally mediated inhibition of interneurons within human primary motor cortex. Exp Brain Res 112:381-391

Schubert M, Curt A, Jensen L, Dietz V (1997) Corticospinal input in human gait: modulation of magnetically evoked motor responses. Exp Brain Res 115:234-246

Sohn YH, Jung HY, Kaelin-Lang A, Hallett M (2003) Excitability of the ipsilateral motor cortex during phasic voluntary hand movement. Exp Brain Res 148:176-185

Stedman A, Davey NJ, Ellaway PH (1998) Facilitation of human first dorsal interosseous muscle responses to transcranial magnetic stimulation during voluntary contraction of the contralateral homonymous muscle. Muscle & Nerve 21:1033-1039

Stinear JW (2003) Coordinated movement induces changes in corticomotor function post-stroke. Doctoral thesis, University of Auckland, New Zealand pp 181-197

Stinear JW, Byblow WD (submitted) Rhythmic bilateral movement training enhances upper limb motricity post-stroke.

Stinear JW, Byblow WD (2001) Phase transitions and postural deviations during bimanual kinesthetic tracking. Exp Brain Res 137:467-477

Stinear JW, Byblow WD (2002) Disinhibition in the human motor cortex is enhanced by synchronous upper limb movements. J Physiol 543:307-316

Stinear CM, Walker KS, Byblow WD (2001) Symmetric facilitation between motor cortices during contraction of ipsilateral hand muscles. Exp Brain Res 139:101.105

Whittal J, Waller S, Silver KHC, Macko RF (2000) Repetitive bilateral arm training with rhythmic auditory cueing improves motor function in chronic hemiparetic stroke. Stroke 31:2390-2395

Ziemann U, Rothwell JC, Ridding MC (1996) Interaction between intracortical inhibition and facilitation in human motor cortex. J Physiol 496:873-881

Ziemann U, Corwell B, Cohen LG (1998) Modulation of plasticity in human motor cortex after forearm ischemic nerve block. J Neurosci 18:1115-1123

Ziemann U, Hallett M (2001) Hemispheric asymmetry of ipsilateral motor cortex activation during unimanual motor tasks: further evidence for motor dominance. Clin Neurophysiol 112:107-113

Ziemann U, Muellbacher W, Hallett M, Cohen LG (2001) Modulation of practice-dependent plasticity in human motor cortex. Brain 124:1171-1181

Chapter 8

NEURAL NETWORKS INVOLVED IN CYCLICAL INTERLIMB COORDINATION AS REVEALED BY MEDICAL IMAGING TECHNIQUES

Nicole Wenderoth, Filiep Debaere & Stephan P. Swinnen
Motor Control Laboratory, Department of Kinesiology, Group Biomedical Sciences, K.U.Leuven, Belgium

Abstract: During the past years, several studies have addressed the neural basis of interlimb coordination by means of imaging techniques, such as positron emission tomography (PET) and functional magnetic resonance imaging (fMRI). The general picture emerging from these studies is that a network consisting of the cerebellum, SMA and dorsal premotor cortex becomes particularly activated during demanding interlimb coordination tasks. Additionally, other regions such as Broca's area, ventral premotor cortex as well as secondary sensory areas appear to become involved when rhythmic interlimb tasks require increased monitoring of the individual limb motions performed in accordance to an imposed rhythm.

Key words: coordination, fMRI, PET, SMA, cerebellum, premotor cortex

1. INTRODUCTION

Everyday activities such as tying a shoelace, riding a car, or playing a musical instrument are characterized by moving several limbs in parallel. This requires the central nervous system (CNS) to coordinate the motions of the involved limbs in accordance to a task-specific, spatio-temporal pattern.

While some of these coordination tasks can be performed easily (e.g., clapping the hands or bimanual pulling or pushing activities), others result in a substantial decrease in performance as compared to the single limb task levels. This can be experienced when, for example, circle drawing has to be performed with the right arm and right leg simultaneously, such that the arm circles clockwise, while the leg circles counterclockwise. This shows that the control strategies identified for single limb movements can not unequivocally be extrapolated to multilimb task conditions. In the past years, behavioral work has resulted in the identification of different principles constraining multilimb movements. Some of these principles are briefly described in Box 8-1 (for a detailed overview see Swinnen, 2002). However, in this chapter we will focus on encoding principles in the CNS with respect to patterns of interlimb coordination. From this viewpoint, earlier behavioral results can be reformulated in such a way that multilimb coordination patterns are performed very successfully as long as the CNS can encode the same movement parameters for both limbs. By contrast, when the CNS has to specify different movement parameters in parallel, coordination performance decreases markedly, unless sufficient practice is provided.

To identify the neural basis of this phenomenon, recent medical imaging techniques such as functional magnetic resonance imaging (fMRI) and positron emission tomography (PET) have revealed the patterns of brain activation during the execution of coordination tasks. Both techniques indirectly assess changes in the neurons' metabolism, which reflect the degree of synaptic activity (see Box 8-2).

This chapter will review current PET and fMRI studies (Table 8-1) and discuss the reported results obtained in healthy humans and patients. To draw some general conclusions, activation maxima reported in the reviewed studies will be displayed against an exemplary anatomical image (single subject T1 MPI, acquired by the McConnell Brain Imaging Center of the Montreal Neurological Institute). However, it has to be kept in mind that the normalization methods as well as the standard brains used, can differ between studies. Therefore, the locations shown in this chapter are tentative[1].

Table 8-1, Overview of the reviewed coordination studies.

	Reference	Meth.	Task	Freq. [Hz]
1	Debaere et al., 2001	fMRI	iso- vs. anti-directional vs. single limb wrist-ankle movements	1
2	Debaere et al., 2003	fMRI	learning a bimanual 90° out-of-phase movement	1
3	Debaere et al. (unpub. data)	fMRI	cyclical bimanual in-, anti-, and 90° out-of-phase movement	0.9-1.8
4	de Jong et al., 2002	PET	cyclical hand-hand and foot-foot anti-phase vs, in-phase movements	0.5

5	Ehrsson et al., 2000a	PET	iso- vs. anti-directional vs. single limb wrist-ankle movements	0.5
6	Ehrsson et al., 2002	fMRI	synergistic vs. non-synergistic finger movements of the right hand	0.5
7	Goerres et al., 1998	PET	bimanual symmetric vs. asymmetric vs unimanual finger movements	dis-crete
8	Immisch et al., 2001	fMRI	bimanual in- vs. anti-phase finger movements	1
9	Jancke et al., 2000a	fMRI	bimanual 2:1 tapping vs. unimanual movements	1.3 : 2.6
10	Meyer-Lindenberg et al., 2002	PET	in-/anti-phase finger movements with increasing mov. frequency	1-2
11	Nair et al., 2003	fMRI	uni- vs. symmetrical bimanual finger-thumb opposition sequences	max.
12	Sadato et al., 1997	PET	bimanual symmetric vs. asymmetric vs. unimanual finger thumb opposition sequences / bimanual anti- vs. in-phase finger mov.	1.0
13	Stephan et al., 1999a	fMRI	bimanual in- vs. anti-phase vs. unimanual index finger-thumb opposition task	0.35-0.4
14	Stephan et al., 1999b	fMRI	bimanual in- vs. anti-phase vs. unimanual index finger-thumb opposition task	1.0
15	Toyokura et al., 1999	fMRI	bimanual in- vs. anti-phase vs. unimanual closing/opening of fist	0.8
16	Toyokura et al., 2001	fMRI	bimanual symmetrical vs. unimanual finger-thumb opposition sequences	1 Hz
17	Tracy et al., 2001	fMRI	bimanual anti-phase vs. unimanual rotation movements of the forearms	1
18	Ullen et al., 2003	fMRI	in-phase vs. anti-phase vs. 2:3 index finger tapping	1.33

Box 8-1: Coordination difficulty

Some interlimb coordination patterns are executed quite easily, while others seem to be rather demanding. This experienced difference in coordination difficulty can be quantified by determining performance accuracy and consistency (often measured as the relative phase ϕ between the limbs) or the attentional costs. Coordination difficulty is probably influenced by different factors. The first one is the spatio-temporal complexity of the coordination pattern itself: For rhythmical movements, some general coordination principles have been discovered describing the preferred coordination modes, which emerge spontaneously within biological systems.

1. The magnet effect. Van Holst was the first one showing that simultaneous limb movements are performed most accurately and consistently, when both limbs move with the same frequency (1:1 coupling) than with integer frequency ratios (1:2, 1:3 etc.) or non-integer polyrhythms (2:3, 3:5).

2. The egocentric constraint: muscle homology. Coordination patterns, requiring bilateral hand-hand or foot-foot movements, are performed most accurately and stable, when limbs can be moved in-phase (upper panel), that means in a mirror-symmetric way, requiring the simultaneous activation of homologous muscle groups than anti-phase, requiring the simultaneous activation of non-homologous muscles (lower panel).

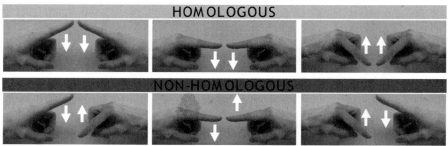

Accordingly, bimanual multi-finger movements, are coordinated more easily, when synergistic digits are moved in parallel (e.g. index finger left together with index finger right), while it is more demanding to move non-synergistic digits simultaneously (little finger left together with index finger right).

3. The allocentric constraint: iso-directionality. Moving upper and lower limbs simultaneously, coordination is more successful, when both limbs are moved iso-directional, (same direction in external space, upper panel) than anti-directional (opposite directions in external space, lower panel).

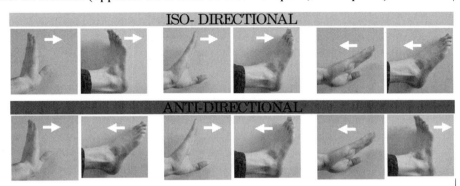

All aforementioned principles focus on the relation between the trajectories of the moving limbs. In this chapter we will summarize the above principles under the terminology **compatible limb trajectories** (i.e. 1:1 frequency ratio, in-phase, iso-directional or synergistic movements) **versus incompatible limb trajectories** (i.e. 2:3 frequency ratio, anti-phase, anti-directional or non synergistic movements).

Secondly, coordination difficulty depends on the required **cycling frequency**. Thus, coordination demands rise when e.g. anti-phase has to be produced with higher speed (see also fig. 8-2).

Finally, coordination difficulty is probably also related to the **limb-combinations** used to perform the coordination pattern (e.g. hand-hand versus hand-foot, see also fig. 8-2) and the **number of limbs** involved (Meesen et al., 2003).

Box 8-2: Functional imaging techniques: PET and fMRI

The most widely and commonly used brain-imaging techniques are positron emission tomography (PET) and functional magnetic resonance imaging (fMRI). Both techniques are capable to detect metabolic changes that occur in the brain while the subject is engaged in cognitive or motor tasks and therefore represent an indirect measurement of neural activity. For PET, radioactive isotopes are injected in the blood stream. The most common isotope used in neuroimaging experiments is ^{15}O (half-life, 2 min), injected in the form of water (H_2O). This isotope rapidly decays by emitting a positron from its atomic nucleus and when a positron collides with an electron, a gamma ray is created witch can be detected by the PET scanner. Because the tracers are in the blood stream and accumulate in regions with high metabolic demands, increased radiation (sites where positron-electron collisions occur) will be found in those regions of the brain exhibiting an increased blood flow as a consequence of the cognitive-motor tasks presented. A reconstructed PET image therefore indicates a relative activation state by showing the distribution of blood flow. For fMRI, the most commonly used method is blood-oxygenation-level-dependent imaging (BOLD). For this method, haemoglobin serves as an endogenous contrast agent. The magnetization properties differ between oxy- and deoxy-haemoglobin and form the basis of the fMRI signal (electromagnetic signal). In regions with increased neural activity, increased oxygen usage will rapidly (1-2 sec.) be followed by a larger fractional increase of blood flow and blood volume, which results in a local net decrease of deoxygenated heamoglobin. The decreased ratio of deoxy-HB/oxy-HB produces magnetic inhomogenities and less proton spin dephasing around the blood vessels, which increases the dectectable fMRI signal. The exact neural correlate for the fMRI BOLD signal, however, still remains speculative. Linear relationships between neuronal activity and haemodynamic responses have been found in visual and somatosensory cortices (Arthurs et al., 2000; Heeger et al., 2000; Logothetis et al., 2001; Rees et al., 2000), suggesting that the fMRI signal is proportional to neural firing rates. Logothethis et al. (2001), on the other hand, simultaneously recorded electrical and fMRI data in primates and found a significant relationship of the haemodynamic response with local field potential measurements. These findings indicate that BOLD contrast rather reflects the input and intracortical processing of a neural population than its spiking output (Logothetis et al., 2001).

2. THE GENERAL NETWORK ACTIVATED DURING INTERLIMB COORDINATION

Studies on interlimb coordination have typically made use of rhythmical tasks, requiring either bilateral finger (Ehrsson et al., 2002; Goerres et al., 1998; Immisch et al., 2001; Jancke et al., 2000a; Nair et al., 2003; Sadato et al., 1997; Stephan et al., 1999a; Stephan et al., 1999b; Toyokura et al., 2001; Ullen et al., 2003) and forearm movements (Toyokura et al., 1999; Tracy et al., 2001), or ipsilateral wrist and foot movements (Debaere et al., 2001; Ehrsson et al., 2000a). These tasks have consistently resulted in the activation of a general network, predominantly containing areas within primary sensorimotor cortex (SM1), premotor cortex (PM), supplementary motor area (SMA), cingulate motor cortex (CMC), superior parietal cortex as well as some subcortical structures, such as basal ganglia (BG), thalamus and cerebellum. These areas are part of a general sensorimotor network that is typically activated during the execution of motor tasks (e.g. Mattay and Weinberger, 1999). The important conclusion that can be drawn from these observations is that a specialized structure outside the sensorimotor system acting as "coordination controller", does not appear to exist.

2.1 Single limb activation versus interlimb coordination

During physical practice, one experiences that even simple tasks such as "patting the head" can become difficult when they are combined with "rubbing the stomach". This effect can be easily quantified by comparing the performance level obtained during coordination to that of the single limb subtasks performed in isolation. Accordingly, experimental results have shown that, depending on the particular task combination, producing coordinated behavior is often associated with lower performance levels and higher attentional costs as compared to single limb task levels. Thus, integrating multiple effectors into a common spatio-temporal pattern seems to require an additional *coordination effort* from the nervous system that exceeds the sum of the single effector demands. Several studies addressed this question by scanning brain activity during the execution of hand- (Toyokura et al., 1999; Toyokura et al., 2001; Tracy et al., 2001) or hand-foot coordination tasks (Debaere et al., 2001; Ehrsson et al., 2000a). With substantial consistency, these studies revealed that the primary motor cortex, premotor cortex, and SMA (Debaere et al., 2001; Ehrsson et al., 2000a; Toyokura et al., 1999; Toyokura et al., 2001) as well as the cerebellar hemisphere and vermis (Debaere et al., 2001; Ehrsson et al., 2000a; Tracy et al., 2001) were significantly more activated during the coordination task than during one of the single tasks. However, to identify areas reflecting the coordination effort more specifically, a stricter comparison is needed in which one has to establish which regions respond more strongly to the coordination task, than would be expected from *summing up* the responses evoked by the single limb subtasks. So far, only two studies performed this comparison and revealed contradictory results. Whereas Debaere et al. (2001) identified mainly the same areas as described above, Ehrson et al. (2001) did not confirm these results. However, this incongruency may have arisen as a result of differences in the required cycling frequency, so that coordination difficulty differed substantially between both studies (see Box 8-1 and fig. 8-1).

2.2   Simple versus complex coordination patterns

An alternative approach to identify areas specifically involved in interlimb coordination is to compare coordination tasks that differ in degree of *coordination difficulty* (see Box 8-1). A substantial amount of behavioral data has shown that coordination is more difficult and requires higher efforts when the coordination pattern has to be executed with *high* versus *low* cycling frequency or when the limbs have to be moved simultaneously along *incompatible* versus *compatible* trajectories. Hereby, "compatibility" implies that the CNS is enabled to encode identical spatial and temporal movement parameters for all the limbs involved in the task. This is, for example, the case during in-phase movements or 1:1 tapping. By contrast, incompatible coordination tasks require the CNS to encode different movement parameter in parallel, such as during anti-phase movements or 3:2 tapping (for further explanation see Box 8-1).

Coordination difficulty is influenced by both cycling frequency and compatibility as well as their interaction, such that the difference between compatible and incompatible coordination patterns becomes more pronounced when the cycling frequency is increased. This has been established most convincingly for in-phase versus anti-phase movements, as shown in figure 8-1.

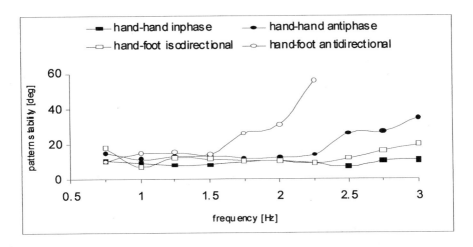

Figure 8-1. The influence of cycling frequency on the stability of in- and anti-phase movements during bimanual hand-hand (squares) as well as ipsilateral hand-foot (circles) coordination. For hand-hand coordination, it can be observed that differences between in- (filled symbols) and anti-phase (open symbols) movements become more pronounced when cycling frequency increases. The effect becomes even stronger when hand-foot coordination patterns are performed. These exemplary data were derived from a single subject, who was highly trained. However, the substantial influence of cycling frequency is still clearly present, indicating that stability changes are inherent to the motor system. For untrained subjects, marked stability differences between in- and anti-phase movements become already evident at lower frequencies.

The majority of imaging experiments have varied coordination difficulty by using spatially compatible versus incompatible tasks. Accordingly, subjects were required to produce rhythmical bimanual movements, which were either in- or anti-(Immisch et al., 2001; Meyer-Lindenberg et al., 2002; Stephan et al., 1999a; Stephan et al., 1999b; Toyokura et al., 1999; Ullen et al., 2003). The areas reported as being more activated during anti- as compared to in-phase movements have revealed differences among studies. However, one has to bear in mind that the aforementioned experiments differed substantially with respect to the required cycling frequencies (see Table 8-1 and fig. 8-2). The low frequency (< 1 Hz) studies predominantly obtained an increased activity of the supplementary motor area (SMA), sometimes extending into the cingulate motor cortex. These results support the traditional view that the SMA plays a critically important role in bimanual coordination. Interestingly, an SMA involvement has also been revealed in studies comparing ipsilateral coordination tasks between arms and legs, i.e. anti-directional versus iso-directional hand-foot coordination

(Debaere et al., 2001), as well as non-synergistic versus synergistic finger movements of the right hand (Ehrsson et al., 2002). Thus, the SMA is not only active during bimanual movements in particular, but also contributes to the coordination of more than one limb or limb segment in general. However, other studies using higher movement frequencies (1 Hz) have shown that interlimb coordination is not exclusively controlled by SMA. In particular, the premotor cortex and the cerebellum have frequently been shown to play an important role in coordination tasks, constituting a cortico-subcortical network together with the SMA. This network further extends during the production of rhythmic high frequency tasks (>1 Hz), evoking additional activation within Broca's area as well as the secondary auditory/somatosensory cortex (fig. 8-2). Moreover, Meyer-Lindenberg et al. (2002) tested the interaction between pattern compatability and cycling frequency directly, by scanning subjects while performing in- and anti-phase movements at four frequency levels of 1, 1.5, 1.7, and 2 Hz. Using a parametric design, they showed that SMA, premotor cortex, the cerebellum and, additionally, spots within Broca's areas and the secondary auditory/somatosensory cortex became increasingly activated during anti- as compared to in-phase movements when cycling frequency rose.

It has to be born in mind, however, that in- and anti-phase patterns are considered to be part of the human's daily movement repertoire and that these tasks are spontaneously performed with high accuracy and consistency. By contrast, other interlimb coordination tasks such as playing the piano or flying a helicopter are experienced as much more demanding. This results from higher degrees of incompatability between the single limb subtasks. Particularly the temporal structure of left and right finger actions differs dramatically between executing in- or anti-phase movements versus playing, for example, Johan Sebastian Bach's "Well-Tempered Clavier". Ullén et al. (2003) addressed this question experimentally by making use of a 3:2 polyrhythm tapping task in addition to in- and anti-phase coordination patterns. In accordance with previous results, they identified higher activation levels in the SMA, premotor cortex, the cerebellum and the secondary auditory cortex when comparing 3:2 tapping to in-phase movements.

Taken together, the aforementioned studies indicate that SMA, premotor cortex, and the cerebellum, play a crucial role in interlimb coordination. Additionally, Broca's area as well as the secondary sensory areas become involved when high frequency tasks are performed. Furthermore, the activation level within this network appears to be highly correlated to the coordination difficulty of the task. In the following paragraphs, we will review and discuss recent findings pertaining to these areas in further detail and try to describe their presumed contribution to interlimb coordination more specifically.

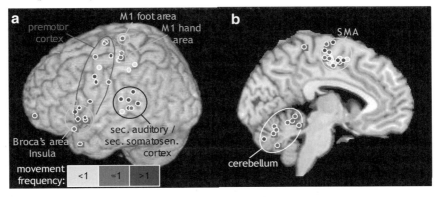

Figure 8-2. Overview of activation maxima that are specific to interlimb coordination, as reported by the reviewed publications. Most studies compared incompatible to compatible interlimb coordination tasks. However, some spots (semitransparent symbols) reflect increased activation during interlimb coordination as compared to single limb movements. The different colors indicate the adopted cycling frequency, which is strongly related to perceived coordination difficulty. The orange line marks the central sulcus. References are numbered in accordance to Table 8-1. *(See color plate section.)*

2.3 SMA and the cingulate motor cortex – sequencing motor actions

The majority of the reported imaging studies have mentioned a specific role of the medial wall areas for interlimb coordination. Activation of this region was either more pronounced during the coordination of the limbs as compared to unimanual or single limb conditions (Debaere et al., 2001; Goerres et al., 1998; Jancke et al., 2000a; Nair et al., 2003; Toyokura et al., 1999; Toyokura et al., 2001), or higher for incompatible compared to compatible coordination tasks (Debaere et al., 2001; Ehrsson et al., 2002; Goerres et al., 1998; Immisch et al., 2001; Sadato et al., 1997; Stephan et al., 1999a; Stephan et al., 1999b; Toyokura et al., 1999; Ullen et al., 2003). In most cases, when coordination conditions were related to unimanual conditions, a higher involvement of the SMA was only evident for the comparison with conditions involving incompatible patterns but not for those involving compatible patterns (Goerres et al., 1998; Stephan et al., 1999b; Stephan et al., 1999a; Toyokura et al., 1999). However, higher SMA activity was observed when movement tasks were sufficiently demanding, i.e. if tasks comprised for example 'more complex' finger sequences relative to compatible mirror-symmetric patterns (Nair et al., 2003; Toyokura et al., 2001). During ipsilateral hand and foot coordination, both the compatible (iso-directional) and incompatible (anti-directional) movement conditions were found to be associated with higher SMA activity (Debaere et al., 2001). This suggests that the specificity of the SMA activation with respect to coordination of limbs is especially related to the degree of difficulty of the interlimb coupling, irrespective of which effector combination is used. This hypothesis is clearly supported by the overall difference in SMA activity that has been obtained when comparing incompatible with compatible coordination conditions (Debaere et al., 2001; Ehrsson et al., 2002; Goerres et al., 1998; Immisch et al., 2001; Sadato et al., 1997; Stephan et al., 1999a; Stephan et al., 1999b; Toyokura et al., 1999; Ullen et al., 2003). Further, using a parametric approach, it has been shown that activity in the SMA systematically increases when pattern difficulty increases as defined by increasing incompatibility (e.g. from in-phase to anti-phase to 90° out-of-phase movements) (Debaere et al. unpublished data) or by increasing cycling frequency (Meyer-Lindenberg et al., 2002). Both movement parameters also seem to interact.

Figure 8-3 represents a midline section of the brain and provides an overview of the reported activation maxima within the supplementary motor area, as specifically related to coordination of the limbs. The reported maxima mainly result from the comparison of incompatible versus compatible coordination modes in bimanual finger movements (blue), hand-foot coordination (green) and unimanual conditions (yellow). Maxima showing a parametric response to increasing pattern difficulty are also depicted (red). Across studies, it can be observed that the activation within the SMA related to interlimb coordination is almost exclusively located within the dorsal part of this region behind the anterior commissure landmark (VCA line), corresponding to the SMA-proper (Picard and Strick, 1996; Picard and Strick, 2001). Furthermore, activations are often close to or around the cingulate sulcus, sometimes extending into the dorsal cingulate motor area (Debaere et al., 2001; Ehrsson et al., 2002; Immisch et al., 2001; Stephan et al., 1999b; Stephan et al., 1999a). Both regions are strongly interconnected and, due to considerable variability in the sulcal anatomy of the medial wall, dissociation between both regions is difficult (Immisch et al., 2001). Additionally, it can be noted that reported maxima for hand-foot coordination are located somewhat more caudal than those reported for hand or finger coordination, which is in agreement with the reported somatotopical organization of the SMA-proper (Fontaine et al., 2002; Mitz and Wise, 1987).

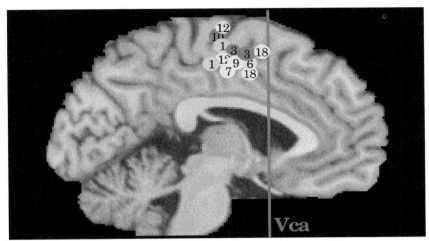

	Limbs	Comparison	MNI-coordinates		
12	bimanual	antiphase vs inphase finger-thumb opposition	-12	-13	68
12		parallel vs mirror finger abduction-adduction	14	-9	52
7	bimanual	assymetric vs symmetric finger tapping	4	-9	51
9	bimanual	assymetric vs unimanual finger tapping	-4	-7	52
18	bimanual	antiphase vs inphase finger tapping	-16	0	60
18		2:3 vs inphase finger tapping	-8	-4	52
3	bimanual	param. response to inc. incompatibility	4	-4	54
3			-4	-10	56
10	bimanual	param. response to inc. frequency in anti-phase	2	-13	66
1	hand-foot	non-isodirectional vs isodirectional coordination	6	-12	63
1			9	-15	54
6	unimanual	non-synergistic vs synergystic finger coordination	-4	-4	52

Figure 8-3. Summary of reported activation maxima within SMA. Locations are reported in MRI coordinates. The plane of the anterior commissure (VCA) is indicated in red. References are numbered in accordance to Table 8-1. *(See color plate section.)*

The SMA and the cingulate motor cortex have direct connections to the primary motor cortex and to the spinal cord (Dum and Strick, 1991; He et al., 1995). Therefore, the function of these regions has mainly been related to executive aspects of motor behavior. This notion has been supported by various imaging studies reporting activation in these regions for simple motor tasks (Picard and Strick, 1996; Picard and Strick, 2001), or by studies reporting a correlation of the activation with the variation of basic movement output parameters such as force and movement frequency (Blinkenberg et al., 1996; Dettmers et al., 1995; Jenkins et al., 1997; Sadato et al., 1996a). However, a specific activation related to sequential movements as compared to simple repetitive movements, has been reported frequently (Catalan et al., 1998; Deiber et al., 1991; Deiber et al., 1996; Rao et al., 1993; Shibasaki et al., 1993; Van Oostende et al., 1997). This might relate well to the specific activation found during the coordination of movements, as spatio-temporal sequencing of the limb movements is a critical component of such activities. The existence of strong reciprocal homotopic and heterotopic interhemispheric connections between primary and secondary motor areas, has further led to the suggestion that the SMA might be in the position of regulating interhemispheric interactions (Rouiller et al., 1994; Wiesendanger et al., 1996). The regulation of such interactions is especially important for bilateral limb control, and probably even more important when bilateral interlimb patterns become more complex (Cardoso De Oliveira, 2002; Serrien and Brown, 2002).

Additional convincing evidence that the SMA has a specific role in interlimb coordination, and more particular bimanual coordination, has also been advanced by lesion studies and TMS-work. These studies add to the imaging work in supporting the hypothesis that the SMA is 'crucial' for the successful performance of interlimb coordination. In three patients with a unilateral excision of the SMA, Laplane et al. (1977) reported deficits in alternating bilateral movements and an inability of performing reciprocal coordination. Similarly, a tendency for simultaneous bilateral or 'mirror' movements has been observed in patients with infarction of the right SMA and anterior cingulate (Chan and Ross, 1988)or with a midline tumor, mainly comprising the right cingulate motor area (Stephan et al., 1999b). Other reports of patients with infarction in the anterior cerebral artery territory (involving the SMA) has further underscored the importance of the SMA in bimanual coordination as such movements were typically found to be severely impaired (Dick et al., 1986; McNabb et al., 1988). The associated coordination deficits of right hemispheric SMA and cingulate lesions (Chan and Ross, 1988; Dick et al., 1986; Stephan et al., 1999b) have led to the hypothesis that the SMA may be responsible for transformation of motor programs originating in the left hemisphere before execution by the primary motor area in the right hemisphere (Chan and Ross, 1988). More generally, the observed deficits have pointed to the SMA as an area which mainly contributes to general integrative aspects of interhemispheric interactions (Stephan et al., 1999b).

Applying transcranial magnetic stimulation (TMS), 'virtual lesions' can be elicited that are somewhat better localized than the lesions in patients, and can therefore provide even more specific information on the parameters controlled by the stimulated region. Very recently, a handful of studies applied TMS to the SMA region during bimanual coordination (Meyer-Lindenberg et al., 2002; Obhi et al., 2002; Serrien et al., 2002; Steyvers et al., 2003). The general observation was that bimanual coordination is disturbed by interfering with the ongoing neuronal processing of the SMA. More specifically, repetitive TMS on the SMA region has been shown to result in increased relative phasing error between the hands. This effect was evident during anti-phase movements but not during in-phase coordination (Serrien et al., 2002; Steyvers et al., 2003). Furthermore, the disturbance occurred immediately after stimulation and did not interfere with the basic movement components of the single limb tasks, such as cycle duration and amplitude (Steyvers et al., 2003). In another study, using double pulse TMS, it was demonstrated that TMS over the SMA induced phase transitions from the behaviorally less stable anti-phase mode to the more stable in-phase mode, whereas the in-phase pattern was not affected (Meyer-Lindenberg et al., 2002). Steyvers et al. (2003) also demonstrated a spatial specificity for the observed effects. The most effective site for eliciting disturbance of bimanual coordination was at a stimulation site 4.0 and 6.0 cm anterior to the M1 leg area. Presumably, this involves a region at the border of the SMA-proper with the pre-SMA, which seems to be somewhat more anterior to the area that has been generally found in most imaging studies.

With respect to a specific function of the SMA in coordinative movement, the present studies suggest that SMA might be critical for simultaneously coding different actions of two or more limbs. Additionally, the TMS experiments have shown that SMA is probably also involved in the suppression of mirror movements. Both these functions can be considered largely compatible with each other.

2.4 The dorso-lateral premotor cortex - suppressing mirror movements

Many bimanual coordination tasks requiring movements along incompatible trajectories, have elicited a higher haemodynamic response in the lateral premotor area than compatible coordination movements (fig 8-4). By contrast, no ipsilateral coordination tasks, such as anti-directional versus iso-directional hand-foot movements, have revealed increased activation of this area. This suggests that the lateral premotor cortex may play a specific role in *interhemispheric* coordination. The premotor cortex lays anterior to the central sulcus, mainly within Broadman area (BA) 6 and can be subdivided in a dorsal part (PMd) and a ventral part (PMv), even though the exact anatomical borders for the human brain are still a matter of debate (Kollias et al., 2001; Lang et al., 1990; Steyvers et al., 2003). All activation maxima within PMd (fig. 8-4, blue dots) were located within its caudal portion (around 10 mm rostral to the M1 hand area), which was recently termed the PMd-proper, corresponding to area F2 in the monkey (Fink et al., 1997; Picard and Strick, 2001). PMd-proper is known to show movement related activation. Most of its neurons project directly to M1 and the spinal cord, and can immediately influence motor execution. Accordingly, functional imaging studies in humans have shown that PMd-proper activity is evoked by voluntary actions, even if the movements are simple (for an overview see Kollias et al., 2001; Picard and Strick, 2001). More specifically, fMRI studies have revealed that the PMd-proper is particularly involved in the preparation of movements (Richter et al., 1997; Thoenissen et al., 2002), confirming earlier results from single-cell recording studies in monkeys (Crammond and Kalaska, 2000; Weinrich et al., 1984).

Interestingly, the premotor areas of both hemispheres are highly interconnected via the SMA (Liu et al., 2002), with the number of these interhemispheric projections exceeding by far those found between the primary motor areas. Thus, the specific anatomy of PMd-proper points already to its important role in the execution of bimanual movements. Nearly all interlimb coordination studies have revealed similar activation spots around the precentral sulcus (close to the human hand area), sometimes bilaterally but mostly within the right hemisphere (fig. 8-4).

This area has been frequently identified in other imaging studies requiring the preparation and execution of finger or arm movements (Kollias et al., 2001; Richter et al., 1997; Thoenissen et al., 2002). Additionally, several experiments have identified nearly the same area during the execution of movement sequences. Together with the SMA, it is considered part of a network that exhibits increasing activation when the movement sequences become more complex (Deiber et al., 1991; Harrington et al., 2000; Kawashima et al., 1998a; Sadato et al., 1996b; Van Oostende et al., 1997).

Additionally, PMd-proper receives strong projections from superior parietal cortices (for an overview, see Colby and Goldberg, 1999). Many studies in humans and monkeys have shown that the superior parietal cortex encodes mainly spatial aspects of an intended movement, which are then forwarded to PMd-proper. Consequently, parietal as well as premotor neurons encode the intended direction of voluntary movement before it is executed. However, recent studies have shown that parietal cortex is probably more strongly involved in the planning of a whole set of alternative motor actions, while PMd-proper activity appears to be more related to selecting a response fitting the environmental constraints (Snyder et al., 2000; Thoenissen et al., 2002). In this respect, PMd-proper is perhaps not only involved in initiating a suitable motor response, but also in inhibiting unwanted motor actions (Thoenissen et al., 2002). Further evidence was yielded by a TMS experiment, showing that disrupting PMd activity led to a disinhibition of automated stimulus-responses combinations (Praamstra et al., 1999). In agreement with this result a recent bimanual study demonstrated that TMS stimulation over the PMd during the execution of fast bimanual anti-phase movements, induced a transition to in-phase movements (Meyer-Lindenberg et al., 2002). Even though similar effects were observed during SMA stimulation, the strongest disturbance of anti-phase coordination was evoked by disrupting the PMd activity of the right hemisphere. Together, these findings indicate that PMd may not only be involved in the selection and execution of an appropriate movement, but also in the suppression of unwanted or disturbing automated behavior.

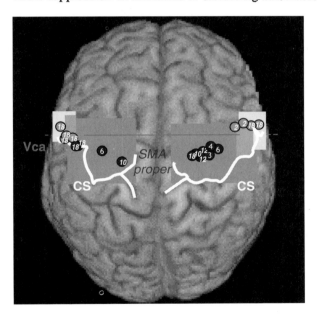

Figure 8-4. Summary of reported activation maxima within premotor cortex. The central sulcus (CS) is indicated in white, the plane of the anterior commissure (VCA) in red. Tentative anatomical locations of the caudal dorsal (PMd-proper) and ventral premotor (PMv) cortex are marked in orange and yellow, respectively. Note that the exact subdivisions of human's premotor cortex are still a matter of debate. References are numbered in accordance to Table 8-1. (*See color plate section.*)

Many of the aforementioned findings were obtained during the study of unimanual movements. Taken together, however, they tentatively suggest that PMd-proper may serve two functions during bimanual coordination. First, it may be involved in integrating both limbs into one sequence of appropriate muscle contractions. Secondly, PMd-proper may perhaps also be involved in the suppression of automated (mirror) movements.

For more demanding coordination tasks, such as 3:2 tapping (Ullen et al., 2003) or during early learning of performing 90° out-of-phase movements (Debaere et al., 2003), additional activation within ventrolateral BA 6 has been reported. This area corresponds most likely to the the ventral premotor cortex (PMv). Anatomical studies have shown that the PMv has dense connections to areas within M1 and the spinal cord, involved in the control of finger movements (Dum and Strick, 2002). Additionally, it receives visual information from the inferior parietal cortex, proprioceptive information from the secondary somatosensory cortex, and timing information from the cerebellum (Middleton and Strick, 1997; Rizzolatti et al., 1998).

In monkeys it has been shown that caudal PMv (corresponding to area F4) is particularly involved in grasping, by transforming the object location into appropriate reaching movements (Rizzolatti et al., 2002). Its function in humans, however, is still uncertain, since the caudal PMv has shown an involvement in many different tasks (see e.g. Schubotz and von Cramon, 2001) without exhibiting a clear functional pattern. Activation spots similar to those observed in the bimanual coordination studies, have been reported for tasks requiring the coordination of fine finger movements, such as during a two-ball rotation task (Kawashima et al., 1998b) or precision grip formation (Ehrsson et al., 2000b; Ehrsson et al., 2001; Kuhtz-Buschbeck et al., 2001). Additionally, it has been shown that nearly the same PMv location is activated in conjunction with the posterior cerebellum (lobule VI/Crus I) during responses to unpredictable auditory stimuli, imposing some temporal uncertainty on movement initiation (Sakai et al., 2000). Sakai et al. hypothesized that the caudal PMv may play a specific role in adjusting temporal aspects of the movement to the environmental requirements. A similar view was also suggested by others, assuming that PMv might become increasingly involved when tasks require a continuous matching of the produced movements with sensory information or environmental constraints (Schubotz and von Cramon, 2001), particularly including prediction, anticipation or imagination (Lutz et al., 2000). Taken together, the role of PMv in movement organization in general and interlimb coordination in particular is not yet clear. So far, only two coordination studies identified caudal PMv activation (Debaere et al., 2003; Ullen et al., 2003). Since both tasks required well-timed but non-synchronized movements, it is possible that this activation reflects larger efforts to monitor and adopt the exact timing of movement initiation between the limbs. Whether this additional effort arises from the fact that the coordination patterns were not sufficiently overlearned or from the temporally incompatible nature of the tasks is not clear.

2.5   The cerebellum - the timing organ

The cerebellum is traditionally believed to have an important function in the coordination of voluntary movements (Holmes, 1939). Therefore, it is not surprising that multilimb coordination tasks also evoke an increased haemodynamic response in several cerebellar regions. Figures 8-5 and 8-6 provide a detailed overview of activation maxima within the cerebellar hemispheres and the vermis, respectively, which are related to a high coordination effort during interlimb coordination as compared to bimanual in-phase or single limb movements. The observed maxima constitute three clusters that mainly differ with respect to their anterior-posterior location for the hemispheres and to their superior-inferior location for the vermis.

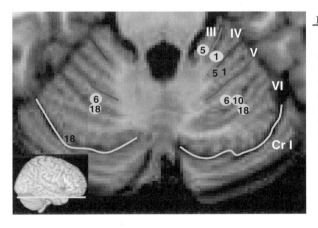

Ref	Lobule	MNI-xyz
5	III/IV	12 -35 -23
1		18 -39 -30
5	IV/V	16 -45 -21
1		21 -45 -27
10		26 -56 -36
6		24 -56 -24
6		-24 -56 -24
18		28 -60 -32
18		-24 -60 -28
18	VI/CrusI	-32 -68 -32

Figure 8-5. Summary of reported activation maxima within the cerebellar hemisphere. Locations are reported in MRI coordinates. The anatomic labeling corresponds to (Schmahmann et al., 2000). References are numbered in accordance to Table 8-1. (*See color plate section.*)

The most anterior-superiorly located activation spots have been found within lobule III/IV (fig. 8-5, light-green) and lobule IV/V (fig. 8-5, dark-green) of the hemispheres as well as lobule III/IV of the vermis (fig. 8-6, dark-green). These areas are also involved in the control of simple, single limb movements of the ipsilateral body side (Desmurget et al., 2001; Rijntjes et al., 1999). In particular, the activation spots observed within the right anterior cerebellar hemisphere lay within a somatotopically organized region, representing the foot and hand area, respectively. Interestingly, during wrist-foot coordination tasks, activation was found to be higher when moving both limbs according to an iso- or anti-directional mode than when moving each limb in isolation (Debaere et al., 2001; Ehrsson et al., 2000a). Similar results were also found for coordinating finger and arm movements within the right limb, showing that this region is more strongly activated by the coordination task than by movements of each segment alone (Ramnani et al., 2001). The anterior cerebellar hemisphere is part of the spinal cerebellum, receiving proprioceptive and cutaneous information over mossy fibers as well as olivary climbing fibers. Although these projections are organized in a grossly somatotopic way, imaging studies have shown that foot and hand representations partly overlap (Rijntjes et al., 1999). This may account for the higher activation levels when both limbs are moved simultaneously. However, Debaere et al. (2001) found that the activation level obtained during the coordination task even exceeded the sum of the single limb activations, pointing to a coordinative function. Interestingly, none of the studies using bimanual tasks reported activation within this region of the hemispheres, indicating that the hemispheric lobule III-V might be specifically involved in *ipsilateral* coordination of different effectors. Similar to the anterior hemispheres, the anterior vermis is also part of the spinocerebellum, receiving projections from the upper limbs mainly in lobule V, and from lower limbs mainly in lobule II (Jasmin and Courville, 1987; Leicht and Schmidt, 1977). Thus, coordination related activation found within the anterior vermis, probably reflects increased processing demands with respect to proprioceptive information from the moving limbs.

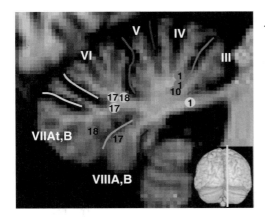

Ref	Lobule	MNI-xyz		
1	II-IV	9	-42	-27
1		6	-51	-18
1		6	-51	-15
10		0	-52	-20
18	VI	8	-64	-24
17		10	-68	-25
17		-2	-69	-24
17	VII/VIII	0	-68	-34
18		0	-76	-32

Figure 8-6. Summary of reported activation maxima within the cerebellar vermis. Locations are reported in MRI coordinates. The anatomic labeling corresponds to (Schmahmann et al., 2000). References are numbered in accordance to Table 8-1. (*See color plate section.*)

More posteriorly, within lobule VI (light blue), several studies reported coordination related activation either bilaterally (Ullen et al., 2003) or only within the right hemisphere (Meyer-Lindenberg et al., 2002) which spread occasionally to the vermis (Tracy et al., 2001; Ullen et al., 2003). Earlier studies have revealed ample evidence that this area within Lobule VI is involved in the timing of movement. With respect to unimanual movements, the lateral hemisphere and occasionally the vermis have revealed activation when subjects had to produce movements according to a predictable rhythm, which was externally (Dreher and Grafman, 2002; Kawashima et al., 2000; Lutz et al., 2000; Rao et al., 1997) or internally triggered (Kawashima et al., 2000; Rao et al., 1997; van Mier and Petersen, 2002). These imaging findings suggest that the cerebellar hemispheres and particularly lobule VI, play a crucial role in the timing of voluntary movements. However, recent results have made it questionable whether this area is primarily related to motor execution per se. Instead, it has been shown that activity within lobule VI, mostly at strikingly similar spots to those identified during interlimb coordination, is specifically increased during perception and discrimination of time intervals (Belin et al., 2002; Parsons, 2001). Studies on patients with lesions of the lateral cerebellar hemisphere support this view, showing that the time perception ability of patients, as revealed by a duration discrimination task, is substantially impaired (Ivry and Keele, 1989). The discrimination of temporal patterns requires the extraction of a time interval from a stream of sensory information, which is subsequently compared to a memorized target interval. Similar mechanisms can also be used to detect or anticipate the coincidence of sensory events, which is an important function for coordination. Thus, even though lobule VI might be endowed with a primarily sensory function, this information is likely to be crucial for controlling the timing of voluntary actions. Accordingly, cerebellar patients are not only impaired with respect to their time perception abilities, but also show a marked increase of temporal variability during a rhythmical tapping task (for an overview see Ivry, 1996, see also Chapter 10). Additionally, a more recent study has revealed that cerebellar patients exhibit a larger desynchronisation between movement onsets of the limbs than healthy subjects, when performing a goal-directed bimanual task (Serrien and Wiesendanger, 2000). However, desynchronization between the limbs did not further increase or was even reduced at the goal, suggesting that mechanisms using proprioceptive information for online-adjustments were still intact. The aforementioned studies indicate that cerebellar patients have difficulties to integrate the sensory information of several internal or external sources into a common temporal framework. Taken together, neuroimaging and lesion studies suggest that the increased activation of cerebellar lobule VI during interlimb coordination tasks probably reflects the larger effort required to synchronize the single limb actions in accordance with a common internally or externally represented temporal pattern.

Comparing a 3:2 tapping task to in-phase tapping (Ullen et al., 2003), additional activation was found in a more posterior and caudal area of the left cerebellar hemisphere, located around the posterior superior fissure, serving as the border between lobule VI and Crus I (fig. 8-5, pink) as well as the posterior vermis (fig. 8-6, pink). Both areas have been shown to be activated for movements in accordance to either a random or new stimulus sequence (Lutz et al., 2000; Penhune et al., 1998; Ramnani and Passingham, 2001) as well as for tasks involving time perception (Rao et al., 2001). However, activation of the lateral cerebellar hemisphere at lobule VI/Crus I is also typically seen during the early phase of motor learning of finger sequences or visuomotor transformations (Doyon et al., 2003). Therefore, a potential explanation for the observed additional activation in this region might be that the 3:2 tapping task was not sufficiently overlearned to reach the same level of automatization as in-phase tapping. Consistent with this assumption, the majority of motor learning studies have reported a decreasing activity in this area with ongoing training. Conversely, a recent fMRI study investigating learning-related changes during the acquisition of a bimanual 90° out-of-phase pattern reported increasing activation at a very similar spot (Debaere et al., 2003). This may point to a more specific function of this region, which is common to the early phase of motor learning as well as to complex interlimb coordination. In a PET experiment,

Blakemore et al. (2001) instructed subjects to move a robot arm with their right hand, which resulted in a tactile stimulation of the left palm. Varying the delay between the right hand action and left hand sensation, they found a positive correlation between the imposed delay and the haemodynamic response of the right lateral hemisphere at the border between lobule VI and Crus I. This experiment suggests that this region represents the time interval between a motor action of the one hand and a sensory event of the other hand. More generally, the lateral cerebellar hemisphere might be involved in predicting the sensory consequences of action in accordance to an internal forward model. In turn, this part of the cerebellum could be used to monitor the discrepancy between the predicted and the actual sensory event structure (Blakemore et al., 2001). A 3:2 tapping task but also a 90° out-of-phase pattern requires continuous control of the temporal relation between the limb trajectories. Therefore, the increased activation of lateral lobule VI/Crus I might indicate a higher level of monitoring and on-line correction of the spatial-temporal relationship between the limbs to preserve the required rhythm. However, this interpretation is highly tentative and can only serve as a working hypothesis for further research.

Taken together, the reported results suggest that the anterior cerebellum may assist the primary motor cortex in executing movements of the ipsilateral limbs in parallel. However, the more crucial function of the cerebellum most likely pertains to controlling the timing of coordinated behavior. More specifically, areas within lobule VI of the cerebellar vermis and hemisphere are activated when movements are performed in synchrony to a predictable rhythm, as observed in the majority of metronome-paced interlimb coordination tasks. By contrast, more posterior regions (hemisphere lobuleVI/Crus I, vermis lobule VII/VIII) seem to become additionally involved when limbs are not moved in synchrony, but when an exact temporal delay between the limb movements has to be maintained.

2.6 Broca's area and secondary sensory cortices – moving to the beat

Several coordination studies have reported an increased activation during difficult coordination patterns within Broca's area, sometimes extending to the neighboring insula as well as for the posterior secondary auditory cortex (BA 22/39) and secondary somatosensory cortex (BA 40). Broca's area and the secondary auditory cortex have traditionally been associated with language-related skills, since Broca's area is tightly linked to speech production (Price, 2000), while the auditory area is activated by language perception (Samson et al., 2001). However, imaging studies have shown that both areas are also activated by non-language tasks. It has been shown that caudal part of Broca's area (BA 44/6) and the insula of the left hemisphere, are activated during rhythm perception (Platel et al., 1997). Additionally, BA 44 has been found to be involved in the execution (Binkofski et al., 1999b; Binkofski et al., 1999a; Ward and Frackowiak, 2003), imagination (Binkofski et al., 2000; Grafton et al., 1996) and imitation (Heiser et al., 2003; Iacoboni et al., 1999; Nishitani and Hari, 2000) of finger movements, suggesting a higher-order control function. With respect to the secondary auditory cortex, Calvert (2001) suggested that it plays an important function in integrating sensory information from multiple modalities, when auditory stimuli are involved. In particular, the posterior auditory cortex around the junction of BA 22, BA 39 and BA 40 has been shown to integrate auditory and somatosensory information (Foxe et al., 2002). This area is close to the activation spots within the secondary auditory/somatosensory cortex, as evoked by difficult interlimb coordination tasks.

Increased activation of the insula has been reported by Ullen et al (2003). Note that in this study, subject's movements were not driven by a metronome but were self-paced. These results are consistent with the notion that the insula is involved in short-term memory of auditory information (Platel et al., 1997). Additionally, it has been suggested that the insula might be involved in detecting coincidence of sensory information from different sources (Calvert, 2001).

Taken together, Broca's area and the secondary sensory cortices could be accounted for by controlling the execution of finger movements in synchrony with a predetermined rhythm. However, it has to be kept in mind, that all these areas were found to be more strongly activated during incompatible than compatible coordination patterns. Thus, these areas are perhaps not only involved in producing rhythmic behavior per se, but may also play a specific role in monitoring whether the motor output of the different limbs matches the temporal requirements.

3.   LEARNING

Both the in- and anti-phase coordination patterns, which formed the primary focus of discussion above, represent coordination modes that are intrinsic to human's motor repertoire and that can usually be performed spontaneously without learning. In spite of this high degree of familiarity, differences between the in- and anti-phase coordination modes are evident at behavioral as well as brain activation level. However, other coordination patterns that deviate from these 'pre-existing' modes and that require more complex spatial or temporal coordination between the limbs, often can not be performed spontaneously, but require extensive practice before stability can be obtained. One can think of tasks such as playing musical instruments or flying a helicopter that require weeks or even years of practice before sufficient accuracy and consistency is manifest. Imaging work addressing the patterns of activation associated with learning new or complex coordination tasks is very scarce. Ullen et al. (2003) investigated the activation patterns associated with a complex temporal coordination task that required prior training (3:2 bimanual polyrhythm tapping). Results showed that polyrhythmic coordination (after training) engaged largely the same neural control circuitry as observed during anti-phase coordination, which is characterized by extensive fronto-parieto-temporal activations. Unfortunately, the evolution of brain activations during the learning of this complex tapping task was not addressed. Using fMRI, Debaere et al. (2003) studied learning-related changes in brain activation during the acquisition of a new bimanual coordination pattern, requiring a complex spatio-temporal relationship between the limbs (90° out-of-phase pattern). Cerebral activation patterns were assessed during initial learning and at an advanced stage following practice, when the pattern was performed with a high degree of stability and consistency. Learning-related changes were addressed relative to a control group who completed equal amounts of practice of in-phase movements that are intrinsic to the motor system and do not require learning. An overview of the regions that changed their activation during learning the new bimanual pattern is provided in figure 8-7. Regions with increasing activity are marked in yellow, regions with decreasing activity are marked in red. Results showed a higher involvement of right prefrontal-premotor-parietal and left cerebellar lobule VI during the early learning phase, whereas, bilateral primary motor, bilateral SMA (partly extending to cingulate motor cortex), left premotor, basal ganglia, and cerebellar dentate nuclei, became more involved as the coordination pattern was well acquired. These results indicate that differential cortico-subcortical routes are preferentially involved, depending on the stage of learning. In other words, a shift in activation from one set of cortico-cerebellar regions to another set of cortico-cerebellar-striatal regions was evident as the new bimanual coordination pattern was acquired. Additionally, an important asymmetry in cortical activations was noted, i.e., the right hemisphere made a more important contribution during early learning (see fig. 8-7).

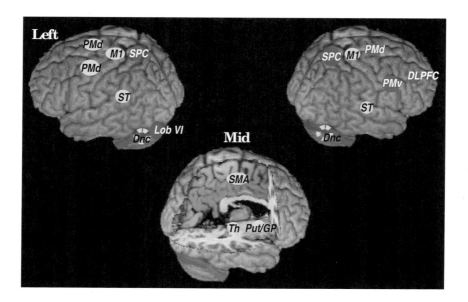

Figure 8-7. Decreases (uncircled) and increases (circled) in activation while learning a 90°
out-of-phase bimanual coordination pattern. M1 = primary motor cortex, SPC = superior
parietal cortex, PMd = dorsal premotor cortex, PMv = ventral premotor cortex, SMA =
supplementary motor area, DLPFC = dorsolateral prefrontal cortex, ST = superior temporal,
Th = Thalamus, Put/GP = Putamen/Globus Pallidus, Dnc = Cerebellum Dentate Nucleus, Lob
VI = Cerebellum Lobule VI.

These changes during bimanual learning bear some similarities with those
reported in various imaging studies on unimanual sequence learning, for
which theoretical models of the changing neural architecture have been
proposed (Doyon et al., 2003; Hikosaka et al., 1999).

It is important to add that Debaere et al. addressed changes in activation
from initial learning to a stage in which the subjects had reached stable
levels of performance on the new bimanual coordination task. It still remains
to be established whether these activation patterns would further change with
more extensive practice. Imaging work in professional pianists has hinted at
plastic changes in the brain that might occur as a result of long-term
bimanual training. Generally, these studies have shown that performing
complex uni- and bimanual finger sequence tasks required substantially less
activation in the primary motor, premotor and supplementary motor areas in
musicians compared to non-musicians (Jancke et al., 2000b; Krings et al.,
2000), with the lower activation being most evident in the secondary motor

cortices such as pre-SMA and cingulate motor cortex (Jancke et al., 2000b). However, when pianists have to play unfamiliar musical pieces, the pre-SMA and SMA seem to become significantly engaged (Sergent, 1993). These findings clearly indicate that long-term bimanual training in professional pianists results in more efficient control of the movements and this gain in efficiency is reflected by a recruitment of smaller neural networks necessary to perform a motor task. Additionally, when highly skilled pianists and non-musicians are exposed to a novel unimanual tapping task, a rapid increase of activation in the primary motor cortex, together with a diminished involvement of SMA, premotor and cerebellar regions is observed to occur in musicians as compared to non-musicians (Hund-Georgiadis and von Cramon, 1999). The effect of rapidly (within minutes) increasing activity in the primary motor cortex for musicians bears strong similarities with the effects observed during slow learning (4 weeks) in normal subjects (Karni, 1995). It has been hypothesized that this effect results from pre-practice experience and may again indicate a higher efficiency in neural recruitment.

Finally, additional insights into the neural processes underlying bimanual skill acquisition have been obtained from studies using an electroencephalogram (EEG). The main observation emerging from these studies is an enhanced interaction between both hemispheres (as quantified by an increased task-related coherence effect, especially with respect to primary motor and frontomesial cortices) during the early stage of bimanual learning, whereas the interhemispheric interaction decreases again following training (Andres et al., 1999). The interpretation of these results is that the dynamic changes in interhemispheric communication reflect the establishment of efficient bimanual 'motor routines'.

4. SUMMARY AND CONCLUSIONS

In this chapter, we have reviewed recent fMRI and PET studies addressing patterns of brain activation during interlimb coordination tasks. Nearly all studies used rhythmical tasks and compared brain activity during coordinated behavior (e.g. bimanual anti-phase movements) to activity during single limb movements or more simple coordination tasks (e.g. bimanual in-phase movements). Using these imaging techniques, it has become evident that the experimental conditions (i.e. the interlimb coordination tasks used) must differ sufficiently from each other to evoke detectable changes in haemodynamic response. Furthermore, the adopted cycling frequency should be sufficiently high to discover reliable differences between compatible and incompatible coordination tasks. Accordingly, studies using low frequencies (<1 Hz) have primarily revealed an increased involvement of SMA only, whereas studies using higher movement frequencies (>= 1 Hz) or more complicated tasks (e.g. 3:2 tapping) have reported additional activation in the premotor cortex, cerebellum, Broca's area/Insula and the superior temporal gyrus. Thus, extending the traditional view that bimanual coordination is mainly controlled by the SMA, it has been shown that different coordination tasks rely on a common network involving several areas, each serving a specific function. The interplay between the areas involved in this network is not yet clear. Moreover, it has to be kept in mind that this review focuses on areas *specifically* involved in inter-limb coordination. Other regions such as the basal ganglia, that are known to be involved in motor control, but that have not exhibited an increased response to coordination demands, are not included. However, a general picture has started to emerge, as summarized in figure 8-8.

For each movement cycle during an interlimb coordination task, different movement parameters for each limb have to be coded in parallel. This is likely to cause larger efforts at the executive level, which is reflected by higher activity of cerebellar and premotor areas and SMA. Hereby, the cerebellar hemispheres (lobule VI) probably serve as an "inner clock", predicting the upcoming temporal event. Thus, it is feasible that the cerebellum provides the general timing structure of rhythmical interlimb coordination tasks (*when to move*), which is forwarded to cortical motor areas. At these higher cortical levels, the SMA perhaps encodes a sequence of actions about how to move next with each limb. This '*how to move next*' information from the SMA, together with spatial information from parietal areas and additional input from contralateral premotor cortex, cerebellum and basal ganglia, are gated to the dorsal premotor area to select the appropriate motor action. Additionally, the premotor areas are also likely to be involved in suppressing mirror movements, which represent the "default coupling" between the hemispheres. This suppression of in-phase movements is particularly important during bilateral coordination to avoid involuntary transitions to naturally preferred coordination patterns. Finally, the motor action initiated by the dorsal premotor area becomes executed under supervision of the primary motor cortex and the spinocerebellum.

In parallel to the increased effort at the executive level, interlimb coordination is likely to cause higher demands on movement monitoring. The majority of interlimb coordination patterns require each limb to move according to a specific spatio-temporal relation with respect to the other limb. This necessitates a continuous monitoring and matching of the limbs' actions. Even though the neural locus of this monitoring function is still unsettled, the reviewed studies suggest that Broca's area, Insula and the ventral premotor cortex, together with the secondary auditory and somatosensory cortex and the cerebellum, may be involved in monitoring interlimb coordination.

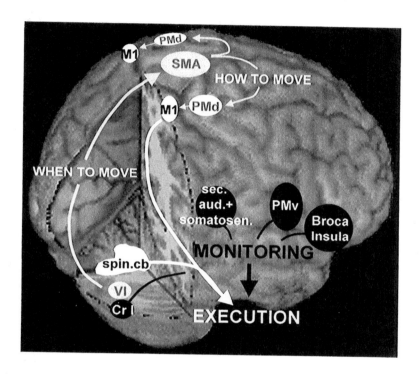

Figure 8-8. Putative cortical networks involved in different aspects of interlimb coordination. The network comprises supplementary motor area (SMA), left and right dorsal premotor cortex (PMd), left and right primary motor cortex (M1), cerebellar hemisphere (Lobule VI/Crus I) and the spinocerebellum as well as Broca's area, Insula, ventral premotor Cortex (PMv) and the secondary auditory/somatosensory cortex.

These areas seem to be particularly activated when tasks are performed rhythmically, as during metronome pacing. The presented global picture of functional assignments of brain areas to interlimb coordination, is highly tentative at this point. More research is needed to better understand the neural basis of interlimb coordination and to discover its remaining mysteries.

NOTES

Note that all locations are reported in MNI coordinates. Transformation from Talairach coordinates were performed in accordance to the procedures developed by Mathew Brett, http://www.mrc-cbu.cam.ac.uk.

ACKNOWLEDGEMENTS

Support for the present study was provided through a grant from the Research Council of K.U. Leuven, Belgium (Contract No. OT/03/61) and the Research Programme of the Fund for Scientific Research - Flanders (Belgium) (F.W.O.-Vlaanderen, G.0105.00 & G.0460.04). F. Debaere was supported by a scholarship from F.W.O.-Vlaanderen.

REFERENCES

Andres FG, Mima T, Schulman AE, Dichgans J, Hallett M, Gerloff C (1999) Functional coupling of human cortical sensorimotor areas during bimanual skill acquisition. Brain 122 (Pt 5): 855-870

Arthurs OJ, Williams EJ, Carpenter TA, Pickard JD, Boniface SJ (2000) Linear coupling between functional magnetic resonance imaging and evoked potential amplitude in human somatosensory cortex. Neuroscience 101: 803-806

Belin P, McAdams S, Thivard L, Smith B, Savel S, Zilbovicius M, Samson S, Samson Y (2002) The neuroanatomical substrate of sound duration discrimination. Neuropsychologia 40: 1956-1964

Binkofski F, Amunts K, Stephan KM, Posse S, Schormann T, Freund HJ, Zilles K, Seitz RJ (2000) Broca's region subserves imagery of motion: a combined cytoarchitectonic and fMRI study. Hum.Brain Mapp. 11: 273-285

Binkofski F, Buccino G, Posse S, Seitz RJ, Rizzolatti G, Freund H (1999a) A fronto-parietal circuit for object manipulation in man: evidence from an fMRI-study. Eur.J Neurosci. 11: 3276-3286

Binkofski F, Buccino G, Stephan KM, Rizzolatti G, Seitz RJ, Freund HJ (1999b) A parieto-premotor network for object manipulation: evidence from neuroimaging. Exp Brain Res 128: 210-213

Blakemore SJ, Frith CD, Wolpert DM (2001) The cerebellum is involved in predicting the sensory consequences of action. NeuroReport 12: 1879-1884

Blinkenberg M, Bonde C, Holm S, Svarer C, Andersen J, Paulson OB, Law I (1996) Rate dependence of regional cerebral activation during performance of a repetitive motor task: a PET study. J Cereb.Blood Flow Metab 16: 794-803

Calvert GA (2001) Crossmodal processing in the human brain: insights from functional neuroimaging studies. Cereb.Cortex 11: 1110-1123

Cardoso De Oliveira S (2002) The neuronal basis of bimanual coordination: recent neurophysiological evidence and functional models. Acta Psychol (Amst) 110: 139-159

Catalan MJ, Honda M, Weeks RA, Cohen LG, Hallett M (1998) The functional neuroanatomy of simple and complex sequential finger movements: a PET study. Brain 121 (Pt 2): 253-264

Chan JL, Ross ED (1988) Left-handed mirror writing following right anterior cerebral artery infarction: evidence for nonmirror transformation of motor programs by right supplementary motor area. Neurology 38: 59-63

Colby CL, Goldberg ME (1999) Space and attention in parietal cortex. Annu.Rev.Neurosci. 22: 319-349

Crammond DJ, Kalaska JF (2000) Prior information in motor and premotor cortex: activity during the delay period and effect on pre-movement activity. J Neurophysiol. 84: 986-1005

de Jong BM, Leenders KL, Paans AM (2002) Right parieto-premotor activation related to limb-independent antiphase movement. Cereb.Cortex 12: 1213-1217

Debaere F, Swinnen SP, Beatse E, Sunaert S, van Hecke P, Duysens J (2001) Brain areas involved in interlimb coordination: a distributed network. NeuroImage 14: 947-958

Debaere F, Wenderoth N, Sunaert S, van Hecke P, Swinnen SP (2003) Changes in brain activation during the acquisition of a new bimanual coordination task. Neuropsychologia submitted:

Deiber MP, Ibanez V, Sadato N, Hallett M (1996) Cerebral structures participating in motor preparation in humans: a positron emission tomography study. J Neurophysiol. 75: 233-247

Deiber MP, Passingham RE, Colebatch JG, Friston KJ, Nixon PD, Frackowiak RS (1991) Cortical areas and the selection of movement: a study with positron emission tomography. Exp Brain Res 84: 393-402

Desmurget M, Gréa H, Grethe JS, Prablanc C, Alexander GE, Grafton ST (2001) Functional anatomy of nonvisual feedback loops during reaching: a positron emission tomography study. J Neurosci. 21: 2919-2928

Dettmers C, Fink GR, Lemon RN, Stephan KM, Passingham RE, Silbersweig D, Holmes A, Ridding MC, Brooks DJ, Frackowiak RS (1995) Relation between cerebral activity and force in the motor areas of the human brain. J Neurophysiol. 74: 802-815

Dick JP, Benecke R, Rothwell JC, Day BL, Marsden CD (1986) Simple and complex movements in a patient with infarction of the right supplementary motor area. Mov Disord. 1: 255-266

Doyon J, Penhune V, Ungerleider LG (2003) Distinct contribution of the cortico-striatal and cortico-cerebellar systems to motor skill learning. Neuropsychologia 41: 252-262

Dreher JC, Grafman J (2002) The role of the cerebellum and basal ganglia in timing and error prediction. Eur J Neurosci. 16: 1609-1619

Dum RP, Strick PL (1991) The origin of corticospinal projections from the premotor areas in the frontal lobe. J Neurosci. 11: 667-689

Dum RP, Strick PL (2002) Motor areas in the frontal lobe of the primate. Physiol Behav 77: 677-682

Ehrsson HH, Fagergren A, Jonsson T, Westling G, Johansson RS, Forssberg H (2000b) Cortical activity in precision- versus power-grip tasks: an fMRI study. J Neurophysiol. 83: 528-536

Ehrsson HH, Fagergren E, Forssberg H (2001) Differential fronto-parietal activation depending on force used in a precision grip task: an fMRI study. J Neurophysiol. 85: 2613-2623

Ehrsson HH, Kuhtz-Buschbeck JP, Forssberg H (2002) Brain regions controlling nonsynergistic versus synergistic movement of the digits: a functional magnetic resonance imaging study. J Neurosci. 22: 5074-5080

Ehrsson HH, Naito E, Geyer S, Amunts K, Zilles K, Forssberg H, Roland PE (2000a) Simultaneous movements of upper and lower limbs are coordinated by motor representations that are shared by both limbs: a PET study. Eur.J Neurosci. 12: 3385-3398

Fink GR, Frackowiak RS, Pietrzyk U, Passingham RE (1997) Multiple nonprimary motor areas in the human cortex. J Neurophysiol. 77: 2164-2174

Fontaine D, Capelle L, Duffau H (2002) Somatotopy of the supplementary motor area: evidence from correlation of the extent of surgical resection with the clinical patterns of deficit. Neurosurgery 50: 297-303

Foxe JJ, Wylie GR, Martinez A, Schroeder CE, Javitt DC, Guilfoyle D, Ritter W, Murray MM (2002) Auditory-somatosensory multisensory processing in auditory association cortex: an fMRI study. J Neurophysiol. 88: 540-543

Goerres GW, Samuel M, Jenkins IH, Brooks DJ (1998) Cerebral control of unimanual and bimanual movements: an H2(15)O PET study. NeuroReport 9: 3631-3638

Grafton ST, Arbib MA, Fadiga L, Rizzolatti G (1996) Localization of grasp representations in humans by positron emission tomography. 2. Observation compared with imagination. Exp Brain Res 112: 103-111

Harrington DL, Rao SM, Haaland KY, Bobholz JA, Mayer AR, Binder JR, Cox RW (2000) Specialized neural systems underlying representations of sequential movements. J Cogn Neurosci 12: 56-77

He SQ, Dum RP, Strick PL (1995) Topographic organization of corticospinal projections from the frontal lobe: motor areas on the medial surface of the hemisphere. J Neurosci. 15: 3284-3306

Heeger DJ, Huk AC, Geisler WS, Albrecht DG (2000) Spikes versus BOLD: what does neuroimaging tell us about neuronal activity? Nat.Neurosci. 3: 631-633

Heiser M, Iacoboni M, Maeda F, Marcus J, Mazziotta JC (2003) The essential role of Broca's area in imitation. Eur J Neurosci. 17: 1123-1128

Hikosaka O, Nakahara H, Rand MK, Sakai K, Lu X, Nakamura K, Miyachi S, Doya K (1999) Parallel neural networks for learning sequential procedures. Trends Neurosci. 22: 464-471

Holmes G (1939) The cerebellum of man. Brain 62: 1-30

Hund-Georgiadis M, von Cramon DY (1999) Motor-learning-related changes in piano players and non-musicians revealed by functional magnetic-resonance signals. Exp Brain Res 125: 417-425

Iacoboni M, Woods RP, Brass M, Bekkering H, Mazziotta JC, Rizzolatti G (1999) Cortical mechanisms of human imitation. Science 286: 2526-2528

Immisch I, Waldvogel D, van Gelderen P, Hallett M (2001) The role of the medial wall and its anatomical variations for bimanual antiphase and in-phase movements. NeuroImage 14: 674-684

Ivry RB (1996) The representation of temporal information in perception and motor control. Curr.Opin.Neurobiol. 6: 851-857

Ivry RB, Keele S (1989) Timing functions of the cerebellum. J Cogn Neurosci 1: 136-152

Jaencke L, Peters M, Himmelbach M, Nosselt T, Shah J, Steinmetz H (2000a) fMRI study of bimanual coordination. Neuropsychologia 38: 164-174

Jaencke L, Shah NJ, Peters M (2000b) Cortical activations in primary and secondary motor areas for complex bimanual movements in professional pianists. Brain Res Cogn Brain Res 10: 177-183

Jasmin L, Courville J (1987) Distribution of external cuneate nucleus afferents to the cerebellum: II. Topographical distribution and zonal pattern--an experimental study with radioactive tracers in the cat. J Comp Neurol 261: 497-514

Jenkins IH, Passingham RE, Brooks DJ (1997) The effect of movement frequency on cerebral activation: a positron emission tomography study. J Neurol Sci 151: 195-205

Karni A (1995) When practice makes perfect. Lancet 345: 395

Kawashima R, Okuda J, Umetsu A, Sugiura M, Inoue K, Suzuki K, Tabuchi M, Tsukiura T, Narayan SI, Nagasaka T, Yanagawa I, Fujii T, Takahashi S, Fukuda H, Yamadori A (2000) Human cerebellum plays an important role in memory-timed finger movements: an fMRI study. J Neurophysiol. 83: 1079-1087

Kawashima R, Matsumura M, Sadato N, Naito E, Waki A, Nakamura S, Matsunami K, Fukuda H, Yonekura Y (1998b) Regional cerebral blood flow changes in human brain related to ipsilateral and contralateral complex hand movements--a PET study. Eur J Neurosci. 10: 2254-2260

Kawashima R, Tanji J, Okada K, Sugiura M, Sato K, Kinomura S, Inoue K, Ogawa A, Fukuda H (1998a) Oculomotor sequence learning: a positron emission tomography study. Exp Brain Res 122: 1-8

Kollias SS, Alkadhi H, Jaermann T, Crelier G, Hepp-Reymond MC (2001) Identification of multiple nonprimary motor cortical areas with simple movements. Brain Res Brain Res Rev. 36: 185-195

Krings T, Topper R, Foltys H, Erberich S, Sparing R, Willmes K, Thron A (2000) Cortical activation patterns during complex motor tasks in piano players and control subjects. A functional magnetic resonance imaging study. Neurosci.Lett. 278: 189-193

Kuhtz-Buschbeck JP, Ehrsson HH, Forssberg H (2001) Human brain activity in the control of fine static precision grip forces: an fMRI study. Eur J Neurosci. 14: 382-390

Lang W, Obrig H, Lindinger G, Cheyne D, Deecke L (1990) Supplementary motor area activation while tapping bimanually different rhythms in musicians. Exp Brain Res 79: 504-514

Leicht R, Schmidt RF (1977) Somatotopic studies on the vermal cortex of the cerebellar anterior lobe of unanaesthetized cats. Exp Brain Res 27: 479-490

Liu J, Morel A, Wannier T, Rouiller EM (2002) Origins of callosal projections to the supplementary motor area (SMA): a direct comparison between pre-SMA and SMA-proper in macaque monkeys. J Comp Neurol 443: 71-85

Logothetis NK, Pauls J, Augath M, Trinath T, Oeltermann A (2001) Neurophysiological investigation of the basis of the fMRI signal. Nature 412: 150-157

Lutz K, Specht K, Shah NJ, Jaencke L (2000) Tapping movements according to regular and irregular visual timing signals investigated with fMRI. NeuroReport 11: 1301-1306

Mattay VS, Weinberger DR (1999) Organization of the human motor system as studied by functional magnetic resonance imaging. Eur J Radiol. 30: 105-114

McNabb AW, Carroll WM, Mastaglia FL (1988) "Alien hand" and loss of bimanual coordination after dominant anterior cerebral artery territory infarction. J Neurol Neurosurg.Psychiatry 51: 218-222

Meesen R, Levin O, Wenderoth N, Swinnen SP (2003) Head movements destabilize cyclical in-phase but not anti-phase homologous limb coordination in humans. Neurosci.Lett. 340: 229-233

Meyer-Lindenberg A, Ziemann U, Hajak G, Cohen L, Berman KF (2002) Transitions between dynamical states of differening stability in the human brain. Proc Natl Acad Sci U.S.A 99: 10948-10953

Middleton FA, Strick PL (1997) Cerebellar output channels. Int.Rev.Neurobiol. 41: 61-82

Mitz AR, Wise SP (1987) The somatotopic organization of the supplementary motor area: intracortical microstimulation mapping. J Neurosci. 7: 1010-1021

Nair DG, Purcott KL, Fuchs A, Steinberg F, Kelso JA (2003) Cortical and cerebellar activity of the human brain during imagined and executed unimanual and bimanual action sequences: a functional MRI study. Brain Res Cogn Brain Res 15: 250-260

Nishitani N, Hari R (2000) Temporal dynamics of cortical representation for action. Proc Natl Acad Sci U.S.A 97: 913-918

Obhi SS, Haggard P, Taylor J, Pascual-Leone A (2002) rTMS to the supplementary motor area disrupts bimanual coordination. Motor Control 6: 319-332

Parsons LM (2001) Exploring the functional neuroanatomy of music performance, perception, and comprehension. Ann N Y Acad Sci 930: 211-231

Penhune VB, Zatorre RJ, Evans AC (1998) Cerebellar contributions to motor timing: a PET study of auditory and visual rhythm reproduction. J Cogn Neurosci 10: 752-765

Picard N, Strick P (1996) Motor areas of the medial wall: a review of their location and functional activation. Cereb.Cortex 6: 342-353

Picard N, Strick PL (2001) Imaging the premotor areas. Curr.Opin.Neurobiol. 11: 663-672

Platel H, Price C, Baron JC, Wise R, Lambert J, Frackowiak RS, Lechevalier B, Eustache F (1997) The structural components of music perception. A functional anatomical study. Brain 120 (Pt 2): 229-243

Praamstra P, Kleine BU, Schnitzler A (1999) Magnetic stimulation of the dorsal premotor cortex modulates the Simon effect. NeuroReport 10: 3671-3674

Price CJ (2000) The anatomy of language: contributions from functional neuroimaging. J Anat. 197 Pt 3: 335-359

Ramnani N, Passingham RE (2001) Changes in the human brain during rhythm learning. J Cogn Neurosci 13: 952-966

Ramnani N, Toni I, Passingham RE, Haggard P (2001) The cerebellum and parietal cortex play a specific role in coordination: a PET study. NeuroImage 14: 899-911

Rao SM, Binder JR, Bandettini PA, Hammeke TA, Yetkin FZ, Jesmanowicz A, Lisk LM, Morris GL, Mueller WM, Estkowski LD, . (1993) Functional magnetic resonance imaging of complex human movements. Neurology 43: 2311-2318

Rao SM, Harrington DL, Haarland KY, Bobholz JA, Cox RW, Binder JR (1997) Distributed neural systems underlying the timing of movements. J Neurosci. 17: 5528-5535

Rao SM, Mayer AR, Harrington DL (2001) The evolution of brain activation during temporal processing. Nat.Neurosci. 4: 317-323

Rees G, Friston K, Koch C (2000) A direct quantitative relationship between the functional properties of human and macaque V5. Nat.Neurosci. 3: 716-723

Richter W, Andersen PM, Georgopoulos AP, Kim SG (1997) Sequential activity in human motor areas during a delayed cued finger movement task studied by time-resolved fMRI. NeuroReport 8: 1257-1261

Rijntjes M, Buechel C, Kiebel S, Weiller C (1999) Multiple somatotopic representations in the human cerebellum. NeuroReport 10: 3653-3658

Rizzolatti G, Fogassi L, Gallese V (2002) Motor and cognitive functions of the ventral premotor cortex. Curr.Opin.Neurobiol. 12: 149-154

Rizzolatti G, Luppino G, Matelli M (1998) The organization of the cortical motor system: new concepts. Electroencephalogr.Clin.Neurophysiol. 106: 283-296

Rouiller EM, Babalian A, Kazennikov O, Moret V, Yu XH, Wiesendanger M (1994) Transcallosal connections of the distal forelimb representations of the primary and supplementary motor cortical areas in macaque monkeys. Exp Brain Res 102: 227-243

Sadato N, Campbell G, Ibanez V, Deiber M, Hallett M (1996b) Complexity affects regional cerebral blood flow change during sequential finger movements. J Neurosci. 16: 2691-2700

Sadato N, Ibanez V, Deiber MP, Campbell G, Leonardo M, Hallett M (1996a) Frequency-dependent changes of regional cerebral blood flow during finger movements. J Cereb.Blood Flow Metab 16: 23-33

Sadato N, Yonekura Y, Waki A, Yamada H, Ishii Y (1997) Role of the supplementary motor area and the right premotor cortex in the coordination of bimanual finger movements. J Neurosci. 17: 9667-9674

Sakai K, Hikosaka O, Takino R, Miyauchi S, Nielsen M, Tamada T (2000) What and when: parallel and convergent processing in motor control. J Neurosci. 20: 2691-2700

Samson Y, Belin P, Thivard L, Boddaert N, Crozier S, Zilbovicius M (2001) Auditory perception and language: functional imaging of speech sensitive auditory cortex. Rev.Neurol (Paris) 157: 837-846

Schmahmann JD, Doyon J, Toga AW, Petrides M, Evans AC (2000) MRI Atlas of the human cerebellum. Academic Press, San Diego, California, USA

Schubotz RI, von Cramon DY (2001) Functional organization of the lateral premotor cortex: fMRI reveals different regions activated by anticipation of object properties, location and speed. Brain Res Cogn Brain Res 11: 97-112

Sergent J (1993) Music, the brain and Ravel. Trends Neurosci. 16: 168-172

Serrien DJ, Brown P (2002) The functional role of interhemispheric synchronization in the control of bimanual timing tasks. Exp Brain Res 147: 268-272

Serrien DJ, Strens LH, Oliviero A, Brown P (2002) Repetitive transcranial magnetic stimulation of the supplementary motor area (SMA) degrades bimanual movement control in humans. Neurosci.Lett. 328: 89-92

Serrien DJ, Wiesendanger M (2000) Temporal control of a bimanual task in patients with cerebellar dysfunction. Neuropsychologia 38: 558-565

Shibasaki H, Sadato N, Lyshkow H, Yonekura Y, Honda M, Nagamine T, Suwazono S, Magata Y, Ikeda A, Miyazaki M, . (1993) Both primary motor cortex and supplementary motor area play an important role in complex finger movement. Brain 116 (Pt 6): 1387-1398

Snyder LH, Batista AP, Andersen RA (2000) Intention-related activity in the posterior parietal cortex: a review. Vision Res 40: 1433-1441

Stephan KM, Binkofski F, Halsband U, Dohle C, Wunderlich G, Schnitzler A, Tass P, Posse S, Herzog H, Sturm V, Zilles K, Seitz RJ, Freund HJ (1999b) The role of ventral medial wall motor areas in bimanual co-ordination. A combined lesion and activation study. Brain 122 (Pt 2): 351-368

Stephan KM, Binkofski F, Posse S, Seitz RJ, Freund HJ (1999a) Cerebral midline structures in bimanual coordination. Exp Brain Res 128: 243-249

Steyvers M, Etoh S, Sauner D, Levin O, Siebner HR, Swinnen SP, Rothwell JC (2003) High-frequency transcranial magnetic stimulation of the supplementary motor area reduces bimanual coupling during anti-phase but not in-phase movements . Exp Brain Res in press:

Swinnen SP (2002) Intermanual coordination: from behavioural principles to neural-network interactions. Nat.Rev.Neurosci. 3: 348-359

Thoenissen D, Zilles K, Toni I (2002) Differential involvement of parietal and precentral regions in movement preparation and motor intention. J Neurosci. 22: 9024-9034

Toyokura M, Muro I, Komiya T, Obara M (1999) Relation of bimanual coordination to activation in the sensorimotor cortex and supplementary motor area: analysis using functional magnetic resonance imaging. Brain Res Bull. 48: 211-217

Toyokura M, Muro I, Komiya T, Obara M (2001) Activation of pre-supplementary motor area (SMA) and SMA proper during unimanual and bimanual complex sequences: An analysis using functional magnetic resonance imaging. J Neuroimaging 12: 172-178

Tracy JI, Faro SS, Mohammed FB, Pinus AB, Madi SM, Laskas JW (2001) Cerebellar mediation of the complexity of bimanual compared to unimanual movements. Neurology 57: 1862-1869

Ullen F, Forssberg H, Ehrsson HH (2003) Neural networks for the coordination of the hands in time. J Neurophysiol. 89: 1126-1135

van Mier H, Petersen SE (2002) Role of the cerebellum in motor cognition. Ann N Y Acad Sci 978: 334-353

Van Oostende S, van Hecke P, Sunaert S, Nuttin B, Marchal G (1997) FMRI studies of the supplementary motor area and the premotor cortex. NeuroImage 6: 181-190

Ward NS, Frackowiak RS (2003) Age-related changes in the neural correlates of motor performance. Brain 126: 873-888

Weinrich M, Wise SP, Mauritz KH (1984) A neurophysiological study of the premotor cortex in the rhesus monkey. Brain 107: 385-414

Wiesendanger M, Rouiller EM, Kazennikov O, Perrig S (1996) Is the supplementary motor area a bilaterally organized system? Adv.Neurol. 70: 85-93

Chapter 9

BEHAVIORAL PRINCIPLES OF INTERLIMB COORDINATION

Will Spijkers[1], Herbert Heuer[2]

[1]*Institut für Psychologie der RWTH Aachen,* [2]*Institut für Arbeitsphysiologie an der Universität Dortmund, Germany*

Abstract: Interference in interlimb coordination is informative with respect to the underlying mechanisms of coordination. A rather comprehensive functional model of bimanual interlimb coordination, in which it is assumed that interference between the limbs may emerge at different levels of motor control, i.e. motor programming and movement execution, was proposed by Spijkers and Heuer (1995). Central to this model is the hypothesis of transient crosstalk during programming of movement parameters in bimanual coordination tasks. A series of experiments is presented, providing support for the model. These studies are based on three prototypical movement tasks in which several movement variables were examined, such as movement amplitude, isometric force and movement direction. It is concluded that the distinction between an execution level and a programming level may not be sufficient to capture all phenomena of intermanual interactions and, therefore, a third "cognitive" level is needed.

Key words: Crosstalk, transient coupling, motor programming, amplitude, isometric force

1.   INTRODUCTION

In this chapter we will outline a rather comprehensive functional model of interlimb coordination and review the respective evidence across a range of experimental paradigms. The model captures important characteristics of soft structural constraints, which give rise to systematic movement errors and the typical impairments of performance which result from them. The premise of this model of interlimb coordination is that bimanual interference in movement performance results from crosstalk between signals involved in motor control. Two core assumptions on the nature of crosstalk are, first, that it can emerge at two levels, the programming level and the execution level, and, second, that crosstalk can be static or phasic (transient). Static crosstalk persists during motor preparation, but phasic crosstalk fades with increasing preparation time.

This chapter is organized as follows. First the basic idea of the model is outlined and the basic assumptions are explained. The prototypical tasks that have been used in most of our experiments are described in the next section. Moving along the different experimental paradigms, the most relevant and decisive results of the studies are presented. A brief discussion of recent challenges of the model will close the chapter.

2.   CONSTRAINTS ON INTERLIMB COORDINATION

Different from gymnastics, ballet-dancing, sport, and the like, our daily experience does not really make us aware of limits of our bimanual skills. Yet, many tasks require coordinated movements of different limbs. For example, the left and right arm are coordinated differently, but smoothly in any case, when we do the dishes, open a bottle or reach for two objects simultaneously. Such actions appear to be guided by the principle of minimal intrusive interaction, that is, of minimizing costs which accrue from soft structural constraints and the associated errors of action. There may be a simple reason why the limits of bimanual movements are so rarely experienced in everyday life: probably we seldom want to do what we cannot do and therefore limit the range of our coordinated actions voluntarily.

When we are asked to draw a circle with the one hand and simultaneously a rectangle with the other hand, or to tap a rhythm like '- - ...' with the right hand while tapping as rapidly as possible with the left hand, we become aware of the sheer impossibility of these tasks. They thus demonstrate the existence of structural constraints on bimanual coordination. Structural constraints can be of a 'hard' type and make performance of a task impossible as in the examples given above. However, more often structural constraints seem to be 'soft': they do not define strict performance limitations, but determine our preference for performing certain movement patterns or produce errors in the case of movement patterns which deviate from the preferred ones. For example, when movements that have different movement times when performed in isolation are performed concurrently, the movement times become highly similar or even almost identical (e.g., Corcos, 1984; Fowler, Duck, Mosher, & Mathieson, 1991; Kelso, Southard, & Goodman, 1979; Marteniuk & MacKenzie, 1980; Marteniuk, MacKenzie, & Baba, 1984). The identical timing, however, is not a 'hard' constraint, but a 'soft' one. It can be overcome, though perhaps not completely so, when different goal durations of the movements are strictly required (Spijkers, Tachmatzidis, Debus, Fischer, & Kausche, 1994).

Whereas the temporal constraints on discrete bimanual movements are fairly tight, the spatial constraints are weaker. Typically, they result only in a small asymmetric assimilation between the hands when different amplitudes must be produced concurrently: the amplitude of the short movement is somewhat increased when the other hand performs a long-amplitude movement, whereas the long amplitude is hardly or not at all affected by the concurrent short-amplitude movement of the other hand (Heuer, Spijkers, Kleinsorge, van der Loo, & Steglich, 1998; Marteniuk & MacKenzie, 1980; Marteniuk et al., 1984; Sherwood, 1990, 1991, 1994a, 1994b; Sherwood & Nishimura, 1992; Spijkers, Heuer, Kleinsorge, & van der Loo, 1997). Comparatively weak spatial assimilation effects have also been observed when circles and lines are drawn concurrently with the two hands (Franz, Zelaznik, & McCabe, 1991) or circles and ellipses (Walter, Swinnen, & Dounskaia, 2002).

The basically soft nature of structural constraints on coordination is well captured by the dynamic approach, which dominates the field. Particular research efforts have been devoted to bimanual periodic movements. These movements can be modeled in terms of nonlinear oscillators, and the structural constraints can be captured by a coupling such that the state variables of each oscillator enter into the differential equation that governs the behavior of the other oscillator (e.g., Haken, Kelso, & Bunz, 1985; Kay, Kelso, Saltzman, & Schöner, 1987). On a higher level of description, the individual oscillators are neglected, and only their relative phase is analyzed. The relative phase is a collective variable, the behavior of which is again governed by a nonlinear differential equation (coordination dynamics). Most research is related to this level of modeling the structural constraints, which has been proven to offer very powerful descriptive and heuristic tools (e.g., Fuchs and Kelso, 1994). Perhaps one of the most important developments is the distinction of "behavioral information", which captures the effects of information coming from perception, memory, and intention, and the "intrinsic dynamics", which capture the effects of the structural constraints on coordination. The intrinsic dynamics produce the deviations from the intended movements, which are defined by the behavioral information (Schöner & Kelso, 1988).

Regarding the question of where the structural constraints on coordination originate, the dynamic approach is essentially silent; as a matter of fact, the dynamical models abstract from the physical substrate (Kelso, 1994) in basically the same way as when linear systems theory was applied to the analysis of human tracking performance (e.g., Ellson, 1959). Therefore, for a fuller understanding of the structural constraints on coordination, dynamic models need to be supplemented by additional analyses that are concerned with questions about the origin of structural constraints (Heuer, 1995). Such analyses can be theoretical, but also experimental, on both a physiological and a behavioral level of observation. In the present contribution we will consider the origin of structural constraints from a behavioral perspective.

2.1 Crosstalk at the execution-level and the programming level

In functional terms, at least two levels of motor control can be distinguished: the programming level and the execution level. It has been suggested that structural constraints on coordination can originate from crosstalk between control signals at both levels (Marteniuk & Mackenzie, 1980; Marteniuk, MacKenzie, & Baba, 1984; Spijkers & Heuer, 1995).

By the term execution level (or outflow level) we refer primarily to efferent commands and associated signals. Execution-level signals change over time as the movement is produced, and they are essentially absent when the movement has not yet started. Depending on how motor control is modeled, execution-level signals can be thought of as the output of a generalized motor program (Schmidt & Lee, 1999, pp. 157-168), the output of a trajectory-generating structure (e.g., Bullock & Grossberg, 1988), or as the state variables of a nonlinear oscillator (e.g., Kay, Kelso, Saltzman, & Schöner, 1987). In anatomical terms, crosstalk at the execution level can originate, for example, from uncrossed fibers of the pyramidal tract (e.g., Preilowski, 1975) or segmental spinal circuits. Unambiguous evidence for crosstalk at the execution level is provided by the existence of associated movements, in particular of mirror movements that tend to be more frequent and stronger as higher forces are developed (Durwen & Herzog, 1992; Todor & Lazarus, 1986). Consistent herewith Walter and Swinnen (1990) and Swinnen, Walter, Serrien, and Vandendriesche (1992) found stronger contralateral effects of a double-reversal movement on a concurrent aiming movement with the other hand when the force in executing the double-reversal movement was increased by added load or reduced duration.

By the term programming level (or parametric level) we refer primarily to movement parameters like direction and amplitude. Before the start of the movement, programming-level signals may change towards the target values required by the movement goal, but during the movement they will largely remain constant. Depending on how motor control is modeled, programming-level signals can be thought of as the parameters of a generalized motor program, of a trajectory-generating structure, or of a nonlinear oscillator. In anatomical terms, there is evidence that parameters like amplitude and direction are coded by neuronal activity (see Lacquaniti, 1996, for an overview) which may be subject to transcortical interactions. Behavioral methods, like the timed-response procedure, also allow one to trace the time course of parameter specification (Favilla & DeCecco, 1996; Favilla, Gordon, Hening, & Ghez, 1990; Favilla, Hening, & Ghez, 1989; Ghez, Favilla, Ghilardi, Gordon, Bermejo, & Pullman, 1997). Both the behavioral and physiological data show that parameter specification is a time-consuming process with continuous characteristics as long as the target values do not deviate too much from the initial values (Ghez et al., 1997).

2.2 Static and phasic crosstalk

In addition to the different levels of crosstalk, we distinguish static and phasic crosstalk effects. Static crosstalk effects are manifested in conditions in which sufficient preparation time is available so that parameter specifications can reach their final steady states before the movements are initiated. For example, when participants have enough time to specify the amplitudes for bimanual movements, the typical asymmetric amplitude assimilation described above can be observed.

Static crosstalk effects should be distinguished from the phasic (or transient) crosstalk effects that can be observed when bimanual movements are initiated before their respective parameter values have been fully specified. The experimental paradigm in which phasic crosstalk effects can be observed most directly is the timed-response procedure, which was introduced by Schouten and Becker (1967) for the study of the speed-accuracy tradeoff. This paradigm has been applied to the study of the specification of characteristics of isometric contractions and movements by Ghez and coworkers (e.g., Favilla, Gordon, Hening, & Ghez, 1990; Favilla, Hening, & Ghez, 1989; Ghez et al., 1997).

The experimental studies reviewed in the present chapter were mainly concerned with phasic or transient crosstalk at the programming level. The reason is, first, that there is little doubt about the existence of execution-level crosstalk, and second, that strong static crosstalk at the programming level makes the study of bimanual movements essentially impossible. Strong static crosstalk at the programming level has the consequence that complete specification of (different) bimanual movements is impossible, so that only more or less erratic movements will be produced or subjects will simply refuse to try the task (strictly concurrent drawing of a circle and a rectangle would be an example of such a task). With weak static crosstalk, effects will be seen even with long preparation times.

3.   THREE PROTOTYPICAL MOVEMENT TASKS

The model was tested with different types of task involving bimanual simultaneous movements, bimanual sequential movements and unimanual movements. The main independent variables were the relation between the movements of both hands, which could be same or different, and the time available for (advance) specification of movement characteristics. With all types of task we used rapid reversal movements with same and different target amplitudes assigned to the left and right hand. Several tasks were also used with isometric contractions of same and different peak forces. By and large, we found the same crosstalk effects with the movements and the isometric contractions. Finally, bimanual simultaneous movements were used to study crosstalk effects in movements with symmetric, parallel, and different directions.

3.1   Tasks with bimanual simultaneous movements

In these tasks bimanual reversal movements with same and different amplitudes, bimanual contractions with same and different peak forces, and bimanual aiming movements with symmetric, parallel, and different directions were used. All task variants required simultaneous production of isometric contractions or movements of the two hands, but otherwise the variants constitute three different experimental paradigms with different rationales which establish the link between the theoretical notion of transient parametric coupling and observable phenomena.

3.1.1   Bimanual timed-response task

In the bimanual timed-response task participants are instructed to produce bimanual movements in synchrony with the fourth (and last) tone of a regular series. Initially, they should prepare themselves for same (intermediate) amplitudes. However, the correct amplitude (short or long for each hand) is cued at a variable interval in advance of the fourth tone and thus of movement initiation. Thus, the time available for amplitude specification is varied, and the amplitudes of the reversal movements produced allow to trace the gradual specification of movement amplitudes. The comparison of the time courses of amplitude specifications when same and different amplitudes are assigned to the other hand allows one to trace the time course of the effect of crosstalk at the programming level.

3.1.2   Bimanual reaction-time task

In the bimanual reaction-time task participants are instructed to produce bimanual movements as rapidly as possible in response to an imperative signal. Cues which specify the target amplitudes for the hands are presented concurrently with the imperative signal or earlier. The bimanual reaction-time task can be seen to complement the timed-response task when it is assumed that the movement is initiated as soon as the amplitude specifications have been completed. When the target amplitudes for the hands are different, crosstalk should slow down the concurrent processes of amplitude specification. Thus reaction time should be longer with different than with same target amplitudes. However, when the amplitude cues are presented well in advance of the imperative signal, final steady states of amplitude specifications should be reached before reaction times are measured. Thus, the reaction-time difference between conditions with same and different target amplitudes should disappear at sufficiently long precueing intervals.

3.1.3   Sequential bimanual task

In the sequential bimanual-movement task participants are instructed to produce bimanual sequences of reversal movements at a certain rate. By keeping the amplitude constant in one hand, but alternating short and long amplitudes in the other hand, crosstalk related to new response specifications for successive movements of the one hand can be evidenced from involuntary amplitude modulations of the other hand. With this paradigm the time available for the amplitude specifications is varied by manipulating the rate of the movements.

3.2   Bimanual sequential-movement task

In the bimanual sequential-movement task, participants are instructed to produce two unimanual movements, one with each hand, in rapid succession in response to two imperative signals. This task is highly similar to the standard procedure of experiments on the psychological refractory period (PRP). There are two imperative signals in succession, the first one requiring a response of the one hand, and the second one requiring a response of the other hand. By means of varying the interstimulus interval, the temporal overlap of processes involved in specifying left-hand and right-hand amplitudes can be manipulated. Similarly, by means of precues which indicate the amplitude of the first movement, crosstalk between temporally overlapping processes of amplitude specification can be reduced. Of main interest is the reaction time of the second movement, which should be longer when the first movement has a different than when it has the same amplitude. When the interstimulus interval and/or the precueing interval is increased, the reaction-time effect should disappear.

3.3   Unimanual movement task

In the unimanual movement task, participants are instructed to produce a left-hand movement or a right-hand movement in response to a choice signal. The amplitudes of both movements which could be required by the choice signal are precued with a variable precueing interval. The rationale of the paradigm is based on the assumption that both amplitudes are specified concurrently in advance of the choice signal. With short precueing intervals advance specification will be incomplete, so that it must be finalized during the reaction-time interval. Depending on whether same or different amplitudes are specified concurrently, the incompleteness of advance specification at short precueing intervals should be less or more. Correspondingly, completion should take less or more time, so that with short precueing intervals choice-reaction time should be longer for choice between a left-hand and a right-hand movement when different rather than same target amplitudes are precued for the two hands.

4. THE EVIDENCE

4.1 Evidence from bimanual timed-response task

4.1.1 Movement Amplitude

In the first study with the bimanual timed-response task (Heuer, Spijkers, Kleinsorge, van der Loo, & Steglich, 1998) participants were instructed to produce rapid bimanual reversal movements in synchrony with the fourth and last tone of a regular series with stimulus-onset asynchronies of 500 ms. The target amplitude for each hand could be short (10 cm) or long (20 cm). It was indicated on a screen at a variable time before the fourth synchronization tone (cueing interval) by means of a short or long horizontal bar. In a second study (Heuer, Kleinsorge, Spijkers, & Steglich, 2000) we varied the difference between target amplitudes (5-30, 15-30, and 25-30 cm). Again, for each combination of target amplitudes, short and long horizontal bars served as cues. As shown in Figure 9-1 (left column), when target amplitudes were the same for both hands, the mean amplitudes of short and long movements changed gradually from a default setting (which we identified for participants by instruction as a medium amplitude) to their respective target values as the time available for amplitude specification increased. In Figure 9-1 the mean amplitudes as a function of the CRI (cue-response interval) are shown. In agreement with the notion of transient parametric crosstalk, we found that the long amplitude was transiently biased toward the short one when target amplitudes were different. This bias disappeared when preparation time became sufficiently long, that is at a CRI of about 300 ms.

A corresponding transient bias was also found in the short-amplitude movements, but it did not always fully disappear at long cueing intervals. The static asymmetric bias of concurrently performed short and long movements, which can be observed even when ample preparation time is provided (cf. Fig. 9-1, upper graph of the left column), has been found repeatedly before (Marteniuk et al., 1984; Sherwood, 1990, 1991, 1994a, b), but it seems to show up only when the difference between target amplitudes is sufficiently large. In the middle graph of the left column of Figure 9-1 it is evident that the late effect disappears when the difference between target amplitudes becomes smaller. In the lower graph, when the amplitude is further reduced, a contrast effect shows up in that the amplitudes in conditions with different target amplitudes become more different than in conditions with same amplitudes. This contrast can support the production of different amplitudes even at short CRIs. From a methodological point of view it has the effect of masking the transient assimilation as long as the amplitudes of short and long movements are not normalized to the amplitudes at long cueing intervals.

With the timed-response procedure, inferences about the state of amplitude specifications at the start of the movements are based on measures taken during the movements, namely at their reversal, about 200-300 ms after their start. Thus, the observed findings could be affected by feedback. In particular, there could be on-line amendments which overcome static crosstalk effects during movement execution so that they appear transient as those in Figure 9-1. On-line amendments affect primarily antagonist activity and thus the deceleration phase (van der Meulen, Gooskens, Denier van der Gon, Gielen, Wilhelm, 1990; Fowler, Duck, Mosher, & Mathieson, 1991). Therefore, we analyzed movement amplitudes at the time of peak velocity. There was a striking correspondence between the time courses of amplitudes at movement reversals and at peak velocities (Heuer, Spijkers, Kleinsorge, & Steglich, 1998). Thus, the results seem not to depend critically on on-line processing of feedback.

Inferences based on measures taken during the movement cannot unambiguously be related to crosstalk at the programming level, but they can also reflect execution-level crosstalk. However, there are at least two reasons to claim that the transient crosstalk effect at short preparation intervals is related to programming-level crosstalk rather than to execution-level crosstalk. First, the transient effect is already present when the produced amplitudes are hardly different, and it is absent, when the difference between the produced amplitudes is large. This seems to contradict the principle that the strength of crosstalk at the execution level is related to the forces produced (Durwen & Herzog, 1992; Swinnen, Walter, Serrien, & Vandendriesche, 1992; Todor & Lazarus, 1986; Walter & Swinnen 1990). Second, the transient effect at short cueing intervals and the late effect at long cueing intervals can be dissociated when the difference between target amplitudes is reduced. As shown in Figure 9-1 (left column), the asymmetric late effect disappears when target amplitudes become similar (cf. Sherwood, 1994b; 1999). However, the (almost symmetric) transient effect does not disappear. It becomes smaller in absolute terms, but not in relative terms.

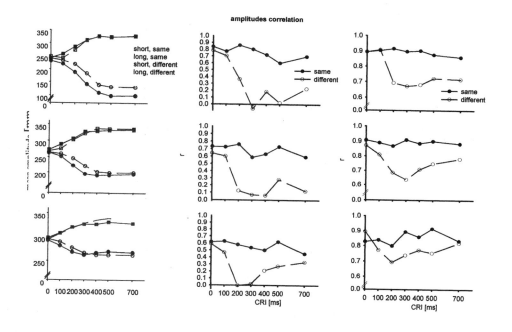

Figure 9-1. Results for rapid reversal movements obtained with the timed-response procedure. Target amplitudes were 30 and 5 cm (upper row of graphs), 30 and 15 cm (middle row), and 30 and 25 cm (bottom row). In the left column mean amplitudes as a function of the cue-response interval (CRI) are shown, in the middle column intermanual amplitude correlations, and in the right column intermanual movement-time correlations. (after Heuer, Kleinsorge, Spijkers, & Steglich, 2000)

The time-course of amplitudes as a function of preparation time is consistent with the hypothesis of a transient parametric coupling, which essentially means an initial coupling which is followed by a de-coupling as preparation time increases. However, the analysis of the mean amplitudes does not allow one to decide whether the de-coupling is mandatory or strategic. A mandatory de-coupling during motor programming should be observed both with different and with same target amplitudes. In contrast, a strategic de-coupling should be observed only when it is functional, that is, when target amplitudes are different. In fact, with same target amplitudes a persistent coupling would be functional. A measure which can be used to assess intermanual coupling separately for conditions with same and different target amplitudes is the intermanual correlation, in particular the correlation computed across series of trials between left-hand and right-hand amplitudes.

Figure 9-1 (middle column) shows the time course of intermanual amplitude correlations: with same target amplitudes they remain high during motor preparation, but with different target amplitudes they decline from an initial high value. With long preparation time the correlation is thus higher with same target amplitudes than with different target amplitudes (cf. Sherwood, 1991). Thus, the de-coupling is restricted to those conditions in which it is functional; it is strategic, but not mandatory. Though we generally assume that the initial coupling even with different target amplitudes is a kind of default, it should be noted that in the timed-response experiments it might also have resulted from the procedure because participants were instructed to prepare for same-amplitude movements and to produce them at the short precueing intervals when there is no chance for amplitude adjustments. Figure 9-1 (middle column) also shows a phenomenon which we have seen repeatedly, namely an increase of the intermanual correlations at the longest cue-response intervals. Perhaps the de-coupling is relaxed as the amplitude specifications have reached a final steady state, and a new coupling emerges even though the amplitudes are different (cf. Heuer, Kleinsorge, Spijkers, & Steglich, 2000).

Rinkenauer, Ulrich, and Wing (2001) found for isometric contractions with different instructed peak forces not only smaller intermanual correlations between peak forces, but also between rise times, even though the mean rise times were not different in those conditions in which the mean peak forces were different. Thus, de-coupling which is induced by a difference with respect to a certain movement characteristic generalizes to other movement characteristics. Figure 9-1 (right column) shows the intermanual correlations between movement times (times to reversal). These are generally higher than intermanual amplitude correlations, but they also decline as the cueing interval becomes longer, provided that different target amplitudes are cued.

4.1.2   Peak force of isometric contractions

Steglich, Heuer, Spijkers, & Kleinsorge (1999) used the timed-response paradigm to investigate the time course of preparation of bimanually performed isometric contractions. The specific goal of the study was to examine whether the hypothesis of transient crosstalk during the parametrization process holds for isometric contractions as well. Participants had to generate weak (20% of maximal voluntary force, MVF) or strong (40% MVF) bimanual isometric forces at the fourth synchronization tone.

Target forces were indicated by visual cues, which were short and long vertical bars presented on a computer monitor. Contractions were produced bimanually with the abducted thumbs while grasping two handles of a specially constructed apparatus (cf. Fig. 9-2b).

Figure 9-2a shows the peak forces; the points represent the observed data, the lines a simple formal description fitted to them. Consistent with earlier studies of isometric-force production, a gradual specification of peak forces with increasing preparation time was observed (e.g., Hening, Favilla, & Ghez, 1988; Ghez, Hening, & Favilla, 1990). These gradual specifications were different when same and different target forces were cued. In particular, corresponding to the observations made with reversal movements of same and different amplitudes, at short cueing intervals there was a peak-force assimilation which disappeared at longer cueing intervals. In contrast to the observations made with reversal movements, there was no asymmetric assimilation at the long CRI. For the intermanual correlations of peak forces again the decline in the course of peak-force specification was found only when different peak forces were cued. When same peak forces were cued, the correlation remained high even at long cueing intervals.

4.1.2.1 Movement direction

Movement direction is a parameter which is known to have well-defined neurophysiological representations in different areas of the cortex (e.g., Georgopoulos, 1991). Nevertheless, intermanual crosstalk effects in movements with different directions have received only limited attention, though recently for repetitive movements crosstalk effects have been clearly shown (Swinnen, Dounskaia, Levin, & Duysens, 2001; Swinnen, Dounskaia, & Duysens, 2002; Swinnen, Puttemans, Vangheluwe, Wenderoth, Levin, & Dounskaia, 2003). Steglich (2002) studied the time-course of direction specifications of bimanual movements aimed at a target (in these experiments no reversal movements, but aiming movements were used). Movements were performed in symmetric or parallel directions. The initial hypothesis was that coupling would support symmetrical movements, but not parallel movements (cf. Swinnen, Jardin, Meulenbroek, Dounskaia, & Hofkens-van den Brandt, 1997).

Participants were required to concurrently perform bimanual aiming movements of 10 cm amplitude from the start positions to different target combinations. Initially target directions of 90° (straight ahead) were cued for both hands, which changed to one of the target combinations 60° - 60° (parallel movements to the right), 60° – 120° (symmetric inward movements), 120° – 60° (symmetric outward movements), and 120° – 120° (parallel movements to the left). The cueing intervals ranged from 0 to

a)

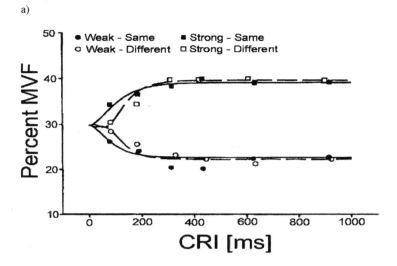

b)

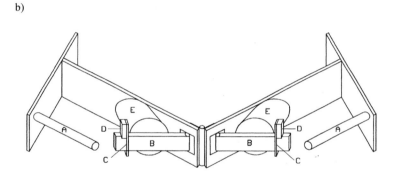

Figure 9-2. (a) Results for isometric contractions obtained with the timed-response procedure in two experiments. Circles and squares represent observed mean peak forces as a function of the cue-response interval (CRI), whereas the lines are from a simple formal representation of the data. Peak forces are expressed as percentages of maximal voluntary force (MVF). (Steglich, Heuer, Spijkers, & Kleinsorge. 1999, Exp. 1) (b) Sketch of the apparatus used for measurements of isometric thumb adduction. Handle (A) were grasped with both hands and thumbs were placed on the levers (B), contacting the steps (C). Positioning of the thumbs was monitored by means of photo electric wake-and-breaks (D). Pressure on the levers was propagated to load cells in the housings (E).

700 ms prior to the fourth (and last) synchronization tone; the cues were the target positions presented on a monitor (cf. Fig. 9-3b). The results fully conformed to expectations based on the hypothesis of transient crosstalk: When parallel directions were specified, the specification process exhibited an initial assimilation which diminished as preparation time increased. The assimilation was predominantly present when movements were directed to outward targets, a finding for which there seems to be no ready explanation.

Figure 9-3a shows the intermanual correlations observed by Steglich (2002). These are correlations between initial directions of the left and right hand. The initial directions were determined from linear regressions fitted to the initial 2 cm of the movement paths. Negative correlations indicate a symmetric coupling; if the one movement is more in a leftward direction, the other hand moves more in a rightward direction. Positive correlations indicate a parallel coupling; if the one movement is more in a leftward direction, the other hand moves more in a leftward direction, too. At short preparation intervals the intermanual correlations were negative. Note that initially participants had to prepare for movements in the 90° direction which are neither symmetric nor parallel (or both symmetric and parallel). With symmetric movements the intermanual correlation stayed negative, but with parallel movements it turned into the positive range. With the longest preparation intervals the intermanual correlations were positive or negative, depending on whether parallel or symmetric target directions were cued. Nevertheless there was a bias toward symmetric coupling as indicated by the initial negative correlations and the final difference between absolute values, negative correlations being (absolutely) larger than positive correlations. As we will discuss below, these findings seem to represent a challenge of our original notion of transient coupling during motor programming.

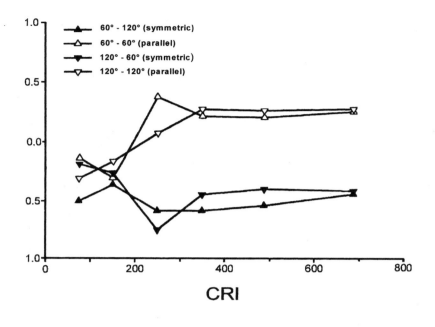

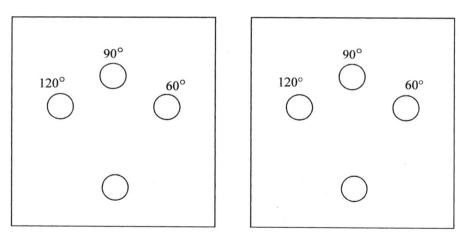

Figure 9-3. (a) Intermanual correlations between initial directions obtained with the timed-response procedure. Initially straight-ahead movements (90°) were prepared; with variable cueing interval target directions 60° - 60°, 60° - 120°, 120° - 60°, or 120° - 120° were cued. (after Steglich, 2002) (b) Sketch of the display with target directions being labeled. In the experiment directions were cued by presenting one of the three targets for each hand.

4.2   Evidence from bimanual reaction-time task

4.2.1   Movement amplitude

In a bimanual reaction-time task, the responses should be initiated as soon as the response-specification processes have more or less reached their final steady states. If this is correct, the hypothesis of transient crosstalk at the programming level predicts a longer reaction time for different than for same responses. This difference should vanish when precues are presented prior to the imperative signal and when sufficient preparation time is available, so that the response specifications can reach their final steady states before the start of the reaction time interval. We tested this hypothesis in a bimanual reaction-time task with same and different amplitudes of reversal movements (Spijkers, Heuer, Kleinsorge, & van der Loo, 1997). Amplitudes were short (10 cm) or long (20 cm).

In Figure 9-4 the results of two experiments are shown with words as precues (left graph) and short and long horizontal bars as precues (right graph) which served to specify the target amplitudes of the concurrent left-hand and right-hand reversal movements. Precues were presented on a monitor; imperative signals were tones of 1333 Hz and 100 ms duration. Reaction time for bimanual movements with different amplitudes was longer than for bimanual movements with same amplitudes. This was true for short precueing intervals. At longer precueing intervals, when the amplitudes could be specified more and more in advance of the imperative signal, the reaction-time difference between conditions with same and different target amplitudes grew smaller. This is the result expected on the basis of the transient-coupling hypothesis. At the longest precueing interval, however, the reaction-time difference did not disappear. Most likely this is due to failures to process the precues in a certain proportion of trials (cf. de Jong, 2000).

A comparison of the two graphs in Figure 9-4 makes the role of the type of precues evident both in terms of overall reaction times and in terms of the differences between conditions with same and different target amplitudes. Thus, there is an obvious alternative to the interpretation of the reaction-time difference at short precueing intervals in terms of transient crosstalk: processing different precues for the two hands could take longer than processing the same precues. To exclude this alternative, we used a control condition in all experiments with the bimanual reaction-time task. In the control condition unimanual responses were required, and the precue which specified the amplitude for the other hand in the bimanual condition

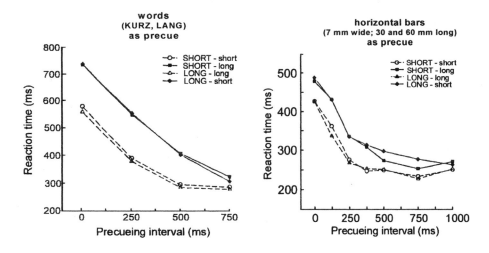

Figure 9-4. Mean reaction times for bimanual reversal movements as a function of the precueing interval. Precues were words (left graph) or short and long bars (right graph). Each curve gives reaction times for movements of a certain amplitude (SHORT, LONG) with a certain amplitude (short, long) of the movement of the other hand. (after Spijkers, Heuer, Kleinsorge, & van der Loo, 1997)

served as a go/no-go signal. Perceptually this control condition is identical to the bimanual condition, but not in terms of processing the precues. In general we also found reaction-time differences in the control condition when precueing intervals were short, and these were not always less than in the bimanual condition. Unfortunately the control condition involves other processes which could give rise to such an effect, so that it is not really satisfactory.

Recently the doubts about the correct interpretation of the findings shown in Figure 9-4 and corresponding findings in other experiments have been nourished by Diedrichsen, Hazeltine, Kennerley, and Ivry (2001) and Hazeltine, Diedrichsen, Kennerley, and Ivry (2003). Basically they replicated the reaction-time difference at short (or zero) precueing intervals with a certain type of cue, but they failed to observe the difference when direct cues were used, that is, when the targets themselves serve as cues. These findings seem to represent a second challenge not only to the usefulness of one or the other of the paradigms we used, but also to the two-level model which embraces an execution level and a programming level only. We shall discuss this issue below.

4.2.2   Peak force of isometric contractions

Using the same reaction-time procedure as for reversal movements with same and different amplitudes, we examined the hypothesis of transient coupling (crosstalk) during the specification of peak forces of isometric contractions produced by the thumbs of the left and right hand (the recording device is shown in Fig. 9-2b). Participants had to produce bimanual contractions with same and different target forces as rapidly as possible in response to an auditory signal. Weak target forces were defined as 20% MVF and strong target forces as 40% MVF. The target forces were cued visually by short and long vertical bars at variable precueing intervals (0 to 750 ms). Consistent with the hypothesis of transient crosstalk between concurrent processes of peak-force specifications, reaction times at short precueing intervals were longer when different peak forces had to be specified than when same peak forces were cued, and this reaction-time difference declined as the precueing interval was increased (Heuer, Spijkers, Steglich, & Kleinsorge, 2002).

4.2.3   Movement direction

The third movement variable that was chosen to examine the hypothesis of transient crosstalk in a bimanual reaction-time task was movement direction (Steglich, 2002). The task consisted of bimanual rapid aiming movements from their respective starting positions at the base of a circle segment for each hand to target positions on the circumference of the circle. Target positions were at 15° and 75° for the right hand, and at 165° and 105° for the left hand. An arrow pointing to the target position served as precue, and the imperative stimulus was a tone. The four target combinations resulted in two symmetric movements: 165° – 15° and 105° – 75°, and two directionally incompatible combinations, for which the directions were neither symmetric nor parallel: 165° – 75° and 105° - 15°. The bimanual reaction times followed exactly the prediction of the transient crosstalk hypothesis: at short precueing intervals (0 and 250ms) it took more time to initiate a response when the targets were in incompatible directions than when movements were symmetric, and this difference decreased as the cueing interval increased.

4.3   Evidence from sequential bimanual task

There is evidence that sequences of movements are not fully pre-programmed, but that some on-line programming occurs (e.g., Garcia-Colera

& Semjen, 1988; Piek, Glencross, Barrett, & Love, 1993), in particular when there is a change of movement characteristics during the sequence. The production of sequences of bimanual responses allows one to test the hypothesis of transient crosstalk during a continuous on-line programming task. With this experimental paradigm participants produce a bimanual sequence of reversal movements of which the tempo is varied across sequences. In one set of conditions, the participants produce constant-amplitude movements with one hand and alternating short- and long-amplitude movements with the other. The main finding is that the voluntary amplitude modulation of the alternating sequence induces a contralateral involuntary amplitude modulation, the size of which increases with tempo.

Figure 9-5 illustrates this main result. In the experiment (Heuer, Spijkers, Kleinsorge, & Steglich, 2000) movements were recorded in the continuation period after an initial period with pacing signal. The voluntary amplitude was changed only every second reversal movement, and the mean amplitudes are shown by the open circles. The filled circles show the amplitudes of the hand for which a constant short or constant long amplitude was instructed. The involuntary amplitude modulation, which accompanies the voluntary amplitude change of the other hand (open circles), is clearly visible from the differences between the first two and the second two amplitudes (filled circles) of the sets of four successive target amplitudes. It is also evident that the size of the involuntary modulation grows smaller as the cycle duration becomes longer. Another feature of the data which we have seen repeatedly is that short-amplitude movements in the constant-amplitude hand are shorter than concurrent short-amplitude movements of the modulated-amplitude hand; mostly long-amplitude movements in the constant-amplitude hand were also longer than concurrent long-amplitude movements in the modulated-amplitude hand, but in this particular experiment this difference is masked by overall shorter amplitudes of the constant-amplitude hand.

The findings shown in Figure 9-5 do not allow one to distinguish between two sources of the involuntary amplitude modulation. First, it could be associated with the amplitude of the concurrent movement of the other hand. Second, it could be associated with the immediately preceding change of the amplitude of the other hand. From the data shown in Figure 9-5 it seems that the amplitude of the concurrent movement is critical and not the amplitude change. The reason is that the involuntary amplitude modulation is seen in two successive movements, although only the first of them is preceded by an amplitude change of the other hand. However, there is other evidence which suggests that the intermanual crosstalk of amplitude specifications induces a new setting of the amplitude parameter which is likely to persist at least for a second movement in a row.

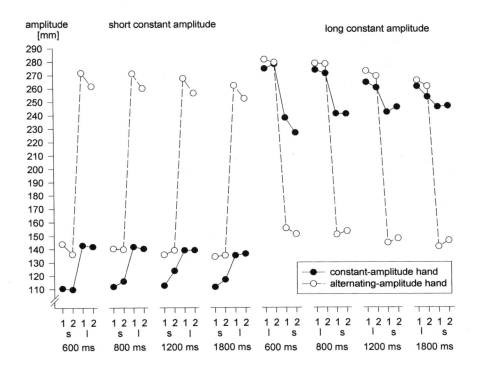

Figure 9-5. Mean amplitudes of rapid bimanual reversal movements produced at different rates (cycle durations of 600, 800, 1200, and 1800 ms). Filled circles show the amplitudes of the hand for which constant short or long amplitudes were instructed. Open circles show the amplitudes of the hand for which alternating runs of two short and two long amplitudes were instructed. Sequences of four successive movements are shown with short (s) and long (l) amplitudes of the hand with voluntary amplitude modulation. (after Heuer, Spijkers, Kleinsorge, & Steglich, 2000)

In a previous study (Spijkers & Heuer, 1995) we had used control conditions with constant amplitudes assigned to the two hands. In that experiment, the participants produced smooth oscillatory movements with different instructed frequencies. Figure 9-6 shows short and long amplitudes as a function of the concurrent amplitude of the other hand and the preceding one. Data with same successive amplitudes of the other hand are from conditions with constant amplitudes for both hands, whereas data with different successive amplitudes of the other hand are from conditions with constant amplitudes for the one hand, but alternating short and long amplitudes for the other hand. It is evident that a long amplitude was longer with a long amplitude of the other hand than with a short one; the long amplitude was even longer, when the long amplitude of the other hand had been preceded by a short amplitude. For the short amplitude again the influence of the amplitude of the other hand was enhanced when the preceding amplitude was different. Thus, in addition to the effect of the concurrent amplitude of the other hand, there is an effect of the amplitude change, that is, of the re-programming of the amplitude. This finding is consistent with the assumption of intermanual crosstalk related to the specification of a new amplitude.

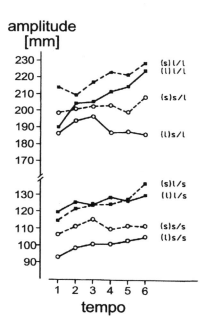

Figure 9-6. Mean amplitudes of movements with instructed short (s) and long (l) constant amplitudes as a function of the instructed tempo. Each curve is for a certain combination of the concurrent and the preceding (in brackets) amplitude of the other hand. Beginning with a comfortable rate (1), tempo was varied by instructions up to maximal tempo (6). (after Spijkers & Heuer, 1995)

In a later study, Heuer, Spijkers, Kleinsorge, and van der Loo (1998) replicated the former basic result of involuntary amplitude modulation when concurrently the other hand performs alternating short- and long-amplitude movements in a continuation paradigm. More important, this study showed that not only physical production, but also the imagination of alternating-amplitude movements with the one hand resulted in period-duration-dependent cross-manual effects, though reduced in size. After a phase of physical performance consisting of a bimanual movement sequence of 10 cycles an imagery phase with a variable length (9 – 12 cycles) followed, which was again followed by a physical-performance phase. At each cycle the pacing tone was presented together with visually presented horizontal bars of either constant or alternating short- and long-lengths in order to provide support for the imagery. The presence of cross-manual involuntary amplitude modulation, when the amplitude-modulated sequence of the other hand is imagined instead of physically produced, can be taken as straightforward evidence of crosstalk at the programming level. At least the effect of imaginary movements cannot be attributed to crosstalk at the execution level.

4.4 Evidence from bimanual sequential-movement task

4.4.1 Movement amplitude

The hypothesis of transient crosstalk for amplitude specification was investigated in two experiments that applied the bimanual sequential-movement procedure (Spijkers, Heuer, Kleinsorge, & Steglich, 2000). Participants performed rapid reversal movements of same or different amplitudes with the left and right hand. The short and long target amplitudes were 10 and 20 cm, respectively. The temporal overlap of the processes of specifying left-hand and right-hand amplitudes was manipulated by varying the delay between the first and the second signal in the first experiment, either relative to presentation of the first signal (stimulus-triggered delays of 100, 200, and 300 ms) or relative to the start of the first response (response-triggered delays of 10, 100, and 500 ms). Conforming to the predictions based on the hypothesis of transient coupling, the main findings were, first, the lengthening of the RT of the second response when its amplitude differed from that of the first response, and second, the disappearance of this lengthening as the delay of the second imperative signal was increased.

In the second experiment the time for advance specification of the amplitude of the first response was manipulated by precueing its amplitude. Precues were again short and long horizontal bars presented on a monitor. Precueing intervals were 0, 125, 250, 375, and 500 ms, and the second imperative signal was delayed relative to the first signal by 100, 300, and 750 ms. The main results are shown in Figure 9-7. As in the first experiment, there was a longer RT of the second response when its amplitude differed from the amplitude of the first response, and this increase disappeared when the delay of the second signal was increased. In addition, the data of

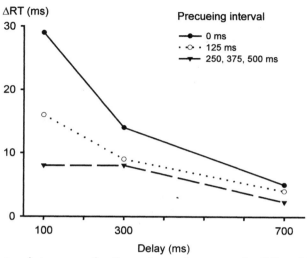

Figure 9-7. Differences between reaction times of second responses that followed first responses with different and same amplitudes in a bimanual sequential-movement task. The reaction-time increase in conditions with different amplitudes is shown as a function of the delay between imperative signals, separately for precueing intervals of the amplitude of the first response of 0 ms, 125 ms, and 250 ms and longer (means of data for precueing intervals of 250, 375, and 500 ms). (after Spijkers, Heuer, Kleinsorge, & Steglich, 2000)

Figure 9-7 show that the lengthening of the second RT in different-amplitude trials also decreased when the precueing interval became longer. With increasing precueing interval the specification of the first movement's amplitude was progressively advanced; finally it overlapped no longer with the specification of the second movement's amplitude. The findings of both experiments add to the evidence for transient crosstalk between concurrent processes of amplitude specification for the two hands. In addition they provide a straightforward account of the observation that the staggering of movements onsets is used as a strategy to reduce intermanual crosstalk effects (Swinnen, Walter, & Shapiro, 1988).

4.4.2   Peak force of isometric contractions

In order to examine the generality of the results obtained with movements with same and different amplitudes, Heuer, Spijkers, Steglich, and Kleinsorge (2002) applied the bimanual sequential-movement task to isometric contractions. Target forces for the first response were precued with variable cueing intervals (ranging from 0 to 500 ms in steps of 125 ms), while for the second response the cues were presented simultaneously with the second imperative signal. Again vertical short and long bars, presented on a monitior, were used as (pre-)cues. The second imperative signal was delayed by 100, 200, or 750 ms relative to the first one. Reaction time of the second response was longer when target forces for the two successive responses were different rather than same, and this reaction-time difference declined when the delay of the second imperative signal was increased as well as when the precueing interval for the first response became longer. These findings are essentially the same as found in the bimanual sequential-movement task with rapid reversal movements. They are consistent with the hypothesis of a transient crosstalk between concurrent processes of peak-force specifications.

4.5   Evidence from unimanual-movement task

At first glance it may appear strange to study intermanual interactions by way of unimanual movements. Nevertheless, the unimanual-movement task has the advantage that it quite obviously avoids effects which could accrue from crosstalk at the execution level. This motivated us to use the task in two experiments with reversal movements with same and different amplitudes. In these experiments the amount of advance specification of the amplitudes was manipulated by varying the precue intervals (Heuer, Kleinsorge, Spijkers, & Steglich, subm.). Short (10 cm) and long (20 cm) amplitudes of reversal movements of the left and right hand were instructed by means of presenting the words KURZ and LANG on a monitor. Precueing intervals ranged between 0 – 1000 ms (Exp. 1) and between 125 - 750 ms (Exp. 2). The task was a unimanual choice task; the choice signal (visual and auditory in the two experiments) indicated whether a left-hand or right-hand movement had to be produced. Reaction time was faster when same amplitudes were precued than when different amplitudes were precued. However, this difference did not disappear at the long precueing intervals; if anything, it became larger as the precueing interval increased.

Figure 9-8 shows the findings of one of the two experiments. Although reaction time (Fig. 9-8a) is longer when different than when same amplitudes are precued both at short (early effect) and at long (late effect) precueing intervals, these reaction-time differences are likely to be of different origins. The early effect is due to differences in the level of advance specification, but the late effect is due to a change of the process of choosing between hands. Corresponding to this account, the reaction time differences at short and long precueing intervals are associated with amplitude assimilation and a difference in choice-error frequency, respectively. We shall consider both origins of the reaction-time differences in turn.

Transient parametric coupling should result in a slowing of the concurrent specifications of different movement amplitudes as compared with the concurrent specifications of same amplitudes. Thus, with different amplitudes advance specification should be less advanced than with same amplitudes at the time the choice signal is presented, provided the precueing interval is sufficiently short (but, of course, longer than zero). As a consequence, completion of amplitude specification after presentation of the choice signal for the selected hand should take longer. As shown in Figure 9-1, the amplitude specifications are continuous. In addition, trials with same and different precued amplitudes were arranged in a random sequence within each block of trials. Thus, reaction time in trials with same target amplitudes might become somewhat longer and reaction time in trials with different amplitudes somewhat shorter than actually required to complete amplitude specifications. The consequence of the reduced reaction-time difference is the amplitude assimilation shown in Figure 9-8b, which is present only at short precueing intervals.

The reaction-time difference at long precueing intervals we attribute to a change of the process of choosing between hands. One way to think of the process of choosing between hands is in terms of a coupled-accumulator model (Heuer, 1987) or competing-accumulator model (Usher & McClelland, 2001). As illustrated in Figure 9-1, the de-coupling of concurrent processes of movement specifications can generalize to specifications of other movement characteristics which are not instructed to be different. There is limited evidence that it can even generalize to concurrent processes of movement initiation (Heuer, Spijkers, Steglich, & Kleinsorge, 2002). The gradual decline of crosstalk during movement specification could result from increasing gains of forward inhibition of crosstalk lines (cf. Heuer, 1993), and this change could generalize, no matter what the initial setting of the gain is. By this generalization inhibitory crosstalk between central representations of the choice alternatives could be strengthened. Simulations show that as a consequence reaction times become longer and choice errors less frequent. This is likely to be the origin of the late effect, the difference between conditions with same and different precued amplitudes at long precueing intervals. Corresponding to this account trials with incorrect choice of the hand were more frequent when same amplitudes were precued than when different amplitudes were precued, provided the precueing interval was sufficiently long.

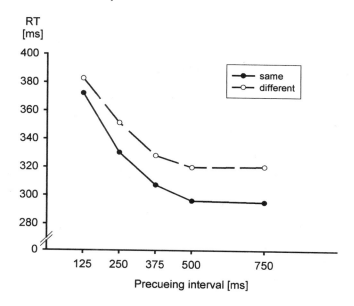

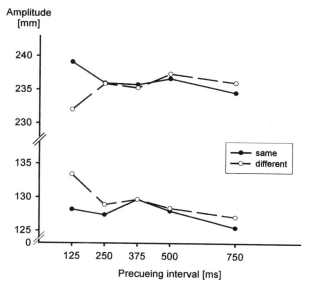

Figure 9-8. (a) Mean reaction times in the unimanual-movement task. (b) Mean amplitudes; at the **shortest** precueing interval there is an amplitude assimilation. (after Heuer, Kleinsorge, Spijkers, & Steglich, submitted)

5.   STATIC (STEADY-STATE) CROSSTALK AND OTHER ORIGINS OF LATE EFFECTS

Transient crosstalk effects at the programming level should no longer be visible when sufficient time is available for the preparation of bimanual movements. However, with long preparation intervals late effects have been observed as well. Here we discuss four such effects, which, as we believe, have different origins. Only two of them seem to be static crosstalk effects.

Late effects can be indirect. An example is the late reaction-time effect discussed in section 3.5. Here, the de-coupling during movement specification, which is an integral part of the notion of transient coupling, generalizes and affects other processes, in particular the process of choosing between responses with the left or right hand.

A second kind of late effect has been shown in Figure 9-1 (bottom graph of left column): With different amplitudes of reversal movements, short and long amplitudes at long preparation times can differ more from each other than short and long amplitudes in same-amplitude trials. The same has been observed for peak forces of isometric contractions (Steglich, Heuer, Spijkers, & Kleinsorge, 1999, Exp. 2). Most likely this contrast effect is strategic. The strategy is chosen because it results in different amplitudes (or peak forces) even when preparation time is short.

In section 3.1.1 we have contrasted the early rather symmetric amplitude assimilation with the well-known asymmetric amplitude assimilation found at long preparation intervals: the latter, but not the former, disappears when the difference between target amplitudes assigned to the left and right hand is reduced (Heuer, Kleinsorge, Spijkers, & Steglich, 2001). There is good reason to believe that the asymmetric amplitude assimilation arises from execution-level crosstalk, provided the difference between target amplitudes is sufficiently large. Overflow of the stronger muscle activation, which is associated with the long amplitude, should be stronger than overflow of the weaker muscle activation, which is associated with the short amplitude. Thus, the assimilation should be asymmetric, as it is. This, then, seems to be a real static crosstalk effect which originates at the execution level.

A fourth kind of late effect seems to be due to a non-negligible static crosstalk at the programming level: in general there is a strong temporal assimilation, which we have found also in the production of reversal movements with different amplitudes (Spijkers, Heuer, Kleinsorge, & an der Loo, 1997). Similarly, the intermanual correlation between movement durations is generally higher than the intermanual correlation between movement amplitudes (Schmidt, Zelaznik, Hawkins, Frank, & Quinn, 1979; Sherwood, 1991, 1994). In addition, at least for isometric contractions the intermanual correlation of rise times does not decline as much as the intermanual correlation of peak forces when different contractions are produced (Rinkenauer, Ulrich, & Wing, 2001). Finally, informal observations do strongly suggest that it is essentially impossible to specify (and thus to produce) bimanual movements with different spatio-temporal patterns concurrently. Perhaps this is possible only to the extent that the (apparently) two temporal patterns can be generated from a single timing structure, that is, as an integrated bimanual pattern (cf. Heuer, 1996, for review), or when different temporal movement patterns go along with same temporal patterns of visual feedback (cf. Mechsner, Kerzel, Knoblich, & Prinz, 2001).

6. CONCLUDING REMARKS: CHALLENGES AND FUTURE DEVELOPMENTS

The series of experiments reviewed in this chapter has been intended to examine the implications of a simple model in a variety of experimental paradigms. The core ingredient of the model which was put to test is the notion of transient coupling at the programming level. Of course, this notion has different implications for different paradigms, and not all implications have been confirmed. The reason is that performance in each single paradigm is also affected by other factors, the effects of which may mask or even override the effects of transient coupling. Notwithstanding the failures, the notion of transient crosstalk at the programming level has remained with us for years. There has only been one major change of the concept: it became clear that the de-coupling during motor programming is not mandatory, but limited to tasks where it is functional. Recent observations suggest that this change must be more radical; in addition they suggest that the distinction of two levels of crosstalk might be insufficient. We shall address both these challenges in turn.

6.1 From "transient crosstalk" to "adaptive modulation of crosstalk"

Transient crosstalk implies a de-coupling of concurrent processes of movement specification during motor preparation. The de-coupling is adaptive because it allows the concurrent production of different movements. To the extent that it is incomplete, that is, to the extent that a static crosstalk remains, concurrent production of different movements becomes impossible. Recent observations by Steglich (2002), which have been confirmed by independent experiments, do strongly suggest that decoupling is just a particular instance for adaptive modulations of crosstalk during motor programming. Depending on whether movements were to be performed with symmetric or with parallel directions, Steglich found negative and positive intermanual correlations which evolved during motor preparation (cf. Fig. 9-3). Whereas there is already considerable evidence that coordination patterns are task-dependent, the data of Steglich show the development of the proper coupling during motor preparation.

The notion of adaptive crosstalk modulation during motor programming implies that the observable crosstalk effects depend on the state of motor programming. It suggests investigators must look not only for the softer or harder constraints on coordination, but also for the dynamics of change and for the conditions that support adaptive modulations. For example, does adaptive modulation depend only on the movements actually to be produced, or does is also depend on the cues that are used to instruct these movements?

6.2   From a "two-level model" to a "three-level model"

The distinction between an execution level and a programming level may not be sufficient to capture all phenomena of intermanual interactions (see also Chapter 10). In addition a higher level seems to be needed, which for lack of a better term we call "cognitive". Diedrichsen, Hazeltine, Kennerley, and Ivry (2001) as well as Hazeltine, Diedrichsen, Kennerley, and Ivry (2003) implicate such an additional level by their account of the findings with the bimanual reaction-time task and the bimanual sequential-movement task. They found the early effect described above only with symbolic cues for the movement targets, but not with direct cues. The effect was attributed to interference in processing different symbolic precues, and the absence of the effect to the absence of such interference with different direct cues. Similarly, Mechsner, Kerzel, Knoblich, and Prinz (2001) claim that major coordination phenomena are perceptual in nature. Crosstalk at the execution level and the programming level seems to play no role in such accounts, and all what matters is crosstalk at some cognitive level, at which perhaps movement goals are represented.

The three-level scheme may not really be well-defined, but it certainly serves to broaden the thinking on motor coordination. When bimanual movements are placed in the context of dual-task performance (cf. Heuer & Wing, 1984), it is evident that crosstalk is not restricted to output-related processes (see Heuer, 1996b, for an overview). However, the fact that interference does exist at a cognitive level or – more generally – that interference is not restricted to execution and programming levels, does not imply that all interference originates at that level. Instead, different types of crosstalk phenomena can arise at different levels of action control. An example has been reported by Klein and Heuer (1999) and Heuer and Klein (2001). They studied the interactions between head rotations or eccentric head positions and rotations of a steering device (handle-bar or steering wheel). They found (a) different kinematics of handle-bar rotations depending on left or right eccentric head position, (b) no effect of eccentric head positions on reaction times for leftward and rightward handle-bar rotations, but an effect of an imminent head rotation, specifically a shorter latency for handle-bar rotations in the same direction, and (c) a bias in the random generation of leftward and rightward handle-bar rotations in the direction of eccentric head positions. These effects have been attributed to execution-level crosstalk, programming-level crosstalk, and cognitive-level crosstalk (response selection in particular), respectively (Heuer & Klein, 2001).

A three-level model of multilimb actions not only broadens the field because it embraces additional types of crosstalk phenomena, but also because it suggests the exploration of interactions between levels. For example, in Section 3.5 we have reported data which strongly suggest that de-coupling at the programming level spreads to crosstalk at the cognitive level and thereby modifies the process of choosing between left-hand and right-hand responses. In fact, the observed data can easily be captured by a simple model with such between-level effects (Heuer, Kleinsorge, Spijkers, & Steglich, subm.). Franz, Zelaznik, Swinnen, and Walter (2001) have suggested what can be understood as between-level effect in the opposite direction: when the traces of bimanual movements can be perceived as a single entity (a circle in that case), cross-manual interference declines. Thus, crosstalk at lower levels may be modulated by the higher levels. Such modulation may also underly the observation of Mechsner, Kerzel, Knoblich, and Prinz (2001) that non-harmonic cycle durations of the two hands in periodic movements become possible when visual-feedback signals have same cycle durations.

Whereas a three-level model broadens the field, it also raises additional issues which may be hard to settle. A core issue is the operational distinction of the various levels. For example, the findings of Diedrichsen, Hazeltine, Kennerley and Ivry (2001) and Hazeltine, Diedrichsen, Kennerley, and Ivry (2003) (see Chapter 10) as well as those of Mechsner, Kerzel, Knoblich, and Prinz (2001) challenge previous accounts of various crosstalk phenomena which had been attributed to the programming level or the outflow level by way of ascribing them to the cognitive level. Settling such issues may become difficult not only for methodological reasons, but also for conceptual reasons. For example, the distinction between motor programming and response selection is not clear. The ambiguity is evident since the introduction of Rosenbaum's (1980) movement precueing technique and the subsequent discussion of the confounding of precued movement features and the number of choice alternatives (e.g., Zelaznik, Shapiro, & Carter, 1982). It might become necessary to settle this and other conceptual issues to answer questions on the origins of intermanual crosstalk. On the other hand, exploring the origins of intermanual crosstalk phenomena might also help to sharpen the conceptions of functional levels involved in motor control.

ACKNOWLEDGEMENTS

A large part of this chapter is based on experimental results obtained in a series of experiments which have been published previously. The research was supported by grants He 1187/9-1 and He 1187/9-2 of the Deutsche Forschungsgemeinschaft to H. Heuer and W. Spijkers. We would like to thank two anonymous reviewers for their valuable comments on an earlier version of this chapter.

REFERENCES

Bullock D, Grossberg S (1988) Neural dynamics of planned arm movements: Emergent invariants and speed-accuracy properties during trajectory formation. Psychol Rev 95:49-90

Corcos DM (1984) Two-handed movement control. Res Q Exerc Sport 55:117-122

De Jong R (2000). An intention-activation account of residual switch costs. In: Monsell S, Driver J (ed) Attention and performance XVIII: Control of cognitive processes, MIT Press, Cambridge, MA pp 357-376

Diedrichsen J, Hazeltine E, Kennerley S, Ivry RB (2001) Moving to directly cued locations abolishes spatial interference during bimanual actions. Psychol Sci 12:493-498

Durwen HF Herzog AG (1992) Electromyographic investigation of mirror movements in normal adults: Variation of frequency with site, effort, and repetition of movement. Brain Dysfunction, 5:310-318

Ellson DG (1959). Linear frequency theory as a behavior theory. In: Koch S (ed) Psychology: a study of a science. Vol. 2: General systematic formulations, learning, and special processes, McGraw-Hill, New York, pp 637 – 662

Favilla M, De Cecco E (1996) Parallel direction and extent specification of planar reaching arm movements in humans. Neuropsychology 34:609-613

Favilla M, Gordon J, Hening W, Ghez C (1990) Trajectory control in targeted force impulses. VII. Independent setting of amplitude and direction in response preparation. Exp Brain Res 79:530-538

Favilla M, Henin W, Ghez C (1989) Trajectory control in targeted force impulses. VI. Independent specification of response amplitude and direction. Exp Brain Res 75:280-294

Fowler B, Duck T, Mosher M, Mathieson B (1991) The coordination of bimanual aiming movements: Evidence for progressive desynchronization. Q J Exp Psychol A 43A:205-221

Franz EA, Zelaznik HN, McCabe G (1991) Spatial topological constraints in a bimanual task. Acta Psychol 77:137-151

Franz EA, Zelaznik HN, Swinnen SP, Walter CB (2001) Spatial conceptual influences on the coordination of bimanual actions: when a dual task becomes a single task. J Mot Behav 33:103-112

Fuchs A, Kelso JAS (1994) A theoretical note on models of interlimb coordination. J Exp Psychol Human Percept Perform 20:1088-1097

Garcia-Colera A, Semjen A (1988) Distributed planning of movement sequences. J Mot Behav 20:341-367

Georgopoulos AP (1991) Higher order motor control. Annu Rev Neurosci 14:361-377

Ghez C, Favilla M, Ghilardi MF, Gordon J, Bermejo R, Pullman S (1997) Discrete and continuous planning of hand movements and isometric force trajectories. Exp Brain Res 115:217-233

Ghez C, Hening W, Favilla M (1990) Parallel interacting channels in the initiation and specification of motor response features. In: Jeannerod M (ed) Attention and performance XIII, Erlbaum, Hillsdale, NJ, pp 265-293

Haken H, Kelso JAS, Bunz H (1985) A theoretical model of phase transitions in human hand movements. Biol Cybern 51:347-356

Hazeltine E, Diedrichsen J, Kennerley SW, Ivry RB (2003) Bimanual cross-talk during reaching movements is primarily related to response selection, not the specification of motor parameters. Psychol Res 67:56-70

Hening W, Favilla M, Ghez C (1988) Trajectory control in targeted force impulses. V. Gradual specification of response amplitude. Exp Brain Res 71:116-128

Heuer H (1987) Visual discrimination and response programming. Psychol Res 49:91-98

Heuer H (1993) Structural constraints on bimanual movements. Psychol Res 55:83-98

Heuer H (1995) Modelle motorischer Koordination. Psychologische Beiträge 37:396-452

Heuer H (1996a) Coordination. In: Heuer H, Keele SW (ed) Handbook of Perception and Action. Vol. 2: Motor skills, Academic Press, London, pp 121-180

Heuer H (1996b) Dual-task performance. In: Neumann O, Sanders AF (ed) Handbook of Perception and Action. Vol. 3: Attention, Academic Press, London, pp 113-153

Heuer H, Klein W (2001) Eccentric head positions bias random generation of leftward and rightward handle-bar rotations. Acta Psychol 106:23-49

Heuer H, Kleinsorge T, Spijkers W, Steglich C (2001) Static and phasic cross-talk effects in discrete bimanual reversal movements. J Mot Behav 33:67-85

Heuer H, Kleinsorge T, Spijkers W, Steglich C (submitted) Intermanual cross-talk effects in unimanual movements

Heuer H, Spijkers W, Kleinsorge T, Steglich C (2000) Parametrische Kopplung bei Folgen beidhändiger Umkehrbewegungen mit gleichen und unterschiedlichen Weiten. Zeitschrift für experimentelle Psychologie 47:34-49

Heuer H, Spijkers W, Kleinsorge T, van der Loo H (1998) Period duration of physical and imaginary movement sequences affects contralateral amplitude modulation. Q J Exp Psychol A 51A:755-779

Heuer H, Spijkers W, Kleinsorge T, van der Loo H, Steglich C (1998) The time course of cross-talk during the simultaneous specification of bimanual movement amplitudes. Exp Brain Res 118:381-392

Heuer H, Spijkers W, Steglich C, Kleinsorge T (2002) Parametric coupling and generalized decoupling revealed by concurrent and successive isometric contractions of distal muscles. Acta Psychol 111:205-242

Heuer H, Wing AM (1984) Doing two things at once: Process limitations and interactions. In: Smyth MM, Wing AM (ed) The psychology of human movement, Academic Press, London, pp 183-213

Kay BA, Kelso JAS, Saltzman E, Schöner G (1987) Space-time behavior of single and bimanual rhythmical movements: Data and limit cycle model. J Exp Psychol Human Percept Perform 13:178-192

Kelso JAS (1994) Elementary coordination dynamics. In: Swinnen SP, Heuer H, Massion J, Casaer P (ed) Interlimb coordination: Neural, dynamical, and cognitive constraints. Academic Press, San Diego pp 301-318

Kelso JAS, Southard DL, Goodman D (1979) On the coordination of two-handed movements. J Exp Psychol Human Percept Perform 5:229-238

Klein W, Heuer H (1999) The effects of eccentric head positions on leftward and rightward turns of a handle-bar. Acta Psychol 103:311-329

Lacquaniti F (1996) Control of movement in three-dimensional space. In: Lacquaniti F, Viviani P (ed) Neural bases of motor behaviour. Kluwer, Dordrecht, pp 1-40

Marteniuk RG, MacKenzie CL (1980) A preliminary theory of two-hand co-ordinated control. In: Stelmach GE, Requin J (ed), Tutorials in motor behavior, North-Holland, Amsterdam, pp 185-197

Marteniuk RG, MacKenzie CL, Baba DM (1984) Bimanual movement control: Information processing and interaction effects. Q J Exp Psychol A 36A:335-365

Mechsner F, Kerzel D, Knoblich G, Prinz W (2001) Perceptual basis of bimanual coordination. Nature 414:69-73

Piek JP, Glencross DJ, Barrett NC, Love GL (1993) The effect of temporal and force changes on the patterning of sequential movements. Psychol Res/ Psychologische Forschung 53:116-123

Preilowski B (1975) Bilateral motor interaction: Perceptual-motor performance of partial and complete "split-brain" patients. In: Zülch KJ, Creutzfeldt O, Galbraith GG (ed), Cerebral localization. Springer, Berlin, pp 115-132

Rinkenauer G, Ulrich R, Wing AM (2001) Brief bimanual force pulses: Correlations between the hands in force and time. J Exp Psychol Human Percept Perform 27:1485-1497

Rosenbaum DA (1980) Human movement initiation: Specification of arm, direction, and extent. J Exp Psychol Gen 109:444-474

Schmidt RA ,Lee T (1999) Motor control and learning: A behavioral emphasis (3rd edition). Human Kinetics Publishers, Champaign, Ill

Schmidt RA, Zelaznik HN, Hawkins B, Frank JS, Quinn JT (1979) Motor-output variability: A theory for the accuracy of rapid motor acts. Psychol Rev 86:415-451

Schöner G, Kelso JAS (1988) A synergetic theory of environmentally-specified and learned patterns of movement coordination. II: Component oscillator dynamics. Biol Cybern 58:81-89

Schouten JF, Becker JAM (1967) Reaction time and accuracy. Acta Psychol 27:143-153

Sherwood DE (1990) Practice and assimilation effects in a multilimb aiming task. J Mot Behav 22:267-291

Sherwood DE (1991) Distance and location assimilation in rapid bimanual movement. Res Q Exerc Sport 62:302-308

Sherwood DE (1994a) Interlimb amplitude differences, spatial assimilations, and the temporal structure of rapid bimanual movements. Hum Mov Sci 13:841-860

Sherwood DE (1994b) Hand preference, practice order, and spatial assimilations in rapid bimanual movement. J Mot Behav 26:123-134

Sherwood DE, Nishimura KM (1992) EMG amplitude and spatial assimilation effects in rapid bimanual movement. Res Q Exerc Sport 63:284-291

Spijkers W, Heuer H (1995) Structural constraints on the performance of symmetrical bimanual movements with different amplitudes. Q J Exp Psychol A 48A:716-740

Spijkers W, Heuer H, Kleinsorge T, Steglich C (2000) The specification of movement amplitudes for the left and right hand: evidence for transient parametric coupling from overlapping-task performance. J Exp Psychol Human Percept Perform 26:1091-1105

Spijkers W, Heuer H, Kleinsorge T, van der Loo H (1997) Preparation of bimanual movements with same and different amplitudes: Specification interference as revealed by reaction time. Acta Psychol 96:207-227

Spijkers W, Tachmatzidis K, Debus G, Fischer M, Kausche I (1994) Temporal coordination of alternative and simultaneous aiming movements of constrained timing structure. Psychol Res 57:20-29

Steglich C (2002) Experimentelle Untersuchungen zur bimanuellen Koordination als transiente Kopplung bei der Richtungsspezifikation. Unpublished Dissertation, RWTH Aachen

Steglich C, Heuer H, Spijkers W, Kleinsorge T (1999) Bimanual coupling during the specification of isometric forces. Exp Brain Res 129:302-316

Swinnen SP, Dounskaia N, Duysens J (2002) Patterns of bimanual interference reveal movement encoding within a radial egocentric reference frame. J Cogn Neurosci 14:463-471

Swinnen SP, Dounskaia N, Levin O, Duysens J (2001) Constraints during bimanual coordination: the role of direction in relation to amplitude and force requirements. Behav Brain Res 123:201-218

Swinnen SP, Jardin K, Meulenbroek R, Dounskaia N, Hofkens-van den Brandt M (1997) Egocentric and allocentric constraints in the expression of patterns of interlimb coordination. J Cogn Neurosci 9:348-377

Swinnen SP, Puttemans V, Vangheluwe S, Wenderoth N., Levin O, Dounskaia N (2003) Directional interference during bimanual coordination: is interlimb coupling mediated by afferent or efferent processes. Behav Brain Res 139:177-195

Swinnen SP, Walter CB, Serrien DJ, Vandendriessche C (1992) The effect of movement speed on upper-limb coupling strength. Hum Mov Sci 11:615-636

Swinnen SP, Walter CB, Shapiro DC (1988) The coordination of limb movements with different kinematic patterns. Brain Cogn 8:326-347

Todor JI, Lazarus JC (1986) Exertion level and the intensity of associated movements. Dev Med Child Neurol 28:205-212

Usher M, McClelland JL (2001) The time course of perceptual choice: The leaky, competing accumulator model. Psychol Rev 108:550-592

Van der Meulen JHP, Gooskens RHJM, Denier van der Gon JJ, Gielen CCAM, Wilhelm K (1990) Mechanisms underlying accuracy in fast goal- directed arm movements in man. J Mot Behav 22:67-84

Walter CB, Swinnen SP (1990) Kinetic attraction during bimanual coordination. J Mot Behav 22:451-473

Walter CB, Swinnen SP, Dounskaia NV (2002) Generation of bimanual trajectories of disparate eccentricity: Levels of interference and spontaneous changes over practice. J Mot Behav 34:183-195

Zelaznik HN, Shapiro DC, Carter MC (1982) The specification of digit and duration during motor programming: a new method of precueing. J Mot Behav 14:57-68

Chapter 10

A COGNITIVE NEUROSCIENCE PERSPECTIVE ON BIMANUAL COORDINATION AND INTERFERENCE

Richard Ivry[1], Jörn Diedrichsen[1], Rebecca Spencer[1], Eliot Hazeltine[2], Andras Semjen[3]

[1]Department of Psychology and Helen Wills Neuroscience Institute, University of California, Berkeley,USA, [2]Department of Psychology, University of Iowa, USA , [3]CNRS Marseille, France

Abstract: We argue that bimanual coordination and interference depends critically on how these actions are represented on a cognitive level. We first review the literature on spatial interactions, focusing on the difference between movements directed at visual targets and movements cued symbolically. Interactions manifest during response planning are limited to the latter condition. These results suggest that interactions in the formation of the trajectories of the two hands are associated with processes involved in response selection, rather than interactions in the motor system. Neuropsychological studies involving callosotomy patients argue that these interactions arise from transcallosal interactions between cortically-based spatial codes. The second half of the chapter examines temporal constraints observed in bimanual movements. We propose that most bimanual movements are marked by a common event structure, an explicit representation that ensures temporal coordination of the movements. The translation of an abstract event structure into a movement with a particular timing pattern is associated with cerebellar function, although the resulting temporal coupling during bimanual movements may be due to the operation of other subcortical mechanisms. For rhythmic movements that do not entail an event structure, timing may be an emergent property. Under such conditions, both spatial and temporal coupling can be absent. The emphasis on abstract levels of constraint makes clear that limitations in bimanual coordination overlap to a considerable degree with those observed in other domains of cognition.

Key words: bimanual coordination, spatial coupling, temporal coupling, response selection, event timing, callosotomy, cerebellum, neuropsychology

1.   INTRODUCTION

A cardinal feature of human behavior is the generative capacity we have for using our upper limbs in the production of voluntary actions. With practice, we master the most complex skills-- the elegant scripts of the calligrapher, the lightning quick movements of the concert pianist, the life saving maneuvers of the heart surgeon. Even those of us who claim to be "all thumbs" are vastly superior to all other species in our ability to produce purposeful, manipulative actions.

We frequently speak of a person as being either left- or right-handed, implying that the dominant hand is more skilled than the non-dominant. Yet casual observation convincingly demonstrates that most actions are bimanual: Typing, using a fork and knife, and buttoning a shirt all require the integrated actions of the two hands. Thus, handedness may be more concisely thought of as describing the typical role-assignment of the hands (Guiard 1987). Of the skills that are included in assessments of handedness, many have a bimanual component (Oldfield 1971). For example, writing or cutting paper with scissors are essentially bimanual actions, with the non-dominant hand serving an essential support, or postural role. In sum, evolution is exploitive: Bipedalism has liberated our upper extremities and we take full advantage of this in our interactions with the world.

While the coordination of our two limbs in most tasks feels effortless and beneficial, much of the motor control research has focused on limitations, or constraints on coordination. By determining these constraints, fundamental principles governing the coordination of actions can be identified. This approach may also provide insight into the degrees of freedom problem articulated in the classic writings of Bernstein (see Whiting 1984), namely, how efficient control is achieved given the redundancy inherent in the motor system.

Limitations in the ability to coordinate bimanual movements have been widely studied with tasks requiring rhythmic, repetitive movements. One appeal of this approach is that tasks bear a similarity, at least superficially, to the most fundamental of multi-limb coordination tasks, locomotion. For example, it is assumed that our ability to produce rhythmic movements with the two upper limbs likely shares some of the constraints defining stable modes of locomotion, and indeed, may reflect the operation of similar neural mechanisms. The preference to move the two limbs in either an in-phase relationship (with the left and right arms moving in the same direction at the same time) or an anti-phase relationship (with the left and right arms moving in opposite directions at a given time) may stem from the fact that locomotion typically involves similar coordination modes. Given that these phase relationships are typically maintained between homologous effectors, researchers looking for the neural correlates of coordination have focused their attention on interactions along the motor neuroaxis (e.g., interneurons in the spinal cord or callosal connections between homologous cortical motor regions).

While recognizing the appeal of evolutionary arguments that attempt to establish common principles shared by locomotion and bimanual coordination, we believe that the two phenomena reflect fundamentally different forms of interactions between limbs (see also, Peters 1994; Semjen 2002). Whereas the motions of the limbs during locomotion are integrated to produce rhythmic, stereotyped movement patterns, the pattern of coordination in two-handed activities can be much more complex. In bimanual actions, the hands often perform distinct movements, whose relationship only becomes apparent when one considers the external goal of this action. For example, when tying shoelaces each hand follows a complex spatiotemporal pattern such that the movement onsets and trajectories for the two hands have no immediate symmetry relationship, but are nonetheless highly coordinated.

Recognition of this difference has led us to re-examine the constraints associated with bimanual coordination. Traditional studies of bimanual coordination have used tasks that mimic the rhythmicity and phase-relationships of locomotion. These approaches have led to a formulation of a set of constraints on bimanual movements arising from the interaction of the two movement patterns. In this chapter, we will review evidence demonstrating that many of these constraints have little to do with the motor system per se. Rather, they reflect limitations associated with processing at abstract, conceptual levels of the cognitive architecture. Our intent is not to discount the relevance of other sources of constraint. However, we believe that these more conceptual sources of constraint have been neglected in the literature on bimanual coordination. Appreciating the fact that many limitations of motor behavior reflect more general features of our cognitive architecture can also help us understand and explain our extraordinary ability to perform complex manual actions involving multiple effector systems.

2.   MODEL TASKS FOR EXPLORING SOURCES OF CONSTRAINT ON BIMANUAL MOVEMENTS

As noted above, the study of rhythmic, bimanual movements has been a very productive area of work in the motor control literature; indeed, one could say that the popularization of such tasks in the 1980's represented a true paradigm shift in the field (Kelso 1984). The rich data sets provided by such tasks were refreshing in contrast to the limited movement repertoire (button pressing and highly constrained movements) that characterized traditional studies of motor programming. By studying complex, repetitive actions, the researchers struck on an experimental procedure that was amenable to concepts and analytic tools emerging in other disciplines of the biological and physical sciences. Moreover, this approach held the promise of being applicable to more ecologically valid tasks that demanded the continuous coordination of the two hands for long periods of time.

Consider one variant of these tasks, coordination of wrist flexion and extension movements of both hands. The marked preference for certain phase relationships, the dependency of pattern stability on movement rate, and the asymmetry in the transitions between different stable states are characteristic of certain classes of dynamical systems. The behavior can be formally captured with a component model, in which the movement of each limb is represented by a non-linear oscillator, with its stability described in terms of limit cycle dynamics. Interactions between the limbs arise due to non-linear coupling terms that connect the dynamics of the oscillators (Haken et al. 1985).

While this component model provides an elegant account for the emergent properties of the dynamical system, its formulation is, in essence, abstract. On a representational level it remains unclear whether the component oscillators refer to the position and velocity of a limb, the contraction of agonist and antagonist muscles or more abstract spatial codes. In a similar vein, the theory remains neutral in terms of the neural implementation of its components.

This work emphasized the prominent temporal constraints associated with bimanual movements. People have great difficulty producing movements in which the limbs are not moving at identical or integer ratio frequencies (1:1, 1:2). Even skilled musicians are subject to this constraint, limited in the manner in which they produce complex polyrhythms (e.g., Klapp et al. 1985; Krampe et al. 2000). While these powerful temporal constraints are most evident in repetitive movements, they can also be seen in simpler contexts. A hallmark of bimanual reaching movements is the tendency for the two arms to initiate and terminate at approximately the same point in time, even if the movements span different amplitudes (Kelso et al. 1979).

Other constraints in the production of bimanual movements can be observed on purely spatial measures. Consider the task of simultaneously drawing two shapes, either two lines or a circle with one hand and a line with the other (Fig. 10-1). Spatial assimilation effects are readily observed in the incongruent condition with each shape becoming more elliptical compared to when both hands produce lines or circles (Franz et al. 1991). Similar assimilation effects are observed when people attempt to draw lines of unequal amplitude (Heuer et al. 1998) or produce isometric forces of unequal intensity (Steglich et al. 1999). In all of these conditions, the interference between the two actions occurs on a background of tight temporal coupling between the hands (i.e., similar frequency and stable phase relationship for the repetitive movements).

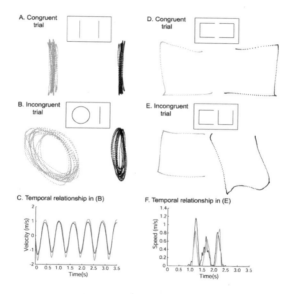

Figure 10-1. Spatial interference in two bimanual drawing tasks. A, B: In the line-circle task, participants are instructed to simultaneously draw two lines (congruent movements) or a circle and line (incongruent movements). The trajectories reveal greater variability in the incongruent condition. C: movement velocity along the y-direction for the left hand (gray) and right hand (black). D, E: In the three-sided figure task participants are instructed to simultaneously draw three-sided figures that are either congruent (symmetric) or incongruent (successive sides are orthogonal). Neurologically healthy participants exhibit interference in the incongruent condition. F: Temporal coupling is strong during the incongruent three-sided drawing task. The data are plotted as the tangential velocity of each hand over time. Adapted from Franz et al. 1991, 1996.

It has been hypothesized that these temporal and spatial interactions arise on separable levels of the control hierarchy (e.g., Heuer 1993). A study involving split-brain patients also suggests that these sources may be separable in terms of their neural implementation (Franz et al. 1996). These patients have undergone resection of the corpus callosum as part of a radical treatment for severe, chronic epilepsy. The procedure eliminates the primary pathway of communication between the two cerebral hemispheres. In this study, the patients and controls were required to simultaneously draw two three-sided boxes, one with each hand. The target shape for the left hand was presented in the left visual field while the target shape for the right hand was presented in the right visual field. The critical comparison was between conditions in which the orientation of the two shapes was either congruent or incongruent. For the congruent conditions the two shapes were mirror symmetric, for the incongruent condition, one shape was rotated by 90 degrees with respect to the other (Fig. 10-1d,e).

For the control participants, the incongruent condition was quite taxing. Compared to the congruent condition, reaction times and movement times were inflated, and spatial assimilation effects were frequently observed. In contrast, the split-brain patients performed similarly on the congruent and incongruent conditions. There was no evidence of any spatial interactions in the productions of the two hands. Interestingly, the patients' bimanual movements remained temporally coupled. Similar to the control participants, the patients initiated each of the three sub-movements in close synchrony. Thus, the patients exhibited persistent temporal coupling in the face of complete spatial uncoupling.

The lack of spatial interactions indicates that spatial crosstalk arises from interactions that involve the corpus callosum. Two neurological accounts have been offered to explain the persistent temporal coupling in the face of spatial uncoupling. First, a single hemisphere may control movement initiation for both hands (Stucchi and Viviani 1993). Alternatively, a subcortical mechanism with access to both effectors may gate the implementation of cortically-generated motor commands (Ivry and Hazeltine 1999; Ivry and Richardson 2002).

The study of split-brain patients by Franz et al. (1996) demonstrates that, at least under certain conditions, constraints associated with trajectory formation can be dissociated from those associated with temporal coordination. This should not be taken to imply that the spatial and temporal aspects of movement can always be decomposed; indeed, later we will argue that for certain types of movement, temporal constraints are an integral part of spatial constraints. However, the dissociation does make clear that constraints on bimanual coordination have multiple sources. In the following sections, we explore these constraints, focusing first on an account of the spatial interactions observed during bimanual movements and then returning to the issue of temporal coupling.

3. REPRESENTATIONAL BASIS OF SPATIAL CONSTRAINTS

Why do bimanual movements exhibit profound spatial interactions in normal participants? The performance of split-brain patients indicates that these interactions result from interhemispheric communication across the corpus callosum. But between which cortical areas and at which level of representation do these interactions occur? The neural locus was examined by Eliassen et al. (1999) who tested a patient on the three-sided figure task over the course of several months. During this period, the patient underwent two successive operations, the first involving resection of the anterior region of the corpus callosum and the second in which the remaining callosal fibers were cut. It was only after the second operation that the patient became spatially uncoupled. This led the authors to suggest that the critical spatial interactions are a reflection of communication between parietal regions, that is, between regions that play a role in the planning, rather than in the motor execution, of spatial trajectories (see also Serrien et al. 2001). Single cell recordings in primates indicate that neural coding of movement in the parietal cortex is best described in terms of spatial direction, rather than in terms of dynamical properties such as force (Kalaska et al. 1990). Thus, evidence from split-brain studies speaks against the possibility that interactions occur between regions associated with activation of homologous muscles.

This conclusion is further supported by studies that have tried to distinguish between symmetry defined in terms of muscular activation and symmetry defined in terms of movement direction. One of the most robust phenomena in rhythmic studies is that symmetric movement patterns are more stable than asymmetric patterns. For example, with the forearms pronated, wrist flexion/extension is more stable when the movements are symmetric. However, in this situation the symmetric pattern involves both symmetric movement directions and homologous muscle activation. If one hand is oriented with the palm facing down and the other hand with the palm facing up, these two factors can be dissociated. In this condition, performance is stable when the hands move up and down together, even though one wrist is flexing while the other is extending. A more compelling preference for common directional coding occurs when the effector combination involves an arm and a leg (Baldissera et al. 1982, 1991). Thus, crosstalk can occur at a level in which movement direction is represented rather than patterns of muscular activation (Swinnen et al. 2002; but see Riek et al. 1992).

3.1   Direct reaching

The observed bias towards movements that are symmetric with respect to the body axis seems counterintuitive when considering how we typically use our limbs. Consider someone clearing the dinner table after a meal, using the right hand to pick up a glass and the left hand to pick up a plate. The movements that bring the hands towards the objects are likely to be asymmetric as the objects are located in different directions and at different distances. Two different grasps have to be shaped and very different grip and lift-forces have to be applied to the objects. If the actions of each hand were subject to strong assimilation effects, we might expect to see that one or both objects would be missed, or the hand shapes would be inappropriately formed. However, we seem to be able to perform this task effortlessly.

These considerations led us to explore spatial interactions for bimanual movements under different movement cueing conditions (Diedrichsen et al. 2001). In these experiments, people were instructed to make two reaching movements on each trial, one with the left hand and one with the right hand. The movement amplitudes could be either short or long. Thus, the bimanual combination could be classified as congruent (i.e., both long or both short) or incongruent (i.e., one short and one long). The critical manipulation centered on the manner in which the movement directions were cued (Fig. 10-2). In the symbolic cueing condition, the four possible target locations (two end locations for each hand) were visible at all times, and the letters "S" and "L" were used to indicate the target locations. One letter was presented in the left visual field to indicate the left-hand movement and the other letter in the right visual field to indicate the right-hand movement. In the direct cueing condition, the target locations were cued by the onset of the target circles, one appearing on each side.

Dramatic differences were observed between the two cueing conditions. In the symbolic condition, congruent responses were initiated much faster than incongruent responses. This result is consistent with previous findings of a preference for symmetric bimanual movements. However, when the movements were directly cued, people were much faster to initiate their movements and, more importantly, there were no differences in reaction time. A similar dissociation was found for movements made in mirror-symmetric or orthogonal directions. On congruent trials, the required movements were either both in the lateral or both in the forward direction; on incongruent trials, the movement directions were orthogonal to each other. Again, reaction time costs were completely eliminated when the target directions for each hand were directly cued by the onset of stimuli at the two target locations. In addition, the initial direction of the movement was in the wrong direction on a significant percent of the trials in the symbolic, but not in the direct condition. The absence of any cost in the initiation of asymmetric movements in the direct condition is underscored by the fact that reaction times on the bimanual direct cueing conditions were similar to those observed in a control condition in which only unimanual reaches were performed.

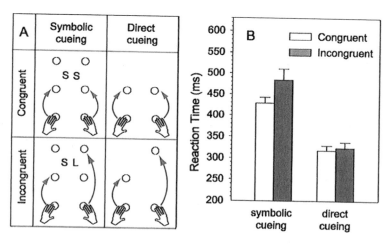

Figure 10-2. Direct reaching experiment. A: In the symbolic condition, the letters indicate the movement amplitude for the hand on the corresponding side (S=short amplitude; L=long amplitude). In the direct condition, the amplitudes were indicated by the appearance of the target circles. Movements could be either congruent (same amplitude) or incongruent (different amplitudes). B: Average reaction times for congruent and incongruent trials for the direct and symbolic conditions, averaged over the two hands. Error bars indicate between-subject standard error. Adapted from Diedrichsen et al. 2001.

The costs observed in the symbolic condition are manifest prior to the initiation of the actions. For both the direct and symbolic conditions, we did not observe assimilation effects in terms of movement amplitude or increased endpoint error in the incongruent condition. Once the movement targets were selected and the movements initiated, even asymmetric movements proceeded with minimal interference. These results stand in drastic contrast to studies that have used fast reversal movements to study amplitude assimilation effects during bimanual movements (e.g., Spijkers and Heuer 1995, Chapter 9). The differences in results between these and our studies may be due to the fact that the movements in our experiment (Diedrichsen et al. 2001) were executed towards visual targets, while in latter studies the movements were produced to match an internally specified amplitude.

3.2   Conflict resulting from the interaction of abstract spatial codes

The contrast in performance between symbolic and direct cues suggests that the constraints associated with spatial interactions during bimanual movements have little to do with the characteristics of the movements per se. The required movements under the symbolic and direct cueing conditions are identical-- participants move from a starting circle to target circles in both conditions. Except for errors in the initial trajectory in the symbolic conditions, the movements themselves are quite similar.   Given these observations, we assume that processes involved in motor programming, defined as the specification of motor commands and motor execution, are highly similar in the two cueing conditions.   If a preference for producing symmetric bimanual actions was a property of the motor system-- for instance, resulting from a bias to activate homologous muscles or plan movement trajectories of a common direction-- then we would have observed congruency effects in both the symbolic and direct cueing conditions.

If the preference for congruent movements is not associated with the motor system, how should we characterize the psychological operations that underlie the spatial interactions during bimanual movements?   One possibility is that the interactions arise at a perceptual level.   Mechsner, Prinz, and their colleagues (Hommel et al. 2001; Mechsner et al. 2001) have favored this interpretation, arguing that the actions are coded in terms of expected sensory consequences.   In a series of elegant experiments, these researchers demonstrated that the two hands can produce stable asymmetric movements when the feedback conditions are altered such that the sensory signals are symmetric. They propose that the coding of the expected sensory consequences plays a critical role in the selection and planning of the movements. Given that the perceptual system is highly sensitive to symmetry, movements resulting in symmetrical visual feedback might be supported by more stable representations.

One perception-based explanation for the costs observed in the symbolically cued bimanual movements centers on processes involved in identifying the various cues. The stimuli are identical in the congruent condition (e.g., "SS"), whereas they are different in the incongruent condition (e.g., "SL"). However, in a follow-up experiment we eliminated the cues altogether and let participants point to colored circles. The color assigned to each hand was constant within each experimental session. Although there was no need to identify a symbolic cue, participants were much slower to select targets of different colors than to select targets of the same color (Diedrichsen et al. 2003). In a different study, arrows and letters were used to cue the movements for the left hand and right hands, respectively. Thus, non-identical stimuli were used to cue the congruent and incongruent conditions. Nonetheless, the reaction time cost for incongruent movements was similar to that observed when the same set of symbolic cues was used for both hands (unpublished observations).

Together, these results suggest that the primary source of spatial interactions during bimanual movements is associated with an intermediate level of processing. On this level the action is represented in relatively abstract terms, without explicit reference to the eliciting stimulus or the execution-related details of the response. Consider a symmetric trial when each hand draws a three-sided square with the open side on the top. Each segment involves the specification of common trajectories for each hand: down, inward, up. Now consider the planning requirements for an orthogonal trial, one in which the open side for the left hand is on top and the open side for the right hand is on the right side. The situation here requires the generation of multiple, spatial codes. The initial movement for the left-hand involves a downward trajectory; for the right hand, a leftward trajectory. For the second segment, the left hand must move rightward and the right hand downward. We hypothesize that the costs observed on orthogonal trials arise from interactions between these various spatial codes. Not only are there conflicts between the component trajectories for each hand, but the spatial trajectories are presented on the left and right sides of the screen and must then be assigned to the left and right hand (Diedrichsen et al. 2003). The overlap between the codes defining the target trajectories and effectors is a ripe source of interference (see Kornblum et al. 1990).

In contrast, action goals for directly cued movements are unlikely to be specified in terms of trajectories or movement paths. Rather, the goals are likely to be related to the endpoint locations. As such, the degree of conceptual overlap is similar for congruent and incongruent movements. Both require the representation of two distinct locations. The lack of a cost on bimanual trials suggests that the representation of multiple locations can be generated and maintained as well as that of a single location.

Support for this hypothesis comes from a recent study in which we compared different types of cueing when performing the three-sided box tasks (Fig. 10-3). In the symbolic condition, the two target shapes were presented above the drawing surface and the participants reproduced the shapes. In the direct reaching condition, two target lights appeared, one on the left and the other on the right. The participants reached to these locations. As soon as their hands entered these target locations, new targets appeared indicating the next locations. The participants were instructed to immediately continue on to the next pair of targets. In this manner, the participants produced the three-sided trajectories, but only by moving from one direct cue to the next. For the tracing condition, the target shapes were presented directly on the drawing surface and the participants were asked to simply trace the two shapes simultaneously.

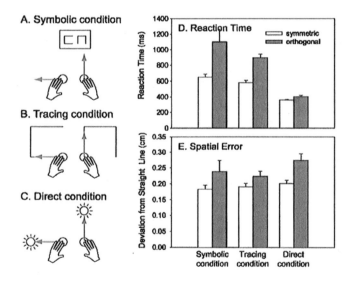

Figure 10-3. Three-sided figure drawing task with three different cueing conditions. A: In the symbolic condition the movements were instructed by small pictures of the target patterns, presented at the top of the table surface. B: In the tracing condition, the target patterns were presented in full size and the participants were instructed to trace these templates. C: In the direct condition, the movements were cued by the successive illumination of each corner location; the complete pattern was never visible. D: Reaction time results for the three conditions. E: Spatial error, calculated as the average deviation of for straight-line trajectories (Diedrichsen, Hazeltine, & Ivry, unpublished data).

As instructed, the participants initiated the movements with each hand in a near-simultaneous fashion. Thus, the effects of bimanual interference are most evident in the initial reaction times, although a similar pattern was evident in the pause between the first and second segment. As can be seen in Figure 10-3, the direct reaching condition was much easier than the other two conditions. Minimal RT differences were observed in the symmetric and orthogonal conditions. In fact, informal observation suggests that the participants were typically unaware of whether a particular trial had required a symmetric or orthogonal stimulus. Most interesting was the performance of the participants in the tracing condition. One might suppose that this condition would be similar to the direct cueing condition since participants simply have to move from one visible target location to the next. However, there was a clear cost on orthogonal trials compared to the symmetric trials. We assume that, by presenting the full shape prior to the initiation of the movements, the participants code the stimuli as target shapes composed of a series of directional vectors. When represented in this manner, interactions between varying spatial codes occurs.

Interestingly, we did not find systematic differences between the conditions on measures of movement accuracy. Figure 10-3e shows the deviation from a straight-line path averaged over the three segments. The spatial costs for the asymmetric shapes were significant in every cueing condition. As these spatial costs persist even after extended preparation time and independent of cue, they likely constitute static, execution-related sources of inference (Heuer 1993; Heuer et al. 2001, Chapter 9).

This experiment demonstrates again that bimanual interference results from interactions of codes on multiple levels (Cardoso de Oliveira 2002) and that the manner in which actions are conceptualized may have a dramatic influence on whether or not spatial interactions are manifest in the preparation of bimanual movements. With direct cues, the actions are specified in terms of target locations. For symbolical cues, the translation of the cues into actions entails a direction-based representation. This distinction touches on a long-debated issue in the motor control literature, namely whether movements are specified in terms of endpoint locations or movement trajectories (e.g., Abrams and Landgraf 1990). Our conjecture is that both of these forms of coding may be relevant, with the form of representation dependent on how the task is conceptualized.

3.3 Neural systems for direct and symbolic actions

It is interesting to consider the relationship between direct and symbolically cued movements and the dichotomy that has been made between the dorsal and ventral visual pathways in the cerebral cortex (reviewed in Goodale and Milner 1992). In simplest form, the dorsal stream across occipital-parietal cortex is hypothesized to be essential for visually guided actions. For example, reaching towards directly cued targets is impaired after lesions to the superior parietal lobe (Perenin and Vighetto 1988; Rushworth et al. 1997). Process-based accounts of the computations subserved by the dorsal stream include the representation of location-based codes and the coordinate transformations required for the translation of sensory information into reference frames useful for action (e.g., Flanders et al. 1992; Cohen and Andersen 2002).

The lack of bimanual interference observed in our reaching studies is in accord with observations that representations within the dorsal pathway are relatively immune to illusions associated with object recognition processes. For example, distortions induced by contrast effects are absent when perceived size is inferred by the aperture of a grasping action or the endpoint location of a pointing response (Goodale and Milner 1992). Correspondingly, the lack of crosstalk between the two limbs in bimanual reaching movements suggests that processing within the dorsal stream of each hemisphere is relatively immune to processing within the other hemisphere, even when these processes are used for the online control of movement.

We have tested this idea in a reaching task in which the target locations were perturbed right around or just after movement onset. Such perturbations are readily accommodated: Adjustments of the movement trajectories occur rapidly and in a smooth fashion. They can typically be detected in the movement kinematics 150-200 ms after the target has been displaced (Goodale et al. 1986; Prablanc and Martin 1992). The high processing speed of the system that allows for closed-loop control even during quick movements (Desmurget and Grafton 2000) has led to its characterization as an "auto-pilot" system that automatically guides the hand towards a visual target without intervention of consciousness. As such, involuntary adjustments can be observed in situations in which a target is displaced but participants are instructed not to adjust their movement (Day and Lyon 2000; Pisella et al. 2000).

In the bimanual version of this task (Nambisan et al. 2002), either one or both of the targets were displaced at the time of movement onset. The results suggest that the reaching movements of each hand are controlled by independent on-line control mechanisms when the targets are directly specified. Performance on trials in which both targets were displaced was very similar to performance on trials in which only one of the targets was displaced and, in fact, similar to that found on unimanual trials. However, there were some small signs of crosstalk between the two hands during the adjustments. Specifically, when the right hand adjusted to a rightward jump, the trajectory of the left hand was also transiently perturbed to the right, a perturbation that was quickly corrected for, before the hand reached the target. Most importantly, this perturbation was in the direction of the displacement of the other target in terms of exocentric coordinates. Spatial interference occurring during non-visually guided movements is usually manifest in egocentric coordinates (Swinnen et al. 2002). Thus the perturbation found in this situation seems to arise from retinal or eye-movement related signals, and may be fundamentally different from the interference underlying the preference for symmetric movements in the context of non-visually guided movements.

While the dorsal stream may be sufficient for directly cued movements, symbolic cues would seem to require the involvement of more ventral visual pathways. In the initial formulation of the dorsal/ventral dichotomy, the ventral stream was considered as part of the perceptual pathways, especially with higher-order object recognition. More recently, the role of such processes in the control of action has been acknowledged. For example, with symbolic cues, ventral areas are likely necessary to identify the stimuli and associate them with the appropriate motor output, perhaps in conjunction with premotor cortex. By this hypothesis, we would assume that symbolically mediated actions entail an additional processing stage, one in which the abstract symbols are mapped onto action codes.

There are a number of reasons why bimanual interference might be observed for actions that engage the more cognitive operations associated with the ventral pathway. Psychologically, the response selection processes required for linking abstract stimuli to intended actions pose a prominent bottleneck in multi-task performance (reviewed in Pashler 1994). Moreover, such interference is likely especially pronounced when the tasks require overlapping representations as we hypothesize is the case for the abstract, trajectory-based codes we associate with symbolically cued actions. On the neural level, we assume that such interactions occur across callosal pathways given the absence of such interference in the split-brain patients (see also, Ivry et al. 1998).

Does this mean that callosal fibers are more prominent for regions within the ventral pathway compared to the dorsal pathway? Indirect support for this conjecture can be found in the physiological literature. While receptive field size increases as one progresses along either the dorsal or visual pathway, a hallmark of inferotemporal cortex is that these neurons respond to stimuli from either visual field. Such neurons must have access to the output from upstream cells in either hemisphere. Alternatively, the lack of interference found with directly cued movements may not reflect a dearth of callosal connections along the dorsal pathway, but rather reduced representational overlap between such actions. As noted above, a location-based code entails two distinct target locations for both congruent and incongruent movements.

A different account of why bimanual interference is restricted to symbolically cued movements comes from recent elaborations of the two visual stream model. It has been proposed that the dorsal and ventral streams sandwich a third stream involving the inferior parietal cortex and that this pathway is highly lateralized (Johnson-Frey in press). Damage to inferior parietal cortex in the left hemisphere in humans leads to the severest forms of apraxia (Leipmann 1907; Heilman et al. 1982) and imaging studies show pronounced activation of this region for actions requiring the representation of complex object properties, for example when interacting appropriately with tools (Johnson et al. 2002). Furthermore, the role of the inferior parietal lobe appears to be most prominent in the planning of actions rather than their on-line control (Glover in press). Taken together, the functions associated with this lateralized region would seem to match those we assume are required in the translation of symbolic cues into actions (see also Schluter et al. 1998; Schluter et al. 2001).

From this perspective, it acknowledges the prominent role of the left inferior parietal cortex whenever actions are planned on the basis of internal goals or symbolic cues; that is, without the affordance of direct targets. Notably, this hypothesis would assume that these operations are required for symbolically cued movements produced with either hand. Interference would be expected to arise when a single processor is trying to plan two incompatible actions. By this hypothesis, bimanual interference for symbolically cued movements reflects a functional hemispheric asymmetry for the mediation of symbolically cued actions. The lack of interference for directly cued movements is attributed to a more symmetric brain organization for regions involved in visually-guided actions.

It is difficult to assess the relative merits of these neural conjectures at present. An appealing feature of the laterality account is that it acknowledges the prominent role for the left hemisphere in the representation of complex, abstract actions. The laterality hypothesis would suggest that split-brain individuals should show a selective impairment in producing symbolically-cued movements with the left hand. While this has not been apparent in our bimanual studies, a recent study reports a left-hand apraxia in some of these individuals (Johnson-Frey, Funell, & Gazzaniga, submitted). Moreover, the apraxia symptoms were especially pronounced for symbolically-mediated actions, for example, when the eliciting cues were pictures rather than real objects. In addition to behavioral tests, physiological studies should prove useful for evaluating the neural hypotheses sketched above. To date, few neuroimaging studies have focused on the manner in which actions are cued, especially with respect to bimanual coordination.

3.4 Spatial constraints revisited

To summarize, we posit that the ease and proficiency with which different bimanual actions are performed is largely determined by the manner in which the tasks are represented and controlled. The issue of task representation has received little attention, yet provides a powerful account of situations that produce spatial interactions during bimanual movements and, as important, situations in which such interactions are essentially absent.

In much of our work we have focused on the preparation phases of the movements. These phasic constraints are clearly influenced by the way the task is cued and conceptualized. However, many constraints on bimanual movements seem to be static in the sense that they do not change with more or less preparation time (Heuer et al. 2001). These have been attributed to lower levels of the motor system, evident during movement execution.

However, recent work has shown that crosstalk during the execution of repetitive movements is also dominated by the manner in which the movement goal is represented. (see Mechsner et al. 2002; Weigelt and Cardoso De Oliveira 2003). Franz et al. (2001) provide a particularly telling example in a study in which each hand traced a semi-circle. In one condition, the two movements started and ended at the same locations, creating an overall shape of a circle. In the other condition, the starting and ending points were spatially displaced, creating an overall shape of two inverted semi-circles that approach each other at the midpoint (Fig. 10-4). Although the two target patterns are both symmetric and involve similar combinations of muscular actions, people were more adept in the former condition. This result was attributed to the fact that the circle is a simpler and more familiar pattern. We note, though, that both combinations of the semi-circles were symbolically cued. Thus, as with our pointing studies, the degree of conflict appears to depend on goal-based representations that govern the action. Swinnen and colleagues (e.g., Swinnen et al. 1997) have made a similar point, demonstrating that the profound interference observed during the production of complex bimanual trajectories can be rapidly overcome when visual feedback requires the participants to focus on an integrated representation of the action goal.

What has been underappreciated in the bimanual literature is the extent to which researchers have relied on movements that are symbolically cued and often executed without a visual external goal. Even when templates are provided, they are used to provide a general trajectory reference and as such, are another form of a symbolic cue. We believe that the reliance on tasks that entail symbolic representations has led to the general impression that interference between the movements always occurs in a muscle-related or egocentric coordinate frame and that our ability to produce asymmetric bimanual movements is highly constrained. Our work with directly cued movements leads us to conclude otherwise, at least in terms of spatial constraints. Interference in this situation may be minimal and often occur in an exocentric reference frame (Nambisan et al. 2002).

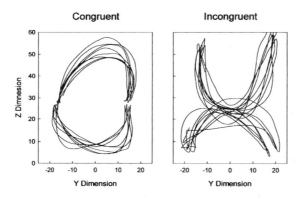

Figure 10-4. Familiarity of target shape influences the extent of bimanual interference. For both conditions, the participants produced two semi-circles in mid-air, one with each hand. In the congruent condition, the circles were aligned such that the hands were closest at the endpoints. In this way, the hands traced a circle. In the incongruent condition, the hands were closest at the midpoint. The patterns were drawn repetitively. Variability is higher for the unfamiliar curved X's. Adapted from Franz et al. 2001.

As noted in the Introduction, studies of bimanual coordination have tended to not use tasks that require the two hands to operate in a synergistic fashion. We would argue that when the actions of two hands are conceptualized as reflecting independent goals, the limitations on performance reflect constraints similar to those identified in the dual-task literature (Duncan 1979; Hazeltine et al. 2003; Pashler 1994) rather than processes that are engaged specifically during bimanual movement.

4. REPRESENTATIONAL BASIS OF TEMPORAL CONSTRAINTS

We now turn to the second major group of constraints, those between temporal features of the movements. When making discrete bimanual reaching movements, people tend to initiate and (approximately) terminate the movements of the two hands in synchrony (Kelso et al. 1979; Marteniuk and Baba 1984). This coupling is even more prevalent during rhythmic movements. We adopt a common frequency for each limb and, without extensive practice, are limited to only two stable phase relationships, in-phase and anti-phase. Even skilled musicians are limited in the flexibility with which they time the movements of their two hands, with their performance generally indicative of an integrated temporal representation rather than a situation in which the timing of each hand is independently controlled (Klapp et al. 1985; Krampe et al. 2000).

Temporal coupling has provided a cornerstone for the dynamic systems approach to the study of motor control. Our movements involve effectors that are physical entities. As such, movements must respect the laws of gravity, inertia, and mechanics (Kugler and Turvey 1987). This approach has produced rigorous formalisms to describe and predict motor behavior across a wide range of situations including unimanual movements, bimanual movements produced by a single person, or the interactions that occur between the movements of different individuals (Kelso 1995). These models have been expressed in terms of abstract dynamics, prompted by the desire to provide a description at a general level. With their focus almost completely on movement trajectories, these general formalisms have failed to provide process models, a description of the neural and psychological representations and processes that might underlie the observable coordination phenomena.

In this section, we review our recent work on this problem. Similar to what was described in the discussion of spatial constraints, we will argue that the manner in which action goals are represented strongly influences temporal constraints (Semjen 2002), as well as the neural correlates driving these phenomena.

4.1 Phase stability and the representation of rhythms

A common formalism for describing temporal constraints associated with rhythmic bimanual movements is that of coupled oscillators. Each limb is described as a limit-cycle oscillator with the interactions between two such oscillators captured by a coupling term (e.g., Haken et al. 1985). When expressed in this manner, a concise description of the dynamics can be obtained. The interactions between the oscillators allow the limbs to maintain a common frequency, even when they have different preferred frequencies (e.g., Turvey 1990). Moreover, the abstract dynamics dictate that certain phase relationships will serve as attractors, with the specific strength of such attractors frequency dependent.

Yamanishi et al. (1980) provided one of the first studies to explore the utility of the coupled oscillator model. They used a simple bimanual finger-tapping task. Each hand was required to tap at 1 Hz. The critical independent variable was the target inter-tap interval (ITI) between successive taps of the two hands. In separate blocks, the ITI ranged from 0 ms to 900 ms in steps of 100 ms. Expressed in terms of relative phase, an ITI of 0 ms corresponds to in-phase tapping and an ITI of 500 ms corresponds to anti-phase tapping. The other target ITI's correspond to more complex target phases (e.g., an ITI of 100 ms is a target phase of 36 degrees).

As predicted, participants were readily able to perform the task when the target ITI was 0 ms or 500 ms. Performance for the other ITI's was less stable and there was a pronounced tendency for the produced phase to be attracted to either the in- or anti-phase pattern. For example, when the target ITI was 400 ms or 600 ms, the participants tended to produce ITI's with mean values closer to 500 ms. On various measures, the coupled oscillator model provided a good account of the data. The model captures the attraction towards the in-phase and anti-phase patterns, as well as the dependency of pattern consistency (e.g., variability of relative phase) as a function of the target phase.

While a coupled oscillator model provides an elegant description of performance in this task, an alternative process model should be considered. When viewed as an integrated pattern, the alternating taps define subintervals that divide the 1000 ms within-hand ITI. These subintervals constitute a rhythmic pattern. The 0 ms and 500 ms ITI conditions create simple rhythms, with subinterval durations of 1000 ms and 500 ms in the in-phase and anti-phase patterns, respectively. The subintervals for the other target ITI's define much more complex rhythmic patterns. For example, for the 600 ms ITI, the successive subintervals are 600 ms and 400 ms, forming a pattern in which the ratio of the longer to shorter interval is 3:2. In the 800 ms ITI, the ratio would be 4:1. Perhaps people represent the temporal goal in this task to create subintervals that match the target ratios. Many studies have shown that people have a strong bias to perceive/reproduce temporal patterns that form simple ratios (Collier and Wright 1995; Essens 1986; Povel 1981). When seen from this perspective, the attraction to in-phase and anti-phase patterns might reflect a bias to the simplest of ratios, the 1:1 ratio created by the isochronous patterns.

To compare the coupled oscillator and simplified rhythm representation hypotheses, Semjen and Ivry (2001) replicated the Yamanishi et al. study with one critical difference; In addition to the bimanual condition, participants were also tested in a unimanual condition. For this condition, a single finger was used to make all of the responses and, the conditions varied in terms of the target durations for the subintervals. The in-phase condition (0 ms ITI) could not, of course, be tested in the unimanual condition.

Consistent with the predictions of the rhythmic representation hypothesis, performance in the unimanual and bimanual conditions was essentially identical (Fig. 10-5a). Regardless of whether the participants tapped with one or two fingers, the deviations from the target intervals were essentially identical. Moreover, when the produced subinterval ratios were calculated, there was a clear attraction towards simple ratios (e.g., 1:1, 2:1, or 3:1). The participants were unable to produce the target durations in the most complex conditions, demonstrating a bias to produce subintervals that yielded relatively simple rhythms.

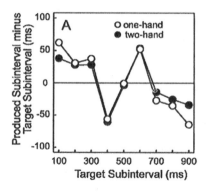

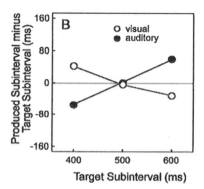

Figure 10-5. A: Participants tapped temporal patterns created by the division of a 1000 ms interval into two subintervals. Successive taps are made with alternating hands (two-hand) or by a single hand (one-hand). The results are plotted as the difference between the produced subinterval and the target subinterval. The produced subintervals tended to conform to simple rhythmic ratios (e.g., 1:1, 2:1), and most important, were the same in the one- and two-hand conditions. B: One-hand tapping was paced by either a visual or auditory metronome. The metronome had a marked effect for the conditions in which the target subintervals were 400 and 600 ms (or 600 and 400 ms). Adapted from Semjen and Ivry (2001).

Interestingly, for one condition, the results of our study appeared at odds with that reported by Yamanishi and colleagues. In the 600:400 (or 400:600 since performance is roughly symmetric), Yamanishi et al. had reported an attraction towards anti-phase tapping: The long interval was shortened and the short interval was lengthened (i.e., bias to produce a subinterval ratio of 1:1). In our study, the bias was in the opposite direction. The long interval tended to be lengthened and the short interval shortened, resulting in a produced ratio close to 2:1.

The key to this discrepancy provides further support for the rhythmic representation hypothesis. Whereas Yamanishi et al. had used a visual metronome to signal the target subintervals, we used an auditory metronome. Temporal acuity is higher in the auditory domain (Allan 1979). As such, participants were likely more attuned to the large difference between the 600 and 400 ms subintervals with the auditory metronome, and due to the bias to simplify the ratio representation, produced subintervals that approximated a 2:1 ratio. To test this hypothesis, visual and auditory metronomes were directly compared in a second experiment. As expected, the 600/400 target subintervals were distorted towards a 1:1 ratio with the visual metronome and towards a 2:1 ratio with the auditory metronome (Fig. 10-5b).

The Semjen and Ivry (2001) study demonstrates how temporal constraints in bimanual movements may reflect the manner in which the task goals are represented. We have emphasized that the key constraint involved the manner in which the target subintervals were represented. Consonant with the music cognition literature, naïve participants exhibited a strong bias towards rhythmic representations involving simple ratios. While the same constraints were operative for visual- and auditory-paced performance, the differential sensitivity of the two modalities resulted in different patterns of distortion.

It is difficult to envision how a coupled oscillator model could be adopted to account for the results of our study. How would one characterize the two oscillators in the unimanual condition? Certainly not in the manner the oscillators are characterized by Yamanishi et al., that is, as two oscillators set to a common frequency of 1 Hz. The modality effect might be accounted for by postulating that coupling strength varies with modality. But even this hypothesis fails to account for the tendency of the perturbations to go in opposite directions for the visual and auditory conditions. We believe that the most parsimonious account of the temporal constraints is given by the rhythm representation hypothesis. Indeed, the key constraints on temporal performance appear to have little to do with the fact that two limbs were used. Rather, these constraints reflect general limitations in our ability to represent complex temporal relationships.

4.2   Probing the event structure of rhythmic movements

The rhythm representation hypothesis can account for the bias people show towards certain phase relationships. In-phase and anti-phase patterns entail especially simple temporal representations given that they result in isochronous subintervals. With more complex ratios, we posit that temporal relationships are organized hierarchically. A fundamental timing unit is established and the hierarchy is used to specify which response to produce and when it should be emitted. These temporal representations define an event structure for the action (Semjen 2002).

To this point, our discussion of temporal constraints has not addressed why in-phase movements are more stable than anti-phase movements. We offered one explanation with respect to spatial constraints; we proposed that in-phase patterns entail more congruent trajectories than anti-phase patterns. However, we also believe that these two patterns may, under certain conditions, be guided by qualitatively different temporal representations, or event structures.

Consider again the example of wrist flexion and extension, taking the situation in which both palms face downwards. When performed repetitively, these movements can be considered as continuous oscillations. However, synchronization studies have shown that certain points in the cycle are more salient than others. If the movements involve contact with an external surface as in table tapping, synchronization with an external metronome will be organized such that the table is contacted coincident with the beat of the metronome (or more accurately, slightly ahead of the metronome, see Vos, Mates and van Kruysbergen 1995). If there is no external surface, most people synchronize with the initiation of each flexion cycle.

Our hypothesis is that the event structure differs for the in-phase and anti-phase patterns. For in-phase movements, a common event defines the cycle initiation point for each hand. In contrast, anti-phase movements entail two events per cycle, one associated with flexion onset of each hand (Fig. 10-6). According to this hypothesis, the event structure for anti-phase movements is more complex than that associated with in-phase movements and stability will be inversely related to complexity.

We have only begun to test the event structure account of the preference for in-phase movements. The goal in our initial studies was to demonstrate that, under conditions of minimal external constraint, people do indeed conceptualize different event structures for in-phase and anti-phase movements. To this end, participants were instructed to perform continuous wrist flexion and extension movements, either in-phase or anti-phase. At the beginning of the first block, participants were instructed to choose their own pace. Once the participant was accurately producing the desired pattern, in subsequent trials they were instructed to "say the word 'BA' repeatedly as you move". We did not give any indication as to when in the cycle the vocalizations should be made nor were any demonstrations provided that might bias performance. In subsequent blocks, the task was repeated but the movements were now made at different paces. Rate-based feedback was given after each training trial (e.g., "Go faster" or "Go slower") until the participant approximated the target rate. At this point, the BA instruction was added.

We expected that the vocalizations would be temporally coupled to "significant" events during the movements. In other words, we used the BA's as a window on the participant's conceptualization of the event structure of each task. The results showed a striking difference between the two tasks. In the in-phase hand movement condition, on 61% of the trials, participants emitted one "BA" for each cycle and the vocalizations tended to occur around the time of flexion onset. In contrast, in the anti-phase hand movement condition, two BA's were vocalized on 100% of the trials, emitted close to the points at which the hands were at maximum flexion and extension.

We had also expected to see that the occurrence of BA's would vary with movement frequency. In particular, we had expected that, as rate increased, there would be a transition in the anti-phase condition from two BA's per cycle to one BA per cycle. However, such transitions were not observed in the anti-phase condition. In the in-phase condition, one participant (17% of the trials) said 2 BA's for every movement. Interestingly, this participant adopted the slowest spontaneous hand movement rate in the first block. The failure to observe a change in the number of BA's per cycle may have been due to the fact that we failed to include rates at which the wrist movements themselves underwent a phase transition. Nonetheless, the study does support the basic tenet of the event structure hypothesis: As measured by the occurrence of BA's, a difference is observed in the temporal representation of in-phase and anti-phase movements. Moreover, if we assume that complexity is related to the number of salient events, anti-phase movements are more complex than in-phase movements (see also, Wimmers et al. 1992).

In a sense, the event structure model is a generalization of the rhythm representation hypothesis. The core idea is that temporal constraints may reflect the manner in which the temporal goals of the task are represented. As these representations become more complex, pattern stability suffers. Complex rhythms such as tapping four against three represent an extreme case. Most of us lack the experience necessary to develop the representations for proficient production of such rhythms. Repetitive movements involving isochronous intervals are, obviously much easier. But, as shown by the BA experiment, differences may exist in the representational structure of even simple rhythms, and these likely have consequences in terms of pattern stability.

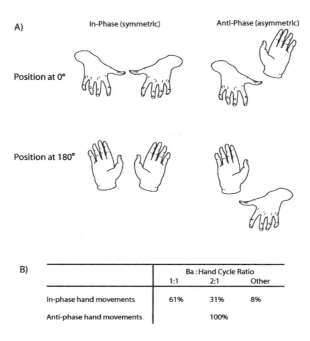

Figure 10-6. Probing the event structure of in- and anti-phase movements. A) Participants were instructed to move in-phase (left column) or anti-phase (right column). Once performance was stabilized, they were required to simultaneously articulate the syllable "BA". B) Although no instructions were given, the vocalizations were usually produced in a fixed temporal relationship with the wrist movements. Participants always vocalized twice per cycle in the anti-phase condition. In contrast, they generally vocalized once per cycle for the in-phase movements, suggesting a simpler event structure for the in-phase pattern. (Spencer, Semjen, & Ivry, unpublished data).

4.3   Neural dissociations in the control of continuous and discontinuous movements

Recently, Zelaznik and colleagues (Robertson et al. 1999; Zelaznik et al. 2000; Zelaznik et al. 2002) presented evidence that the temporal control of unimanual movements may vary as a function of task requirements. Participants were required to produce rhythmic movements, either by finger tapping or circle drawing. On measures of variability, performance across the two tasks was expected to be positively correlated (e.g., Keele et al. 1985). To their surprise, temporal variability in tapping and circle drawing were not correlated (Robertson et al. 1999; Zelaznik et al. 2000). These results suggested that different processes were engaged for controlling the timing of these two tasks.

What could account for these null results? We have proposed that the representation of the task goals, or at least the manner in which these goals are achieved may be fundamentally different for tapping and circle drawing, with these tasks being representative prototypes of two different classes of movements (Ivry et al. 2002; Zelaznik et al. 2002). Repetitive tapping can be conceptualized as the concatenation of a series of discrete events, with discontinuities observed either when the finger contacts the table surface or just prior to the onset of flexion when people typically make a brief pause. The goal for such tasks is to separate each successive event by the target interval; that is, timing is an explicit part of the action goal with an internal timing being used to control the movements of each cycle.

In contrast, circle drawing involves continuous movements. People can, of course, vary the cycle duration of these movements with the same flexibility as for tapping. However, this does not mean that the on-line control of these movements requires the operation of an internal timer. Rather, temporal regularities might be achieved by optimizing some other variable such as maintaining constant angular velocity, perhaps by the continuous modulation of joint stiffness over a targeted range. We refer to this form of timing as emergent, to contrast with the event timing required for tasks that involve discontinuities.

The notion of emergent timing has been espoused previously by proponents of the dynamic systems approach, usually in reference to the idea that there is a preferred frequency for repetitive limb movements (reviewed in Amazeen et al. 1998). We share this view of how temporal regularities may be maintained for continuous movements. However, to account for the fact that these movements are not constrained to a particular frequency, we propose that the control system can determine the mapping between a temporal goal and certain control parameters. In this manner, the goal in a task like circle drawing undergoes a translation. Initially, the goal is of a target cycle duration, similar to the event timing representations for tapping. But because of the continuous nature of the movements and their lack of salient events, the goal can be achieved in an emergent manner by controlling other parameters to optimize performance. Thus, emergent timing tasks have a different form of representation; they lack an event timing structure.

Converging evidence in support of the event/emergent distinction comes from two sets of neuropsychological studies. Various lines of evidence suggest that the cerebellum is essential for tasks that require the precise representation of temporal information, the form of representation that we hypothesize is essential for event timing tasks (reviewed in Ivry et al. 2002). To test this idea, patients with cerebellar damage performed continuous and discontinuous repetitive movements (Spencer et al. 2003). As predicted, the patients exhibited increased temporal variability on various discontinuous movement tasks. These included both tapping tasks and a modified circle drawing task in which the participants were required to insert a pause prior to the onset of each drawing cycle.

Most striking was their performance on the continuous circle drawing task. The movements for this task are considerably more complex than those required for tapping, involving 2-dimensional spatial trajectories that involve multiple joints and interactional torques. Based on conventional neurological thinking concerning the role of the cerebellum in coordinating such movements, one would have expected the patients to be at least, if not more impaired, on the circle drawing task. However, the patients were unimpaired on the continuous circle drawing task: no increase in temporal variability was evident on this task when performing with their impaired limbs. This dissociation is consistent with the idea that the timing of continuous movements does not require continuous control from an internal timing system.

The second neuropsychological dissociation was obtained in studies with callosotomy patients and brings us back to the topic of bimanual coordination. As part of our research on spatial crosstalk, we tested three split-brain patients on the bimanual circle drawing task (Kennerly et al. 2002). The initial goal of the study was to compare their performance when circling in a symmetric mode (one hand clockwise and one hand counterclockwise) versus an asymmetric mode (both clockwise or both counterclockwise). Unimpaired individuals exhibit more stable performance in the symmetric condition (Semjen et al. 1995). The patients failed to exhibit this form of spatial coupling. Their movements were no more accurate in the symmetric condition and phase transitions were observed from the symmetric to the asymmetric mode as often from the asymmetric to the symmetric mode. More striking, however, was that the split-brain patients' movements were frequently temporally uncoupled. On many trials, the hands adopted completely different frequencies, a phenomenon that is never spontaneously observed in normal participants.

The temporal uncoupling during circle drawing was puzzling given previous reports, including our own, that split-brain patients exhibit strong temporal coupling (Franz et al. 1996; Ivry and Hazeltine, 1999; Tuller and Kelso 1989). Interestingly, these previous studies involved tasks associated with an event-based representation. Either the movements were discrete or required tapping-like movements. Thus, we reasoned that the uncoupling during continuous circle drawing may provide another indication that the representational basis of these types of movements is quite different.

To test this idea, we created two hybrid tapping tasks involving flexion-extension movements of the index fingers. In the continuous condition, the participants were instructed to make flexion-extension movements with their index fingers, attempting to move the fingers in a smooth, continuous manner. In the discrete tapping condition, the participants were instructed to insert a brief pause prior to each flexion phase. It is important to note the overall similarity between the two conditions. All movements were made in free space without contacting an external surface, and we did not pace the movements with a metronome or give any instructions regarding synchronization.

Despite this similarity, a dramatic difference was seen in the performance of the split-brain patients (Fig. 10-7). In the air tapping condition, the patients' movements were strongly coupled. A common frequency was adopted for the left and right hand movements, and as measured by the phase difference distribution, the strength of coupling was similar as that found in the control participants. However, in the continuous condition, performance was much more variable. While there were epochs in which the movements were coupled, there were also epochs in which the two hands became temporally uncoupled, similar to what we had observed during the bimanual circle drawing task with these patients. Again, at least parts of the trials, the two hands moved at different frequencies. This dissociation provides converging evidence concerning representational differences between continuous and discontinuous movements, and emphasizes that bimanual coupling arises from a varied set of constraints associated with these representations. First, consider repetitive, continuous movements. We have argued that the event-based representations provided by the cerebellum are not essential for such tasks (Spencer et al. 2003). While normal participants exhibit strong temporal coupling when making continuous movements, this constraint is absent in the split-brain patients. Based on our earlier considerations about spatial coupling, we hypothesize that temporal coupling for continuous, repetitive movements arises from dynamic interactions between time-varying representations of the abstract spatial goals for these actions. For circle drawing, symmetric patterns appear to be more congruent than asymmetric patterns; for one-dimensional movements, congruency is generally associated with movements along the same direction of rotation. When viewed from this perspective, the absence of temporal coupling in the split-brain patients is another manifestation of the fact that interactions between these abstract spatial codes is mediated by communication across the corpus callosum.

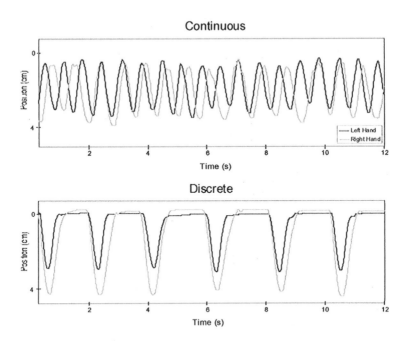

Figure 10-7. Representative trials of temporal coupling in a split-brain patient during continuous and discontinuous movements. Movements involved flexion-extension of the index finger in midair (**flexion** is portrayed upwards). In the discrete condition, the patient was instructed to insert a brief pause prior to each flexion phase. While control participants exhibit strong coupling in both conditions, the coupling is weaker, and occasionally absent for the split-brain patients. (based on Kennerley et al. 2002).

The close relationship between the coupling of spatial and temporal characteristics in the context of continuous movements is also demonstrated by the fact that neurologically healthy individuals show an attenuation of temporal coupling when producing asymmetric (i.e., spatially incompatible) movements at fast rates (Carson et al. 1997; Semjen et al. 1995). Moreover, a complete breakdown of temporal coupling can occur when the spatial overlap between the two tasks is reduced; for example, when the movements are made with non-homologous limbs of very different masses (Jeka and Kelso, 1995).

In contrast, temporal coupling appears to be much more robust for discontinuous movements. We hypothesize that this is because the two movements share a common event structure, a representation of the temporal goals. Stability here is dictated, not in terms of the congruency of spatial relations, but rather in terms of temporal economy. Our movements are biased towards patterns that specify temporal events in a concise and simple manner. In-phase patterns entail a simpler event structure than anti-phase patterns. Both of these patterns are more stable than other phase relations because they entail simple rhythmic representations (e.g., small ratios).

We have argued that the cerebellum is essential for the temporal representations embodied in an event structure (Ivry et al. 2002). We see this structure as one part of a distributed system for controlling actions (and mediating perceptions) that entail complex temporal relationships. For example, a rhythm can be described abstractly-- a drummer can produce a 3:2 bimanual tapping pattern at different speeds. Our expectation is that the abstract level of representation is cortical; the cerebellum is engaged when this abstract pattern is instantiated as a specific action, one that requires the real-time coding of an event structure.

The idea of an event structure for both unimanual and bimanual movements is a powerful tool to understand the temporal coordination of more complicated bimanual actions. For example, when opening a drawer with one hand to grasp an object with the other hand, the timing of the two hands is stereotypically organized (Perrig et al. 1999). However, the phase relationship of the two movements is much more complicated than in simple repetitive tasks. Importantly, the movements in such tasks are part of a generalized motor program, one that specifies the successive events of the two hands to achieve a common goal (Schmidt et al. 1998). We hypothesize that the cerebellum provides the fine temporal resolution necessary for the success of many bimanual actions. Consistent with this idea, cerebellar patients show deficits in the temporal aspects of such a complex bimanual motor tasks (Serrien and Wiesendanger 2000).

Interestingly, split-brain patients are still subject to the temporal constraints imposed by the representation of an event structure. Furthermore, they do not report drastic deficits in many well-learned bimanual skills, even if these require detailed coordination of the hands (Franz et al. 2000; Serrien et al. 2001). We believe this reflects the operation of a mechanism required for the implementation of the actions specified by the event structure. Such a process could ensure that the action codes specified by the cerebral hemispheres are implemented in an efficient manner, an operation that has been likened to a neural gating process. However, we do not wish to imply that this gating process is performed by the cerebellum. At present, our speculations on the localization of such a process are guided by two considerations. First, it should have access to the output from the cerebellum specifying the event structure. Second, it should be capable of initiating actions in a relatively generic manner (e.g., bilaterally). We believe the evidence points to a subcortical locus, perhaps the basal ganglia but it is also possible that a single cerebral hemisphere might meet such requirements (see Ivry and Richardson 2002).

5.   FINAL COMMENTS

The study of bimanual coordination has provided an important tool for exploring the cognitive neuroscience of motor control. Central to this work has been the elucidation of the many ways in which our ability to produce bimanual movements is constrained. We have focused on two primary classes of constraint, those associated with the interactions observed between the two limbs in the spatial and temporal domains, similar to what Semjen (2002) referred to as trajectory-level and event-level constraints. We do not wish to imply that a clean division, either psychologically or neurologically, can always be made between the manner in which the spatial and temporal features of movements are represented and controlled.   Nonetheless, neuropsychological evidence demonstrates that the two types of constraint can be dissociated (Franz et al. 1996). Indeed, there are notable differences in our accounts of these sources of constraint. Spatial interference effects, at least in terms of response planning, are limited to situations in which the movements are symbolically cued, suggesting that the primary constraint arises with response selection rather than motor programming or execution. The neuropsychological evidence points to a cortical locus for such effects, with candidate areas including ventral visual processing pathways and inferior parietal and premotor cortices. Temporal interactions generally reflect the operation of a unified temporal representation, one in which the timing of salient events is explicitly controlled, a process associated with the cerebellum.

A common theme in our analyses, however, is that the way in which the task goal is conceptualized will play a central role in determining patterns of interference between the two movements.   We attribute the difference between symbolically- and directly-cued movements to a difference in task conceptualization, with the former involving goals defined as movement trajectories and the latter involving goals defined as target locations. Similarly, we hypothesize that while an event-based representation is essential for discontinuous movements, this form of representation is not essential for continuous movements.   The emphasis on task conceptualization also leads to the conclusion that many of the constraints underlying bimanual coordination arise at an abstract level, one that can be divorced from processes devoted to motor execution. Acknowledging the limitations imposed by our cognitive architecture should also hold promise for understanding and appreciating the extraordinary flexibility with which humans use their two hands.

ACKNOWLEDGEMENTS

Preparation of this chapter was supported by Grants NS30256, NS40813, NS17778, and NS33504. We are grateful for discussions with Scott Grafton, Scott Johnson, Winston Byblow, and comments on an earlier draft from Steve Keele and Howard Zelaznik. Correspondence should be directed to R. Ivry, Department of Psychology, MC 1650, University of California, Berkeley, CA 94720 USA, or email at ivry@socrates.berkeley.edu.

REFERENCES

Abrams RA, Landgraf JZ (1990) Differential use of distance and location information for spatial localization. Percept Psychophys 47:349-359

Allan LG (1979) The perception of time. Percept Psychophys 26:340-354

Amazeen PG, Amazeen EL, Turvey MT (1998) Dynamics of human intersegmental coordination: Theory and research. In: Rosenbaum DA, Collyer CE (eds), Timing of behavior: neural, computational, and psychological perspectives. MIT Press, Cambridge MA. pp 237-259

Baldissera F, Cavallari P, Civaschi P (1982) Preferential coupling between voluntary movements of ipsilateral limbs. Neurosci Letters 34:95-100

Baldissera F., Cavallari P, Marini G, & Tassone G (1991). Differential control of in-phase and anti-phase coupling of rhythmic movements of ipsilateral hand and foot. Exp Brain Res 83:375-380.

Cardoso de Oliveira S. (2002) The neural basis of bimanual coordination: recent neurophysiological evidence and functional models. Acta Psych 110:139-159

Carson RG, Thomas J, Summers JJ, Walters MR, Semjen A. (1997) The dynamics of bimanual circle drawing. Quart J Exp Psych 50A:664-683

Cohen YE, Andersen RA (2002) A common reference frame for movement plans in the posterior parietal cortex. Nature Neurosci Rev 3:553-562

Collier GL, Wright CE (1995) Temporal rescaling of simple and complex ratios in rhythmic tapping. J Exp Psychol Hum Percept Perform 21:602-627

Day BL, Lyon IN (2000) Voluntary modification of automatic arm movements evoked by motion of a visual target. Exp Brain Res 130:159-168

Desmurget M, Grafton S (2000) Forward modeling allows feedback control for fast reaching movements. Trends Cog Sci 4:423-431

Diedrichsen J, Hazeltine E, Kennerley S, Ivry RB (2001) Moving to directly cued locations abolishes spatial interference during bimanual actions. Psychol Sci 12: 493-498

Diedrichsen J, Ivry RB, Hazeltine E, Kennerley S, Cohen A (2003) Bimanual interference associated with the selection of target locations. J Exp Psychol Hum Percept Perform 29: 64-77

Duncan J (1977) Response selection errors in spatial choice reaction tasks. Quart J Exp Psychol 29: 415-423

Eliassen JC, Baynes K, Gazzaniga MS (1999) Direction information coordinated via the posterior third of the corpus callosum during bimanual movements. Exp Brain Res 128:573-577

Essens PJ (1986) Hierarchical organization of temporal patterns. Percept Psychophys 40: 69-73

Flanders M, Tillery SIH, Soechting JF (1992) Early stages in a sensorimotor transformation. Beh Brain Sci 15:309-362.

Franz EA, Waldie KE, Smith MJ (2000) The effect of callosotomy on novel versus familiar bimanual actions: a neural dissociation between controlled and automatic processes? Psychol Sci 11:82-85

Franz EA, Eliassen JC, Ivry RB, Gazzaniga MS (1996) Dissociation of spatial and temporal coupling in the bimanual movements of callosotomy patients. Psychol Sci 7:306-310

Franz EA, Zelaznik HN, McCabe G (1991) Spatial topological constraints in a bimanual task. Acta Psychol 77: 137-151

Franz EA, Zelaznik HN, Swinnen S, Walter C (2001) Spatial conceptual influences on the coordination of bimanual actions: When a dual task becomes a single task. J Motor Behav 33:103-112

Glover S (in press) Separate visual representations in the planning and control of action. Beh Brain Sci

Goodale MA, Milner AD (1992) Separate visual pathways for perception and action. Trends Neurosci 15:20-25

Goodale MA, Pelisson D, Prablanc C (1986) Large adjustments in visually guided reaching do not depend on vision of the hand or perception of target displacement. Nature 320:748-750

Guiard Y (1987) Asymmetric division of labour in human skilled bimanual action: the cinematic chain as a model. J Motor Beh 19: 86-517

Haken H, Kelso JAS, Bunz H (1985) A theoretical model of phase transitions in human hand movements. Biol Cybern 51:347-356

Hazeltine E, Diedrichsen J, Kennerley SW, Ivry RB (2003) Bimanual cross-talk during reaching movements is primarily related to response selection not the specification of motor parameters. Psychol Res 67:56-70

Heilman KM, Rothi LJ, Valenstein E (1982) Two forms of ideomotor apraxia. Neurology 32:342-346

Heuer H (1993) Structural constraints on bimanual movements. Psychol Res 55:83-98

Heuer H, Kleinsorge T, Spijkers W, Steglich W (2001) Static and phasic cross-talk effects in discrete bimanual reversal movements. J Motor Beh 33:67-85

Heuer H, Spijkers W, Kleinsorge T, van der Loo H, Steglich C (1998) The time course of cross-talk during the simultaneous specification of bimanual movement amplitudes. Exp Brain Res 118:381-392

Hommel B, Musseler J, Aschersleben G, Prinz W (2001) The Theory of Event Coding (TEC): a framework for perception and action planning. Beh Brain Sci 24:849-878

Ivry RB, Franz EA, Kingstone A, Johnston J (1998) The PRP effect following callosotomy: Uncoupling of lateralized response codes. J Exp Psychol Hum Percept Perform 24:463-480

Ivry RB, Hazeltine E (1999) Subcortical locus of temporal coupling in the bimanual movements of a callosotomy patient. Hum Mov Sci 18:345-375

Ivry RB & Richardson T (2002) Temporal control and coordination: The multiple timer model. Brain Cog 48:117-132

Ivry RB, Spencer RM, Zelaznik HN, Diedrichsen J (2002) The cerebellum and event timing. In: Highstein SM, Thach WT (eds) The cerebellum: recent developments in cerebellar research Annals of the New York Academy of Sciences Vol 978 New York Academy of Sciences, NY. pp 302-317

Jeka JJ & Kelso JAS (1995) Manipulating symmetry in the coordination dynamics of human movement J Exp Psychol Hum Percept Perform 21:360-374

Johnson-Frey SH (in press) Cortical mechanisms of human tool use. In Johnson-Frey SH (ed.) Taking action: cognitive neuroscience perspectives on the problem of intentional acts MIT Press, Cambridge MA

Johnson-Frey SH, Funnell MG, Gazzaniga MS A dissociation between tool use skills and hand dominance: Insights from left- and right-handed callosotomy patients. Manuscript under review

Johnson SH, Rotte M, Grafton ST, Hinrichs H, Gazzaniga MS, Heinze HJ (2002) Selective activation of a parietofrontal circuit during implicitly imagined prehension. Neuroimage 17:1693-1704

Kalaska JF, Cohen DA, Prud'homme M, Hyde ML (1990) Parietal area 5 neuronal activity encodes movement kinematics not movement dynamics. Exp Brain Res 80:351-364

Keele SW, Pokorny R, Corcos D, Ivry R (1985) Do perception and motor production share common timing mechanisms? Acta Psychologia 60:173-193

Kelso JAS, Southard DL, Goodman D (1979) On the coordination of two-handed movements. J Exp Psychol Hum Percept Perform 5:229-238

Kelso JAS (1984) Phase transitions and critical behavior in human bimanual coordination. Am J Physio Reg Integ Comp 15:R1000-R1004

Kennerley SW, Diedrichsen J, Hazeltine E, Semjen A, Ivry RB (2002) Callosotomy patients exhibit temporal and spatial uncoupling during continuous bimanual movements. Nature Neuro 5:376-381

Klapp S, Hill MD. Tyler JG, Martin ZE. Jagacinski RJ, Jones MR (1985) On marching to two different drummers: perceptual aspects of the difficulties. J Exp Psychol Hum Percept Perform 11:814-827

Kornblum S, Hasbroucq T, Osman A (1990) Dimensional overlap: Cognitive basis for stimulus-response compatibility: A model and taxonomy. Psychol Rev 97:253-270

Krampe RT, Kliegl R, Mayr U, Engbert R, Vorberg D (2000) The fast and the slow of skilled bimanual rhythm production: Parallel vs integrated timing. J Exp Psychol Hum Percept Perform 26:206-233

Kugler PN, Turvey MT (1987) Information natural law and the self-assembly of rhythmic movement. Lawrence Erlbaum, Hillsdale NJ

Leipmann HMO (1907) Ein Fall von linksseitiger Agraphie unf Apraxie bei rechtsseitiger. Lähmung Monatszeitschrift für Psychiatrie und Neurologie 10:214-227

Marteniuk RG, MacKenzie CL, Baba DM (1984) Bimanual movement control: Information processing and interaction effects. Quart J Exp Psychol 16A:335-365

Mechsner F, Kerzel D, Knoblich G, Prinz W (2001) Perceptual basis of bimanual coordination. Nature 414:69-73

Nambisan R, Diedrichsen J, Ivry RB, Kennerley S (2002) Two autopilots one brain: limitations and interactions during online adjustment of bimanual reaching movements. Paper presented at the annual meeting of the Society for Neuroscience, Orlando FL

Oldfield RC (1971) The assessment and analysis of handedness: the Edinburgh inventory. Neuropsychologia 9:97-113

Pashler H (1994) Dual-task interference in simple tasks: data and theory. Psychol Bul 116:220-244

Perenin MT, Vighetto A (1988) Optic ataxia: a specific disruption in visuomotor mechanisms. I Different aspects of the deficit in reaching for objects. Brain 111:643-674

Perrig S, Kazennikov O, Wiesendanger M (1999) Time structure of a goal-directed bimanual skill and its dependence on task constraints. Behav Brain Res 103:95-104

Peters M (1994) Does handedness play a role in the coordination of bimanual movement? In: Swinnen SP, Heuer H, Massion J, Casaer P (eds) Interlimb coordination: Neural dynamical and cognitive constraints. Academic Press, London, pp 595-615

Pisella L, Grea H, Tilikete C, Vighetto A, Desmurget M, Rode G, Boisson D, Rossetti Y (2000) An 'automatic pilot' for the hand in human posterior parietal cortex: toward reinterpreting optic ataxia. Nature Neuro 3:729-736

Povel D-J (1981) Internal representation of simple temporal patterns. J Exp Psychol Hum Percept Perform 7:3-18

Prablanc C, Martin O (1992) Automatic control during hand reaching at undetected two-dimensional target displacements. J Neurophysio 67:455-469

Riek S, Carson RG, Byblow WD (1992) Spatial and muscle dependencies in bimanual coordination. J Hum Mov Stud 23:251-265

Robertson SD, Zelaznik HN, Lantero DA, Bojczyk KG, Spencer RM, Doffin JG, Schneidt T (1999) Correlations for timing consistency among tapping and drawing tasks: Evidence against a single timing process for motor control. J Exp Psychol Hum Percept Perform 25:1316-1330

Rushworth MF, Nixon PD, Passingham RE (1997) Parietal cortex and movement I Movement selection and reaching. Exp Brain Res 117:292-310

Schluter ND, Krams M, Rushworth MF, Passingham RE (2001) Cerebral dominance for action in the human brain: the selection of actions. Neuropsychologia 39:105-113

Schluter ND, Rushworth MF, Passingham RE, Mills KR (1998) Temporary interference in human lateral premotor cortex suggests dominance for the selection of movements A study using transcranial magnetic stimulation. Brain 121:785-799

Schmidt RA, Heuer H, Ghodsian D, Young DE (1998) Generalized motor programs and units of action in bimanual coordination. In: Latash, ME (ed) Progress in motor control Vol 1: Bernstein's traditions in movement studies. Human Kinetics, Champaign IL, pp 329-360

Semjen A (2002) On the timing basis of bimanual coordination in discrete and continuous tasks. Brain Cog 48:133-148

Semjen A, Ivry RB (2001) The coupled oscillator model of between-hand coordination in alternate-hand tapping: A reappraisal. J Exp Psychol Hum Percept Perform 27:251-265

Semjen A, Summers JJ, Cattaert D (1995) Hand coordination in bimanual circle drawing. J Exp Psychol Hum Percept Perform 21:1139-1157

Serrien DJ, Wiesendanger M (2000) Temporal control of a bimanual task in patients with cerebellar dysfunction. Neuropsychologia 38:558-565

Serrien DJ, Nirkko AC, Lovblad KO, Wiesendanger M (2001) Damage to the parietal lobe impairs bimanual coordination. Neuroreport 12:2721-2724

Spencer RMC, Zelaznik HN, Diedrichsen J, Ivry RB (2003) Disrupted timing of discontinuous but not continuous movements by cerebellar lesions. Science 300:1437-1439

Spijkers W, Heuer H (1995) Structural constraints on the performance of symmetrical bimanual movements with different amplitudes. Quart J Exp Psychol: Human Experimental Psychology 48:716-740

Steglich C, Heuer H, Spijkers W, Kleinsorge T (1999) Bimanual coupling during the specification of isometric forces. Exp Brain Res 129:302-316

Stucchi N, Viviani P (1993) Cerebral dominance and asynchrony between bimanual two-dimensional movements. J Exp Psychol Hum Percept Perform 19:1200-1220

Swinnen SP, Dounskaia N, Duysens J (2002) Patterns of bimanual interference reveal movement encoding within a radial egocentric reference frame. J Cog Neuro 14:463-471

Swinnen SP, Dounskaia N, Walter CB, Serrien D J (1997) Preferred and induced coordination modes during the acquisition of bimanual movements with a 2:1 ratio. J Exp Psychol Hum Percept Perform 23:1087-1110

Tuller B, Kelso JAS (1989) Environmentally-specified patterns of movement coordination in normal and split-brain subjects. Exp Brain Res 75:306-316

Turvey MT (1990) Coordination. Am Psychol 45:938-953

Vos PG, Mates J, Kruysbergen NW (1995) The perceptual centre of a stimulus as the cue for synchronization to a metronome: Evidence from asynchronies. Quart J Exp Psychol 48A:1024-1040

Weigelt C, Cardoso De Oliveira S (2003) Visuomotor transformations affect bimanual coupling. Exp Brain Res 148:439-450

Whiting HTA (Ed) Human motor actions: Bernstein reassessed Amsterdam: North Holland (1984) Advances in Psychology Series Vol 17

Wimmers RH, Beek PJ, Vanwieringen PCW (1992) Phase-transitions in rhythmic tracking movements: a case of unilateral coupling. Hum Mov Sci 11: 217-226

Zelaznik HM, Spencer RM, Doffin J (2000) Temporal precision in tapping and circle drawing movements at preferred rates is not correlated: Further evidence against timing as a general purpose ability. J Motor Beh 32:193-199

Zelaznik HM, Spencer RM, Ivry RB (2002) Dissociation of explicit and implicit timing processes in repetitive tapping and drawing movements. J Exp Psychol Hum Percept Perform 28:575-588

Chapter 11

DYNAMICAL MODELS OF RHYTHMIC INTERLIMB COORDINATION
Relating pattern (in)stability to neural processes and effector properties

C. (Lieke) E. Peper, Andreas Daffertshofer, & Peter J. Beek
Institute for Fundamental and Clinical Human Movement Sciences, Faculty of Human Movement Sciences, Vrije Universiteit, Amsterdam, The Netherlands

Abstract: The dynamical system approach to movement coordination has highlighted the importance of stability and loss of stability of coordination patterns. The empirically observed stability characteristics of rhythmic interlimb coordination have been modeled in terms of gradient dynamics of the (order) parameter that defines the coordination modes. However, the high level of abstraction of these models precludes gaining insight into how the stability features result from underlying processes and system properties. Modeling the limb movements and their interactions as a system of coupled oscillators holds greater promise in this respect. Because the development of such models is mathematically underconstrained, the identified coordination dynamics may be used to determine stability-related aspects of relevant system properties and interaction processes, which may, in turn, constrain the dynamical modeling of rhythmic interlimb coordination.

Key words: dynamical systems, nonlinear oscillators, pattern stability, brain dynamics

1. INTRODUCTION

The types of constraints that play a role in rhythmic bimanual coordination are legion. Broadly speaking, one may distinguish between voluntary and structural constraints, that is, constraints that are brought to bear by the actor to achieve a particular task goal vs. constraints that reflect inherent limitations or coordination tendencies (cf. Heuer, 1996, see also

Chapter 9). Structural constraints vary in the degree to which they are 'hard-wired' or 'soft-wired', that is, the extent to which they may be overcome when performing a certain task. The distinction between voluntary and structural constraints plays a pivotal role in the study of bimanual coordination as it brings along a host of essential questions in the study of motor control and learning, pertaining, among other things, to the conflict between motor freedom and controllability, the ability to cognitively 'override' softly assembled structural constraints (cf. Chapter 10), and the history- and context-dependent nature of motor learning. To be able to address the broader issues in a productive and lasting manner, we deem it necessary to first uncover the structural constraints on coordination. This is already quite a tall order to achieve because the structural constraints that influence bimanual coordination are rich and varied. Heuer (1996) distinguished between temporal coupling, phase coupling, force coupling, and structural constraints during motor preparation, but such taxonomies are necessarily arbitrary (for instance, in Heuer's division spatial coupling appears as a subclass of phase coupling, but one could also argue that temporal and spatial couplings are primordial). In any case, when studying the structural constraints on bimanual coordination, one has to choose a convenient starting point, which, in general, will be a function of both theoretical and methodological considerations. In our research, we have adopted a primary (but not a unique) focus on the temporal coupling between rhythmic (oscillatory) limb movements. The reason for this choice, which will be elaborated in this chapter, is that we believe that pattern stability (and the loss thereof) provides an expedient entry point for studying the interactions between limb movements, especially those that are due to temporal constraints.

A focus on temporal aspects of bimanual coordination implies that all the participating components are viewed as dynamical systems. Since virtually everything that evolves in time is a dynamical system (Newell et al., 2001), any biological system can be formally described in terms of its dynamics. Of course, this approach is only viable when concentrating on essential properties of the system under investigation. In this spirit, one interprets coordination (or coordinative patterns) as the collective behavior of so-called open systems, a term used in thermodynamics to refer to dissipative systems that are in permanent contact with the environment (Haken, 1983). As a rule, such systems are composed of numerous subsystems and may organize themselves into coherent patterns or structures due to interactions between the assembling subsystems, at least under certain circumstances. A plenitude of instances of such systems and corresponding coherent structures have been studied over the years, particularly in the inanimate world (e.g., laser light, sand ripples, cloud formations, etc.). Interestingly, all such coherent structures or patterns have in common that their explicit forms are not prescribed by any external mechanism. Hence, the formation of the patterns is said to be self-organized. Invariably, the driving 'force' for self-organization is the cooperation among the participating microscopic components (photons, molecules, neurons, muscle fibers, etc.), culminating in macroscopic structures that can be characterized by one or few quantities. These quantities reflect ordered states of the entire system and are therefore called order parameters.

Irrespective of the explicit nature of the order parameters, their qualitative changes (their emergence or disappearance, or switches between ordered states) can be described as so-called non-equilibrium phase transitions. Consequently, the system's evolution can be formalized using concepts from theories of self-organized pattern formation, such as Haken's (1983) synergetics: Spontaneous changes of macroscopic patterns, i.e., phase transitions, are cast into the (bifurcation) dynamics of order parameters. To appreciate this rigorous step it is essential to realize that, in the immediate vicinity of phase transitions, order parameters evolve arbitrarily slowly with regard to the generating subsystems, leading to an extremely slow relaxation back to the original state after a perturbation (critical slowing down). The subsystems, however, preserve their individual finite time scales so that, from the perspective of the order parameters, all the subsystems evolve arbitrarily fast. As a consequence, the subsystems adapt (almost) instantaneously to changes of the order parameters, which are therefore considered to prescribe the dynamics of the subsystems (slaving principle). The drastic separation of time scales and the following causal transactions between microscopic components and macroscopic states are referred to as circular causality (Haken, 1983). Thus, the upshot of the slaving principle is that the information necessary to characterize coherent macroscopic structures is significantly less than the information required to describe the accompanying microscopic processes: Qualitative (macroscopic) changes of the system's behavior can be studied via (low-dimensional) order parameter dynamics, provided that the order parameters can be identified.

The aim of the present chapter is to demonstrate how these theoretical concepts have been appropriated to the experimental and theoretical study of bimanual coordination and its neural basis, as well as to discuss the challenges confronted in this enterprise. First we summarize a by now well-established phenomenological approach for studying the relative phasing between limb movements based on gradient dynamics (Section 2). Subsequently, we discuss how this approach may be complemented through explicit modeling of the components and their interactions as nonlinear systems of coupled oscillators (Section 3). Finally, in the second half of the chapter we discuss how the approach can be expanded addressing pertinent constraints (Section 4) and underlying neural processes (Section 5).

2.   GRADIENT DYNAMICS FOR RELATIVE PHASE

2.1   The HKB Potential: The Main Idea

Almost two decades ago, Kelso (1984) experimentally demonstrated the occurrence of phase transitions in rhythmic bimanual coordination: When subjects start out to cycle their index fingers (or hands) rhythmically in antiphase (simultaneous activation of nonhomologous muscle groups) and gradually increase the cycling frequency (as prescribed by a metronome), a spontaneous, involuntary switch to the in-phase pattern (simultaneous activation of homologous muscle groups) occurs at a certain critical frequency. Beyond this critical frequency only the in-phase pattern can be stably performed. In fact, this change in coordination is an example of the kind of qualitative changes described in the introduction, as Kelso already intuited correctly at the time. The subsequent mathematical formulation, the HKB model (Haken et al., 1985), may be considered a landmark paradigm in the field of study now known as coordination dynamics. The model builds on the theory of nonlinearly coupled nonlinear oscillators (see Section 3) yielding a potential to describe the stability properties of relative phase between the two limbs. In brief, the evolution of coordinative patterns is described in terms of gradient dynamics, that is, if the time-dependent variable defining the coordinative patterns of interest (here the relative phase) is denoted by ϕ, then its dynamics receives the form

$$\dot{\phi} = \frac{d\phi}{dt} = -\frac{dV}{d\phi} \tag{1}$$

In words, the evolution of $\phi = \phi(t)$ can be seen as an overdamped motion (of a ball) over the potential landscape defined by $V = V(\phi)$. Put differently, the (gradient of the) potential defines the magnitude and direction of the change of ϕ as a function of its actual value. Evidently, if the time-derivative of ϕ vanishes the corresponding state is stationary. Whenever such a steady situation coincides with a (local) minimum of $V(\phi)$, it is stable, whereas if it reflects a (local) maximum, it is unstable.

The simplest form of $V(\phi)$ that accounts for the phase transition observed by Kelso (1984) proved to be a superposition of two cosine terms, i.e.,

$$V(\phi) = -a\cos\phi - b\cos 2\phi \tag{2}$$

In combination with the definition of gradient dynamics given by eq. (1), this potential results in the following equation of motion for the relative phase between the two fingers (the order parameter equation)

$$\dot{\phi} = -a \sin \phi - 2b \sin 2\phi \qquad (3)$$

In these equations, b/a serves as a so-called control parameter. The minima of $V(\phi)$, $\phi = 0$ and $\phi = \pm\pi$, represent the attractor states of relative phase, that is, the in-phase and antiphase coordination patterns. To account for the observed transition from antiphase to in-phase coordination, the experimental control parameter frequency has to induce changes in the potential landscape. To this end, frequency is assumed to be reciprocally related to b/a: With increasing frequency, b/a decreases, resulting in a differential decrease in stability of the two phase relations. This leads to the observed transition at a critical value of $b/a = ¼$ where the stable solution for antiphase coordination is annihilated (see Fig. 11-1): From a minimum (stable fixed point or 'attractor') it turns into a maximum (unstable fixed point or 'repellor'). In this situation (i.e., for $b/a > ¼$), a slight perturbation is sufficient to cause a transition to the in-phase pattern. Such perturbations are typically modeled by adding random fluctuations, which are always present in empirical systems, to the phase dynamics (see Schöner et al., 1986).

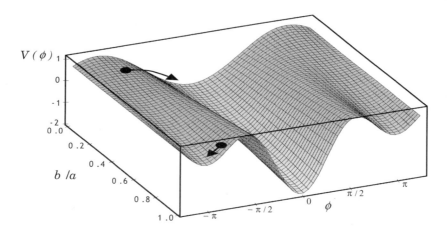

Figure 11-1. HKB potential for different values of the control parameter b/a. While the in-phase solution is always stable (global minimum at $\phi = 0$), the characteristics at the stationary solutions at $\phi = \pm\pi$ (antiphase) change qualitatively: For large values of b/a the antiphase solution is stable since local perturbations from that state are followed by a relaxation back to $\phi = \pm\pi$; for small values of b/a a perturbation results in a switch to the in-phase solution ($\phi = 0$) since the antiphase state has become a maximum.

The addition of random fluctuations to the phase dynamics is not only important for theoretical reasons pertaining to the explicit transition mechanism (e.g., the occurrence of critical fluctuations prior to the transition), but it also allows for an operational study of stability in terms of variability. After all, if the relative phase is in a steady state the form of the corresponding minimum determines the degree of noise-induced variability: In a more shallow minimum the same amount of noise results in larger relative phase variability. As such, the standard deviation of ϕ (SDϕ) provides an index of coordinative stability. However, some caveats are in order here. First, SDϕ may also be affected by the explicit form of the limb trajectories, which need not be harmonic (as was assumed in the derivation of the HKB potential). Second, SDϕ only represents an indirect measure of stability if rather stringent assumptions regarding the phase dynamics apply. It therefore remains important to compare relative phase variability with direct estimates of the form and strength of the attractor, such as may be derived from the response characteristics following a (transient) perturbation. In two recent empirical studies (Post et al., 2000a,b), we demonstrated the existence of subtle differences between such indirect and direct estimates of pattern stability.

2.2   The HKB Potential: Generalizations And Extensions

The HKB potential proved to be generalizable to a variety of situations. The characteristic transition from antiphase to in-phase coordination was, for instance, also observed for coordination between two nonidentical limb segments (e.g., a hand and a foot, cf. Carson et al., 1995), between two segments of a single limb (Buchanan & Kelso, 1993), between a single limb and an environmental stimulus (Kelso et al., 1990; Wimmers et al., 1992), and between the movements of two different persons (Schmidt et al., 1990). These observations revealed that the identified coordination principles are relatively independent of the system's components and the precise mechanism of their interaction, thereby underscoring the relevance of modeling these dynamics at a high level of abstraction.

In subsequent research, several factors were identified that affected the dynamics of rhythmic interlimb coordination. A series of experiments using a dual-task paradigm revealed a relation between pattern stability and attention: More attention was required to perform the less stable antiphase pattern, variations in pattern stability due to changes in movement frequency were associated with similar variations in the required amount of attention, and coordinative stability decreased when attention was primarily directed to another task (cf. Monno et al., 2002; Temprado et al., 1999; see also Carson et al., 1999). This relation between attention and pattern stability indicates that the HKB potential may be affected by attentional processes.

Other factors appeared to affect the exact values of relative phase that allowed for stable performance. Handedness, for instance, induces a slight shift in relative phase: In left-handed persons the left hand is leading in time (e.g., $\phi = \phi_{\text{left}} - \phi_{\text{right}} > \pi$ for intended antiphase coordination), whereas the opposite is **true** for right-handers (Treffner & Turvey, 1995, 1996). To account for this phase shift, Treffner and Turvey proposed that the HKB potential be expanded with two additional terms:

$$V(\phi) = -a\cos\phi - b\cos 2\phi + c\sin\phi + d\sin 2\phi \tag{4}$$

Note that the third term was actually not used in explaining the observed coordination characteristics (i.e., $c = 0$), whereas $d > 0$ for right-handers and $d < 0$ for left-handers. Figure 11-2A shows how nonzero values of d affect the location of the stable fixed points, inducing the empirically observed shifts in relative phase. Thus, without providing an explicit account of the mechanism(s) through which handedness influences interlimb coordination, its effects onto the coordination dynamics were captured phenomenologically by extending the HKB potential. Interestingly, the same potential also provided an adequate account of asymmetries in the coordination dynamics induced by directing attention to one of the coordinated limbs (Amazeen et al., 1997), which may be regarded as support for the suggestion that handedness is related to an inherent asymmetry in the degree of attention devoted to the limbs (cf., Peters, 1994).

In addition, (bio)mechanical differences between the coordinated limbs were shown to affect interlimb coordination. In particular, a large number of studies indicated that a difference in the eigenfrequencies (i.e., natural frequencies) of the coordinated limbs can lead to a variety of coordinative phenomena, ranging from shifts in relative phase (e.g., Rosenblum & Turvey, 1988; Schmidt et al., 1993; Sternad et al., 1992) in combination with reduced stability (cf. Schmidt & Turvey, 1995) to 'relative coordination', in which the behavior is no longer permanently attracted to a stable coordination pattern, although the constantly changing relative phase is still characterized by momentary preferences ('remnants' of the original attractors; cf. Kelso & Jeka, 1992). The observed range of behaviors could be understood by adding a single symmetry-breaking term to the original (symmetric) HKB potential, indicating that they all resulted from the same basic coordination principles (Kelso et al., 1990):

$$V(\phi) = -\Delta\omega\phi - a\cos\phi - b\cos 2\phi \tag{5}$$

where $\Delta\omega$ functions as an additional control parameter that affects pattern stability and may induce transitions between coordination modes (see Fig. 11-2B). On the basis of the above mentioned empirical results this parameter is commonly interpreted as reflecting the eigenfrequency difference between the coordinated limbs.

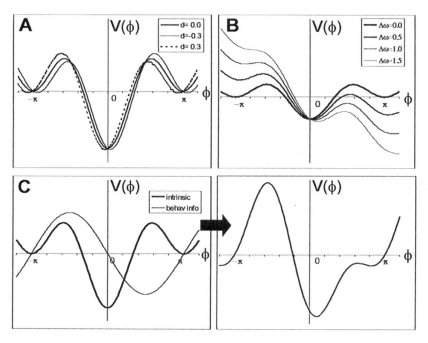

Figure 11-2. Three extensions of the HKB potential. A: Influence of handedness according to eq. (4), for both $d > 0$ and $d < 0$ ($a = b = 1$; $c = 0$). B: Influence of a difference in eigenfrequency ($\Delta\omega$) according to eq. (5). By tilting the potential, $\Delta\omega$ shifts the locations of the minima. For large values of $\Delta\omega$ minima are no longer present (not shown), so that no stable phase relation can be attained. However, the form of the potential still results in momentary attraction to particular phase relations ('relative coordination'); $a = b = 1$. C: The coordination dynamics (right) results from the interplay between the intrinsic dynamics and behavioral information (left).

A more general contribution to the formalization of relative phase dynamics was developed by Schöner and Kelso (1988; see also Schöner et al., 1992), who introduced the distinction between *intrinsic dynamics*, being the dynamics of the system in the absence of external or cognitive influences (presumably captured by the HKB potential), and *behavioral information*, which reflects the influence of a variety of factors that influence the relative phase dynamics (e.g., a phase relation specified in the environment, or an intended or memorized coordination pattern). Since both are specified in terms of the same order parameter, their summation yields the dynamics as observed at the behavioral level (cf. Fig. 11-2C). The resulting dynamics are particularly interesting in the context of learning, where the learning process is regarded as a dynamic process, during which the memorized behavioral information evolves towards the to-be-learned pattern (Schöner, 1989; Schöner et al., 1992). Consequently, mastering a new coordination pattern is conceptualized as a deformation of the initial potential (i.e., the intrinsic dynamics) toward a potential in which the to-be-learned pattern is present as a stable fixed point. Thus, according to this approach learning new coordination patterns affects the coordination dynamics as a whole, rather than resulting solely in improved performance of the new pattern. Furthermore, it indicates that for each individual the learning process is influenced by his or her initial intrinsic dynamics. Empirically, these model predictions may be assessed by examining performance for the full range of values of the order parameter before and after the learning process, so that the changes in the potential as a whole can be visualized (Kelso & Zanone, 2002; Zanone & Kelso, 1992).

In sum, explicit modeling of the relative phase dynamics by means of a potential has instigated research focused on stability-related aspects of rhythmic interlimb coordination, addressing (changes in) the location of stable fixed points and their associated degree of stability. This work has underscored that pattern stability (and loss of stability) is an essential feature of interlimb coordination and suggested that the underlying functional organization of the central nervous system is reflected by the identified coordination dynamics. Moreover, the high level of abstraction has highlighted the generality of these coordination principles, indicating that similar dynamics may be observed for systems that differ with respect to their components and manner of interaction. At the same time, however, due to this level of abstraction the HKB potential does not provide any insight into the origins of coordinative (in)stability. Haken et al. (1985) addressed this latter aspect by including a second level of modeling, in which the rhythmically moving limbs and their interactions were formalized as a nonlinear system of coupled oscillators.

3. MODELS OF COUPLED OSCILLATORS

3.1 The HKB Model: A Brief Summary

The oscillator model proposed by Haken et al. (1985) is based on the view that salient coordination phenomena, such as phase and frequency locking and frequency-induced phase transitions, are (primarily) brought about by specific nonlinear couplings. To model these coupling constraints, one first requires an explicit dynamical account for the oscillating limbs. As a first approximation, the rhythmically moving limbs have been modeled as self-sustaining or 'limit cycle' oscillators with stable periodic orbits and amplitudes due to particular combinations of linear and nonlinear damping terms (Beek et al., 1996; Haken et al., 1985; Kay et al., 1987). To account for the empirically observed spatiotemporal characteristics of rhythmic limb movements, such limit cycle models usually take the form of the so-called 'hybrid' model, which is characterized by a decrease in amplitude and an increase in peak velocity with increasing movement frequency.

For a given limit cycle description, the bimanual system may be cast into the following form

$$\underbrace{\ddot{x}_j + \omega_j^2 x_j}_{\text{oscillation}} + \underbrace{n\left(x_j, \dot{x}_j\right)}_{\substack{\text{amplitude} \\ \text{stabilization}}} = \underbrace{I_{j \leftarrow k}}_{\text{coupling}} \qquad (6)$$

in which ω_j refers to the frequency of oscillator j, $n(\cdots)$ guarantees its stable limit cycle properties, and $I_{j \leftarrow k}$ refers to the coupling with oscillator k. In the HKB model (Haken et al., 1985), this coupling is modeled in terms of a nonlinear function that allows for stable performance of both in- and antiphase coordination at low movement frequencies, whereas at high movement frequencies only the in-phase pattern is stable. Specifically, in the commonly adopted version of the model, the empirically observed (pitchfork) bifurcation structure, with movement frequency as control parameter, is viewed as the result of a cubic amplitude coupling, i.e., the frequency-induced changes in stability are mediated by accompanying changes in movement amplitude (cf. Peper & Beek, 1998a,b, 1999). Incorporation of higher-order nonlinearities yields more complex stability patterns including transitions between different frequency ratios (Haken et al., 1996), which may account for the frequency-induced switches between polyrhythmic coordination patterns as observed by Peper et al. (1991, 1995).

3.2   Two-Level Model Of Coupled Oscillators

Extensive experimentation and model testing has demonstrated that the strength of the HKB model resides predominantly in the properties of the relative phase dynamics, as described by the HKB potential (see Section 2.1). Theoretically, the derivation of two nonlinearly coupled oscillators complementing the HKB potential implied a nontrivial addition because it formally explicated both the dynamics of the limb movements and their coupling (even though there is an arbitrarily large number of coupled oscillators that are consistent with the HKB potential). However, the results of empirical tests suggested that the validity of the proposed model of coupled oscillators is questionable, in terms of both the intrinsic features of the moving limbs as captured by the limit cycle description and the postulated coupling (see Beek et al., 2002 for a review). Specifically, detailed studies of single limb movements revealed various properties that are difficult to reconcile with a simple limit cycle description, such as: (i) non-monotonic frequency-peak velocity relations (Beek et al., 1996); (ii) the occurrence of a phase-dependent phase shift following a transient perturbation (Kay et al., 1991), (iii) the existence of negative correlations between adjacent movement cycle durations (Daffertshofer, 1998), and (iv) the possibly chaotic nature of oscillatory limb movements (Mitra et al., 1997). Furthermore, with regard to the postulated amplitude coupling, several studies have indicated that whereas movement frequency has a distinct effect on pattern stability, this is not the case for movement amplitude (Peper & Beek, 1998a,b, 1999; Post et al. 2000a).

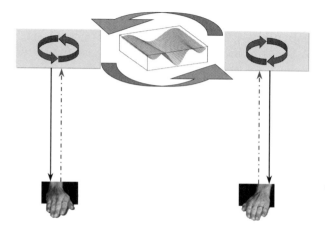

Figure 11-3. The coordination dynamics are decomposed into the dynamics of end-effectors and neural units. The latter interact via bilateral coupling functions that resemble the original HKB coupling and force the end-effectors. The state of the end-effectors, in turn, is fed back to the neural level.

In view of these shortcomings, Beek et al. (2002) formally explicated a dynamical model of interlimb coordination composed of two levels (see Fig. 11-3), referred to as the 'neural' and the 'effector' level, respectively. The movements of each individual limb are modeled by means of a nonlinear, neural oscillator which forces a linear, effector oscillator. This effector oscillator (reflecting the peripheral properties of the end-effector), in turn, influences the neural oscillator on the basis of afferent feedback. Moreover, interactions between the limbs are accounted for by means of a bilateral nonlinear coupling between the neural oscillators. Although this model is still rather abstract, its basic structure allows for relating the coordination dynamics (as observed at the behavioral level) to the functional influences of various system properties, such as biomechanical characteristics of the end-effectors and stability-related aspects of various interaction processes (see Section 4). In this context it is useful to note that the neural level, accounting for the generation of the oscillatory patterns as well as their interactions, may in fact involve various neurophysiological processes at different levels in the nervous system. It is therefore conceivable that in order to capture the particular contributions of distinct processes, the proposed neural level may be further decomposed into additional functional levels in the future. Finally, it should be noted that the model's forcing of the end-effectors and the influence of generated afferent feedback (cf. Fig. 11-3) also reflect neurophysiological processes, although for the sake of functional transparency they are not defined at the neural level of the model.

This basic model construct allows for an adequate account of the mentioned properties of single limb movements, with (i) spatiotemporal characteristics ensuing from resonance phenomena; (ii) perturbations of the end-effector inducing (via afferent signals) modulation of the activity of the neural oscillator; (iii) the characteristic negative autocorrelations resulting from the incorporation of a single noise source; and (iv) the possibility of chaotic behavior due to the increased number of state variables (for more details, see Beek et al., 2002). Since the neural oscillators associated with the two limbs are coupled, the coordination dynamics of rhythmic interlimb coordination can be accounted for by an adequate formulation of the coupling functions. Given the lack of support for the original amplitude coupling (see above), the changes in stability may be realized by incorporating a coupling function through which movement frequency is directly related to pattern stability (rather than its influence being mediated by associated changes in amplitude).

4.   CONSTRAINTS FOR FURTHER MODELING

4.1   The Need For Modeling Constraints

Further elaboration of the two-tiered model sketched in Section 3.2 requires casting the model's components and their interactions into an adequate mathematical form such that the overall dynamics corresponds to the identified coordination dynamics. Although Beek et al. (2002) demonstrated that this is possible in principle, the choice of the formal representation of the system's properties is largely underconstrained. A natural source of constraints may be sought in underlying neurophysiological and biomechanical system properties. However, to date it is still largely unknown which properties and processes contribute to the stability-related aspects of rhythmic interlimb coordination. Therefore, we choose to follow a top-down approach in which the abstract coordination dynamics is used as a guide in the search for stability-related aspects of the underlying system properties and interaction processes.

In this approach, various steps need to be taken. At the highest level of abstraction, the potential provides an account of the dynamics of relative phase. Because these dynamics are critically dependent on the way in which the limbs interact, the potential may be used to determine global requirements for the model's coupling functions. As we are dealing with a many-to-one mapping (i.e., a variety of coupling functions can give rise to the same potential), it is useful to limit the number of options by focusing on qualitative aspects of the relation between interlimb interactions and the overall coordination dynamics. This conceptual step is illustrated in the next paragraph. Because the thus obtained requirements are still rather abstract and defined independently of more structural aspects of the movement system, additional assessments of the way in which particular system properties contribute to coordinative (in)stability may provide further modeling constraints. In this context we may distinguish between influences related to particular (e.g., neuromuscular or biomechanical) limb properties and those arising from various neurophysiological interaction processes. Both are illustrated below.

Establishing the relation between coordinative (in)stability and underlying system properties and processes may cut both ways: The identified coordination dynamics may guide the quest for essential ingredients of the coordination process, while at the same time identification of these ingredients may constrain further development of the proposed model construct, leading to model components that reflect functional aspects of the underlying movement system.

4.2   Global Coupling Requirements: An Example

To illustrate how a potential may be instrumental in uncovering qualitative requirements for modeling the interlimb coupling, we consider the extended potential proposed by Treffner and Turvey (1995, 1996), i.e., eq. (4), which captures the asymmetries in coordination induced by hand dominance. Characteristic of this model is the nonzero coefficient of the second sine term in the potential, which reflects the influence of handedness (see Section 2.2). What implications does this have for the formulation of the coupling between the oscillators? Defining relative phase as $\phi = \phi_L - \phi_R$

(with L and R referring to the left and right oscillator, respectively) implies that $\phi = \phi_L - \phi_R$. Thus, the order parameter equation that corresponds to the proposed potential is directly related to the ways in which the phases $\phi_{L,R}$ of the component oscillators vary over time due to their (mutual) interactions (cf. Haken et al., 1985). (Note that if, for harmonic oscillators, $\dot{\phi}_{L,R} = 0$ the oscillations solely depend on the oscillation frequencies, i.e., the oscillatory movements are not affected by coupling influences exerted by the other oscillator.) Assuming that the functional form of the coupling is identical in both directions, we may suggest (in line with Haken et al., 1985) that

$$\dot{\phi}_L = a_L \sin(\phi_R - \phi_L) + 2\,b_L \sin 2(\phi_R - \phi_L) - 2d_L \cos 2(\phi_R - \phi_L) \qquad (7a)$$

or, equivalently,

$$\dot{\phi}_L = -a_L \sin(\phi_L - \phi_R) - 2b_L \sin 2(\phi_L - \phi_R) - 2d_L \cos 2(\phi_L - \phi_R) \qquad (7b)$$

and

$$\dot{\phi}_R = a_R \sin(\phi_L - \phi_R) + 2b_R \sin 2(\phi_L - \phi_R) - 2d_R \cos 2(\phi_L - \phi_R) \qquad (8)$$

Subtracting eq. (8) from eq. (7b) indeed yields the required order parameter equation (cf. eqs. (1) and (4)):

$$\dot{\phi} = -(a_L + a_R)\sin \phi - 2(b_L + b_R)\sin 2\phi - 2(d_L - d_R)\cos 2\phi \qquad (9)$$

(Note that for the sake of clarity, we have omitted the first additional term proposed by Treffner and Turvey, since its coefficient was set to 0.)

Although Treffner and Turvey (1995, 1996) confined their contribution to the potential, relating it to this lower (but still rather high) level of abstraction reveals an interesting possible source of the coordinative asymmetries due to hand dominance: The compound coefficient of the third term $(d_L - d_R)$ is nonzero only if the coefficients associated with the two hands are not identical. Because these coefficients are related to the strength of the interaction forces exerted by the other oscillator, an imbalance between d_L and d_R reflects an asymmetry in coupling strength. This suggests that the empirically observed phase shifts may be caused by an asymmetry in the degree to which the limbs are influencing each other. Thus, for right-handers the observed shifts in relative phase may indicate that the coupling influence exerted by the right hand onto the left hand (reflected by d_L) is larger than the influence in the opposite direction (reflected by d_R), while the reverse situation may explain the results obtained for left-handers. This is in line with earlier suggestions that the nondominant hand is more strongly influenced by the dominant hand than vice versa (e.g., Byblow et al., 1994).

In this context it is useful to note that the first two terms of eq. (9) reflect the result of interaction processes through which the two oscillators attract each other to a phase relation of 0 or π rad, which implies that asymmetries in coupling strength will not affect the resulting phase relation between the limbs. The third term, however, reflects a competition between two interaction processes that operate in opposite directions, both driving the phase relation away from in-phase and antiphase coordination. The influences of the latter processes cancel when the strengths of the coupling forces are identical, but result in a phase shift in case of an asymmetry in coupling strength (e.g., due to a left-right difference in the degree of inhibition of particular interaction processes).

4.3   Stability-Related Influences of Limb Properties

In a series of studies, Carson and colleagues (e.g., Carson, 1996; Carson & Riek, 1998) demonstrated that for unimanual coordination with an external rhythmic stimulus, synchronization of finger flexion with the stimulus was more stable than synchronization of finger extension (and, in a similar vein, synchronized forearm pronation was more stable than synchronized supination, Byblow et al., 1995). In addition, the degree of stability was affected by the postural orientation of the limb segments in question, which entailed manipulation of the lengths of the muscles involved. Given the relation between a muscle's length and its strength, this may suggest that relative muscular strength is related to coordinative stability (cf. Carson, 1996; Carson & Riek, 1998). Carson et al. (2000) further investigated these phenomena by manipulating the axis of rotation relative to the long axis of the forearm while subjects performed rhythmic pronation-supination movements. These manipulations were shown to affect the stability of synchronizing either peak pronation or peak supination with an external stimulus. Moreover, bimanual coordination was examined for various combinations of these axes of rotation (as applied to the individual arms). Interestingly, a gradual increase in movement frequency typically resulted in a transition to the composite of the two most stable unimanual patterns associated with the two axes of rotation that were used in a particular trial. This indicates that the stability of interlimb coordination is at least partially dependent on (stability-related) neuromuscular properties of the individual limb movements.

Another indication that peripheral limb properties may affect the coordination dynamics is provided by the asymmetries in coordination that are typically observed if the coordinated limb segments differ in eigenfrequency (see Section 2.2). Although many studies have revealed that such situations are characterized by shifts in relative phase and reduced coordinative stability, it remains unclear how such a (bio)mechanical aspect of the individual limbs may affect the coordination between them. Part of the observed phase shifts may directly be explained in terms of the proposed two-level organization of single limb movements (see Section 3.2), because the oscillation phase of a forced oscillator (the end-effector in our model) is delayed in time with respect to the forcing (neural) signal and, for a given forcing frequency, this phase lag increases with decreasing eigenfrequencies of the forced oscillator (e.g., Karnopp & Rosenberg, 1975). Consequently, the relative phasing between two limbs with different eigenfrequencies may be expected to deviate from the relative phase as imposed by the corresponding control (forcing) signals. However, this rather straightforward peripheral effect cannot readily account for the observed changes in coordinative stability and the empirically observed difference between (intended) in-phase and antiphase coordination with respect to the size of the induced phase shifts (cf. Schmidt & Turvey, 1995). This suggests that a difference in eigenfrequency (effector level) also affects the underlying control processes (neural level) (cf. Peper et al., 2003).

In a first attempt to pinpoint relevant variables in this regard, Van Soest et al. (in press) examined the way in which the dynamics of the effector oscillator (which, in their analysis, had the mechanical characteristics of a male lower leg) were affected by mass addition. This analysis showed that not only the effector's eigenfrequency was systematically affected, but also its low-frequency control gain. As this gain reflects the inverse of the muscular torque that is required to perform a particular movement, asymmetrical mass addition does not only induce an eigenfrequency difference between the limbs, but also an imbalance in the required muscular effort. An experiment in which the effects of a difference in eigenfrequency were dissociated from the influences stemming from the co-varying difference in the low-frequency control gains of the limb segments revealed that the latter difference did not affect the interlimb coordination dynamics, whereas a difference in eigenfrequency led to the typically observed shifts in relative phase (Peper et al., 2003). The question remains, however, how such an end-effector property (i.e., a difference in eigenfrequency) affects the (neural) control processes. Future studies in this regard may focus on compensatory changes in EMG timing (e.g., Baldissera et al., 1991, 2000; Mackey et al., 2002) and/or influences of differences in the *actual* levels of muscular torque (which, due to co-contractions, may exceed the *net* required torques as reflected by the low-frequency control gains; cf. Peper et al., 2003)

4.4   Stability-Related Interaction Processes

An essential insight gained by modeling the coordination dynamics at the level of coupled oscillators is that coordinative (in)stability and changes therein are critically dependent on the interactions between the limbs. Therefore, in modeling the dynamics of rhythmic interlimb coordination, the choice of adequate coupling functions is a central issue. A natural source of modeling constraints may be sought in known aspects of interlimb interaction processes. Several sources of interaction have been proposed in the literature, many of them substantiated by empirical results. For instance, activation of the muscles in one limb may affect muscle activity in the contralateral limb due to bilateral activation processes (cf. Cattaert et al., 1999; Swinnen, 2002), probably associated with the co-existence of ipsilateral and contralateral projections from the cortex. On the other hand, passive movements of one limb have been shown to affect the movement patterns of other limbs (e.g., Gunkel, 1962; Swinnen et al., 1995), which underscores the interactive influences of movement-induced afferent signals, possibly through entrainment of spinal reflex pathways (cf. Baldissera et al., 1991). Conversely, interaction between the limbs may also be an inherent consequence of the way in which the desired movement pattern is centrally generated or represented, as is suggested in a variety of models (e.g., Grossberg et al., 1997; Helmuth & Ivry, 1996; Jirsa & Haken, 1996). However, it is still largely unclear to what extent these various sources of interlimb interaction are related to (changes in) coordinative stability. Once more insight is gained into their relative contributions in this regard, their functional characteristics may serve as a basis for formulating adequate coupling functions.

Determining whether the proposed sources of interaction contribute to coordinative (in)stability is not straightforward, in particular due to difficulties in manipulating such processes and in monitoring relevant variables. Nevertheless, experiments along both lines may still provide valuable information. One option is to (attempt to) manipulate proposed interaction processes and examine how this affects the coordination dynamics. Manipulation of afferent signals, for instance, can be performed relatively easily, e.g., by means of passive movements or tendon vibration. Indeed, studies using tendon vibration to distort muscle spindle afferents, (Steyvers et al., 2001; Verschueren et al., 1999) indicated that the control of the relative phasing between two limbs involves proprioceptive information from both limbs. On the basis of adaptations to load perturbations, which were more efficient during antiphase than in-phase coordination, Baldissera et al. (1991) suggested that performance of the former pattern entailed the use of afferent signals in supraspinal processes in order to overcome (unintended) spinal entrainment to in-phase coordination. Moreover, Stinear and Byblow (2001) revealed that coordination between an actively moving hand with a hand that was passively driven ('kinesthetic tracking') adhered to similar coordination principles as active bimanual coordination (viz., differential stability of antiphase and in-phase coordination and frequency-induced transitions between these coordination patterns). This may suggest that also during active interlimb coordination, pattern stability is related to interactions based on afferent signals. In this context, it is useful to note that other sources of information may also be used in the coordination process. In particular, visual information about the interlimb phase relation appears to affect coordinative stability (e.g., Mechsner et al., 2001), possibly due to differential stability of the visual perception of specific relative phases (e.g., Bingham et al., 2001).

Other interesting results in this regard were recently obtained using transcranial magnetic stimulation (TMS). Stinear and Byblow (2002) revealed that during kinesthetic tracking inhibitory processes in the motor cortex associated with the passively moved limb were modulated due to movement-elicited afferences stemming from this limb. This modulation was influenced by the phase relation between the two limbs, resulting in decreased inhibition during in-phase coordination. When the limbs moved $\pi/3$ out of phase, however, intracortical inhibition was maintained, suggesting that this inhibition enhances the ability to generate independent muscle activation during active bimanual coordination. Meyer-Lindenberg et al. (2002) demonstrated that disturbance of the premotor and supplementary motor areas resulted in transitions from antiphase to in-phase coordination at the behavioral level, while applying such disturbances during in-phase coordination did not affect coordinative stability. Moreover, the degree of pattern stability (manipulated through movement frequency) corresponded to the strength of the pulse needed to elicit the phase transition. These results do not only indicate that the specified cortical areas were involved in stabilizing the antiphase pattern, but also that (in)stability at the behavioral level was associated with neural (in)stability.

In addition, experiments with patients in whom essential aspects of a proposed source of interaction are affected may provide valuable insights. For instance, although it has been demonstrated that interlimb interactions in the spatial domain are mediated by interhemispheric communication via the corpus callosum (e.g., Eliassen et al., 1999; Franz et al., 1996) persons with a dissected corpus callosum still showed strong attraction to the in-phase and antiphase coordination modes while performing bimanual 1:1 frequency tasks, indicating that the corresponding temporal coupling was not critically dependent on interhemispheric interactions (Tuller & Kelso, 1989). Recently, this interpretation has been refined by dissociating between temporal interlimb coupling during either repetitive discrete movements, such as tapping, or continuous rhythmic movements, such as fluent flexion-extension oscillations (Kennerley et al., 2002). The performance of the former class of movements was not affected in callosotomy patients, suggesting that in this situation interhemispheric communication is not essential for the temporal coupling between the limbs (which may, alternatively, rely on subcortical interaction mechanisms). However, these patients showed clear temporal uncoupling during continuous rhythmic movements, underscoring the relevance of the corpus callosum for the temporal coupling observed in the healthy control subjects. Thus, these results indicate that discrete and continuous movements may not only involve different timing mechanisms, but also different sources of interlimb interaction (cf. Kennerley et al., 2002)

Although manipulating crucial aspects in the coordination process and examining the coordination dynamics in patients with particular disorders may enhance our understanding of interlimb coordination, care is to be taken in generalizing the thus obtained findings to 'normal' bimanual coordination, in view of possible compensation or adaptation processes. This caveat is less salient when relevant variables are monitored during unmanipulated coordination in healthy subjects (i.e., without interventions aimed at altering the influence of particular aspects of the coordination process). In particular, the relation between the coordination dynamics at the behavioral level and the associated cortical activity has been studied and successfully modeled. In the next section the rationale behind this research and some of the key results are outlined.

5. CORTICAL DYNAMICS

Various research groups have started looking for cerebral contributions to dynamical interlimb interactions by analyzing patterns of temporal encoding across cortical areas during the course of limb movements. Obviously, when studying dynamical processes the experimental recordings should have sufficient temporal resolution, as is usually achieved by means of non-invasive techniques like electro- or magneto-encephalograms. The latter are assumed to reflect temporal changes of dendritic currents within and across neural clusters (for a recent discussion, see Jirsa et al., 2002) and may allow for a reliable 3D localization of active cortical areas even though the corresponding magnetic fields are recorded above the scalp (Tass et al., 2003). To illustrate the kind of insights one may gain through such recordings, we here focus on the relation between cortical signals and behavioral data during bimanual rhythmic tapping.

In general, when studying rhythmic performance one can presume that relevant interactions can be uncovered via phenomena like frequency and phase locking, here, between the cortical signals and behavioral measurements. That is, correlations between different components are manifested in the degree of concurrent spectral power (frequency) and subsequent timing of activity (phase). Indeed, interactions may be expected to cause correlations between the left and right sides at both the cortical and the behavioral level.

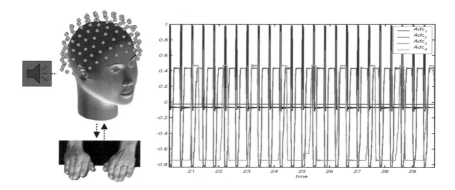

Figure 11-4. Multifrequency performance synchronized to an acoustic stimulus. The subject taps a polyrhythm, that is, the two index fingers move at a fixed frequency ratio, here 3:8. Right panel: Stimulus train specifying the tempo of the fast cadence (top; dark gray line) and taps of the left (slow) finger and of the right (fast) finger (bottom; light grey lines); Left panel: Simultaneous magneto-encephalographic recordings are indicated as sensor arrays above the scalp – cf. Fig. 11-5.

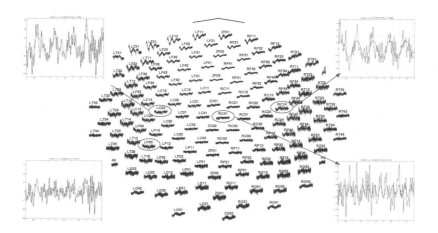

Figure 11-5. Time series recorded via magneto-encephalography during polyrhythmic tapping (3:8 polyrhythm – cf. Fig. 11-4). The location of the sensors is mapped onto a plane, nasion on top. As illustrated in the enlarged figures both movement frequencies can be found at different locations, be they ipsi- or contralateral to the corresponding limb. The according spectral power, however, accumulates contralaterally.

Because encephalographic signals may contain various spectral components corresponding to both limb movements, the discrimination between such couplings is a challenge, even when looking exclusively at the cortical level: Spectral properties associated with both (behavioral) sides are typically distributed in a nontrivial fashion across cortical signals (see, e.g., Daffertshofer et al., 2000; Frank et al., 2000 for 'delocalized' phase distributions during unimanual behavior; or Andres et al., 1999, for coherence distributions in the isofrequency bimanual case). An immediate way for discriminating between cortical correlations with movements of either the left or the right index finger is to study polyrhythmic (or multifrequency) performance, during which the two fingers move at distinct frequencies (see Fig. 11-4; right panel).

As can be seen in Figure 11-5, during such polyrhythmic performance different cortical areas show different frequency components that are not restricted to a specific hemisphere. A detailed decomposition into spatial and temporal components and subsequent cross-spectral analyses revealed that different spatial correlations were discernible within different frequency regimes. Various cortical areas appeared to be strongly frequency-locked with the movements, suggesting, at first blush, that the bimanual activity pattern can be expressed as a superposition of the two corresponding unimanual activity patterns. Moreover, these cortical areas were also phase locked with the motor activity (Daffertshofer et al., 2000). In addition to activity in (primary) motor areas, the mesial, central cortex appeared to be active (Lang et al., 1990), which is consistent with studies on isofrequency bimanual behavior (e.g., Gerloff & Andres, 2002; Jirsa et al., 1998) and related studies using functional imaging (e.g., Debaere et al., 2001; De Jong et al., 1999; Nair et al., 2003; Sadato et al., 1997).

Apart from their intricate spatial distribution, one has to realize that event-related cortical activities are by no means restricted to single frequency components (here: the movement frequencies of the two fingers). Dynamical interaction phenomena between and within hemispheres may be evident in various spectral components, in particular the sub- and/or super-harmonics of the movement frequency (Daffertshofer et al., 2000). To stay focused on the most relevant aspects of cortical activation during bimanual coordination, we refrain here from a more detailed interpretation of different cortical areas or frequency regimes and summarize instead that the dynamics of both the cortical activities and the behavioral components show (multi)frequency and (generalized) phase locking (i.e., for a $n:m$ frequency ratio, $\phi = n\phi_1 - m\phi_2$). When exploring the corresponding dynamics and their more general impact on behavior, the relative phases between the two hands or between hands and cortical areas or across cortical areas become of central importance. That is, we focus on the stability properties of the (generalized) relative phase rather than looking at the differential stability of the frequency locking. Although phase-locked cortical areas do not necessarily coincide with areas of strong (spectral) power, they indicate that overall timing processes like frequency stabilization or error correction are present. This becomes particularly apparent when investigating qualitative macroscopic changes in performance. As mentioned in the introduction, within the vicinity of phase transitions between different polyrhythmic patterns (Peper et al., 1991, 1995) the corresponding spatiotemporal patterns are expected to be reducible to low-dimensional dynamical systems irrespective of their origin (e.g., Haken, 1996). Due to this reduction of dimensionality, one can link the complicated spatially distributed neural activation with the essential features at the behavioral level, which eventually may lead to a closed mathematical description of both the macroscopic pattern formation at the behavioral level (polyrhythmic performance) and the spatiotemporal patterns of the MEG signals (see e.g., Frank et al., 2000; Fuchs et al., 2000; Haken, 2002).

In sum, encephalographic recordings of the cortical activity during bimanual coordination show qualitative properties that are consistent with behavioral (in)stabilities. The cortical dynamics, however, appears to be much more complex, not in the least because of nontrivial spatially distributed activation and a broad range of contributing spectral components. Two or multi-tiered models such as the one briefly summarized in Section 3.2 need to be elaborated further to account for these rich dynamical aspects. In prospect, this integrated approach will help to disentangle the explicit temporal features of bimanual coordination.

6.   CONCLUSION

The dynamical systems approach has highlighted the importance of stability and loss of stability in movement coordination. According to this approach the identified coordination dynamics, which reveals the way in which particular factors influence coordinative stability, is to a large extent the result of interactions between the coordinated limbs. Thus, apart from examining and modeling interlimb coordination in terms of gradient dynamics, sorting out how interlimb interactions lead to pattern (in)stability and establishing the extent to which they contribute to changes in stability (e.g., due to learning or reduced attention) will provide vital information for our understanding of the behavioral coordination dynamics. Because such interactions may stem from various neurophysiological processes, ranging from spinal entrainment to cortical (and transcortical) connections, establishing their relation to coordinative stability is by no means a trivial exercise and will involve a variety of experimental manipulations and settings. The thus obtained insights may be helpful in formulating an encompassing dynamical model of rhythmic interlimb coordination that, in spite of its relatively high level of abstraction, reflects essential features of the underlying system properties and processes.

ACKNOWLEDGMENT

The contribution of Lieke Peper was facilitated by ASPASIA grant 015.001.040 of the Netherlands Organization for Scientific Research (NWO).

REFERENCES

Amazeen EL, Amazeen PG, Treffner PJ, Turvey MT (1997) Attention and handedness in bimanual coordination dynamics. J Exp Psych: Hum Percept Perf 23:1552-1560

Andres FG, Mima T, Schulman AE, Dichgans J, Hallett M, Gerloff C (1999) Functional coupling of human cortical sensorimotor areas during bimanual skill acquisition. Brain 122:855-870.

Baldissera F, Cavallari P, Marini G, Tassone G (1991) Differential control of in-phase and anti-phase coupling of rhythmic movements of ipsilateral hand and foot. Exp Brain Res 83:375-380

Baldissera F, Borroni P, Cavallari P (2000) Neural compensation for mechanical differences between hand and foot during coupled oscillations of the two segments. Exp Brain Res 133:165-177

Beek PJ, Peper CE, Daffertshofer A (2002) Modeling rhythmic interlimb coordination: Beyond the Haken-Kelso-Bunz model. Brain Cogn 48:149-165

Beek PJ, Rikkert WEI, Van Wieringen PCW (1996) Limit cycle properties of rhythmic forearm movements. J Exp Psych: Hum Percept Perf 22:1077-1092

Bingham GP, Zaal FTJM, Shull JA, Collins DR (2001) The effect of frequency on the visual perception of relative phase and phase variability of two oscillating objects. Exp Brain Res 136:543-552

Buchanan JJ, Kelso JAS (1993) Posturally induced transitions in rhythmic multijoint limb movements. Exp Brain Res 94:131-142

Byblow WD, Chua R, Goodman D (1995) Asymmetries in coupling dynamics of perception and action. J Mot Behav 27:123-137

Byblow WD, Carson RG, Goodman D (1994) Expressions of asymmetries and anchoring in bimanual coordination. Hum Mov Sci 13:3-28

Carson RG (1996) Neuromuscular-skeletal constraints upon the dynamics of perception-action coupling. Exp Brain Res 110: 99-110

Carson RG, Chua R, Byblow WD, Smethurst CJ, Poon P (1999) Changes in posture alter the attentional demands of voluntary movement. Proc R Soc Biol Sci 266:853-857

Carson RG, Goodman D, Kelso JAS, Elliott D (1995) Phase transitions and critical fluctuations in rhythmic coordination of ipsilateral hand and foot. J Mot Behav 27:211-224

Carson RG, Riek S (1998) The influence of joint position on the dynamics of perception-action coupling. Exp Brain Res 121:103-114

Carson RG, Riek S, Smethurst CJ, Franscisco Lisón Párraga J, Byblow WD (2000) Neuromuscular-skeletal constraints upon the dynamics of unimanual and bimanual coordination. Exp Brain Res 131:196-214

Cattaert D, Semjen A, Summers JJ (1999) Simulating a neural cross-talk model for between-hand interference during bimanual circle drawing. Biol Cybern 81:343-358

Daffertshofer A (1998) Effects of noise on the phase dynamics of nonlinear oscillators. Phys Rev E 58:327-337

Daffertshofer A, Peper CE, Beek PJ (2000) Spectral analyses of event-related encephalographic signals. Phys Lett A 266:290-302

Daffertshofer A, Peper CE, Frank TD, Beek PJ (2000) Spatio-temporal patterns of encephalographic signals during polyrhythmic tapping. Hum Mov Sci 19:475-498

De Jong BM, Willemsen AT, Paans AM (1999) Brain activation related to the change between bimanual motor programs. Neuroimage 9:290-297

Debaere F, Swinnen SP, Beatse E, Sunaert S, Van Hecke P, Duysens J (2001) Brain areas involved in interlimb coordination: A distributed network. Neuroimage 14: 947-958

Eliassen JC, Baynes K, Gazzaniga MS (1999) Direction information coordinated via the posterior third of the corpus callosum during bimanual movements. Exp Brain Res 128:573-577

Frank TD, Daffertshofer A, Peper CE, Beek PJ, Haken H (2000) Towards a comprehensive theory of brain activity: Coupled oscillator systems under external forces. Physica D 144:62-86

Franz EA, Eliassen JC, Ivry RB, Gazzaniga MS (1996) Dissociation of spatial and temporal coupling in the bimanual movements of callosotomy patients. Psychol Sci 7:306-310

Fuchs A, Jirsa VK, Kelso JAS (2000) Theory of the relation between human brain activity (MEG) and hand movements. Neuroimage 11:359-369

Gerloff C, Andres FG (2002) Bimanual coordination and interhemispheric interaction. Acta Psychol 110: 161-186

Grossberg S, Probe C, Cohen MA (1997) Neural control of interlimb oscillations: I. Human bimanual coordination. Biol Cybern 77:131-140

Gunkel M (1962) Über relative Koordination bei willkürlichen menschlichen Gliedbewegungen (Relative coordination in voluntary limb movements in man). Pflügers Archiv 275:427-477

Haken H (1983) Synergetics: An Introduction. Springer, Berlin

Haken H (1996) Principles of Brain Functioning. Springer, Berlin

Haken H (2002) Brain Dynamics. Springer, Berlin

Haken H, Kelso JAS, Bunz H (1985) A theoretical model of phase transitions in human hand movements. Biol Cybern 51:347-356

Haken H, Peper CE, Beek PJ, Daffertshofer A (1996) A model for phase transitions in human hand movements during multifrequency tapping. Physica D 90:179-196; 92:260 (erratum).

Helmuth LL, Ivry RB (1996) When two hands are better than one: Reduced timing variability during bimanual movements. J Exp Psych: Hum Percept Perf 22:278-293

Heuer H (1996) Coordination. In Keele SW (ed) Handbook of Perception and Action (Vol. 2: Motor Skills). Academic Press, London, pp 121-180

Jirsa VK, Haken H (1996) Field theory of electromagnetic brain activity. Phys Rev Lett 77:960-963

Jirsa VK, Fuchs A, Kelso JAS (1998) Connecting cortical and behavioral dynamics: Bimanual coordination. Neural Comput 10:2019-2045

Jirsa VK, Jantzen KJ, Fuchs A, Kelso JAS (2002) Spatiotemporal forward solution of the EEG and MEG using network modeling. IEEE Trans Med Imaging 21:493-504

Karnopp D, Rosenberg R (1975) System dynamics: A unified approach. John Wiley, New York

Kay BA, Kelso JAS, Saltzman EL, Schöner GS (1987) Space-time behavior of single and bimanual rhythmical movements: Data and limit cycle model. J Exp Psych: Hum Percept Perf 13:178-190

Kay BA, Saltzman EL, Kelso JAS (1991) Steady-state and perturbed rhythmical movements: A dynamical analysis. J Exp Psych: Hum Percept Perf 17:183-197

Kelso JAS (1984) Phase transitions and critical behavior in human bimanual coordination. Am J Physiol 246: R1000-1004

Kelso JAS, DelColle JD, Schöner G (1990) Action-perception as a pattern formation process. In Jeannerod M (ed) Attention and Performance XII. Lawrence Erlbaum, Hillsdale NJ, pp 139-169

Kelso JAS, Jeka JJ (1992). Symmetry breaking dynamics of human multilimb coordination. J Exp Psych: Hum Percept Perf 18:645-668

Kelso JAS, Zanone PG (2002) Coordination dynamics of learning and transfer across different effector systems. J Exp Psych: Hum Percept Perf 28:776-797

Kennerley SW, Diedrichsen J, Hazeltine E, Semjen A, Ivry RB (2002) Callosotomy patients exhibit temporal uncoupling during continuous bimanual movements. Nat Neurosci 5:376-381

Lang W, Obrig H, Lindinger G, Cheyne D, Deecke L (1990) Supplementary motor area activation while tapping bimanually different rhythms in musicians. Exp Brain Res 79: 504-514

Mackey DC, Meichenbaum DP, Shemmell J, Riek S, Carson RG (2002) Neural compensation for compliant loads during rhythmic movement. Exp Brain Res 142:409-417

Mechsner F, Kerzel D, Knoblich G, Prinz W (2001) Perceptual basis of bimanual coordination. Nature 414:69-73

Meyer-Lindenberg A, Ziemann U, Hajak G, Cohen L, Berman KF (2002) Transitions between dynamical states of differing stability in the human brain. Proc Natl Acad Sci USA 99:10948-10953

Mitra S, Riley MA, Turvey MT (1997) Chaos in human rhythmic movement. J Mot Behav 29:195-198

Monno A, Temprado JJ, Zanone PG, Laurent M (2002) The interplay of attention and bimanual coordination dynamics. Acta Psychol 110:187-211

Nair DG, Purcott KL, Fuchs A, Steinberg F, Kelso JAS (2003) Cortical and cerebellar activity of the human brain during imagined and executed unimanual and bimanual action sequences: A functional MRI study. Brain Res Cogn Brain Res 15:250-260

Newell KM, YT Liu, Mayer-Kress G (2001) Time scales in motor learning and development. Psychol Rev 108:75-82

Peper CE, Beek PJ (1998a) Are frequency-induced transitions in rhythmic coordination mediated by a drop in amplitude? Biol Cybern 79:291-300

Peper CE, Beek PJ (1998b) Distinguishing between the effects of frequency and amplitude on interlimb coupling in tapping a 2:3 polyrhythm. Exp Brain Res 11:78-92

Peper CE, Beek PJ (1999) Modeling rhythmic interlimb coordination: The roles of movement amplitude and time delays. Hum Mov Sci 18:263-280

Peper CE, Beek PJ, van Wieringen PCW (1991) Bifurcations in bimanual tapping: In search of Farey principles. In Requin J, Stelmach GE (eds) Tutorials in motor neuroscience Kluwer, Dordrecht, pp 413-431

Peper CE, Beek PJ, Van Wieringen PCW (1995) Frequency-induced transitions in bimanual tapping. Biol Cybern 73:301-309

Peper CE, Nooy SAE, Van Soest AJ (2003) Mass perturbation of a body segment: II. Effects on interlimb coordination. Manuscript submitted for publication

Peters M (1994) Does handedness play a role in the coordination of bimanual movement? In Swinnen S, Heuer H, Massion J, Casear P (eds) Interlimb coordination: Neural, dynamical, and cognitive constraints. Academic Press, San Diego, pp 559-615

Post AA, Peper CE, Beek PJ (2000a) Relative phase dynamics in perturbed interlimb coordination: The effects of frequency and amplitude. Biol Cybern 83:529-542

Post AA, Peper CE, Daffertshofer A, Beek PJ (2000b) Relative phase dynamics in perturbed interlimb coordination: Stability and stochasticity. Biol Cybern 83:443-459

Rosenblum LD, Turvey MT (1988) Maintenance tendency in coordinated rhythmic movements: Relative fluctuations and phase. Neuroscience 27:289-300

Sadato N, Yonekura Y, Waki A, Yamada H, Ishii Y (1997) Role of the supplementary motor area and the right premotor cortex in the coordination of bimanual finger movements. J Neurosci, 17:9667-9674

Schmidt RC, Carello C, Turvey MT (1990) Phase transitions and critical fluctuations in the visual coordination of rhythmic movement between people. J Exp Psych: Hum Percept Perf 16:227-247

Schmidt RC, Shaw BK, Turvey MT (1993) Coupling dynamics in interlimb coordination. J Exp Psych: Hum Percept Perf 19:397-415

Schmidt RC, Turvey MT (1995) Models of interlimb coordination – equilibria, local analysis, and spectral patterning: Comment on Fuchs and Kelso (1994). J Exp Psych: Hum Percept Perf 21:432-443

Schöner G (1989) Learning and recall in a dynamical theory of coordination patterns. Biol Cybern 62:39-54

Schöner G, Haken H, Kelso JAS (1986) A stochastic theory of phase transitions in human hand movement. Biol Cybern 53:247-257

Schöner G, Kelso JAS (1988) A synergetic theory of environmentally-specified and learned patterns of movement coordination: I. Relative phase dynamics. Biol Cybern 58:71-80

Schöner G, Zanone PG, Kelso JAS (1992) Learning as change of coordination dynamics: Theory and experiment. J Mot Behav 24:29-48

Sternad D, Turvey MT, Schmidt RC (1992) Average phase difference theory and 1:1 entrainment in interlimb coordination. Biol Cybern 67:223-231

Steyvers M, Verschueren SMP, Levin O, Ouamer M, Swinnen SP (2001) Proprioceptive control of cyclical bimanual forearm movements across different movement frequencies as revealed by means of tendon vibration. Exp Brain Res 140:326-334

Stinear JW, Byblow WD (2001) Phase transitions and postural deviations during bimanual kinesthetic tracking. Exp Brain Res 137:467-477

Stinear JW, Byblow WD (2002) Disinhibition in the human cortex is enhanced by synchronous upper limb movements. J Physiol 543:307-316

Swinnen SP (2002) Intermanual coordination: From behavioural principles to neural-network interactions. Nat Rev Neurosci 3:350-361

Swinnen SP, Dounskaia N, Verschueren S, Serrien DJ, Daelman A (1995) Relative phase destabilization during interlimb coordination: The disruptive role of kinesthetic afferences induced by passive movement. Exp Brain Res 105:439-454

Temprado JJ, Zanone PG, Monno A, Laurent M (1999) Attentional load associated with performing and stabilizing preferred bimanual patterns. J Exp Psych: Hum Percept Perf 25:1579-1594

Treffner PJ, Turvey MT (1995). Handedness and the asymmetric dynamics of bimanual rhythmic coordination. J Exp Psych: Hum Percept Perf 21:318-333

Treffner PJ, Turvey MT (1996) Symmetry, broken symmetry, and handedness in bimanual coordination dynamics. Exp Brain Res 107:463-478

Van Soest AJ, Peper CE, Selles RW (in press). Mass perturbation of a body segment: I. Effects on segment dynamics. J Mot Behav

Verschueren SMP, Swinnen SP, Cordo PJ, Dounskaia NV (1999) Proprioceptive control of multijoint movement: Bimanual circle drawing. Exp Brain Res 127:182-192

Tass PA, Fieseler T, Dammers J, Dolan K, Morosan P, Majtanik M, Boers F, Muren A, Zilles K, Fink GR (2003) Synchronization tomography: A method for three-dimensional localization of phase synchronized neuronal populations in the human brain using magnetoencephalography. Phys Rev Lett 90:088101

Tuller B, Kelso JAS (1989) Environmentally-specified patterns of movement coordination in normal and split-brain subjects. Exp Brain Res 75:306-316

Wimmers RH, Beek PJ, Van Wieringen PCW (1992) Phase transitions in rhythmic tracking movements: A case of unilateral coupling. Hum Mov Sci 11:217-226

Zanone PG, Kelso JAS (1992) The evolution of behavioral attractors with learning: Nonequilibrium phase transitions. J Exp Psych: Hum Percept Perf 18:403-421

Index

DATE DUE